Geometry Reconstruction from Images (2nd Edition)

Geometry Reconstruction from Images (2nd Edition)

Guest Editor

Daniel Meneveaux

MDPI

Basel • Beijing • Wuhan • Barcelona • Belgrade • Novi Sad • Cluj • Manchester

Guest Editor
Daniel Meneveaux
XLIM Institute, UMR CNRS 7252
University of Poitiers
Poitiers
France

Editorial Office
MDPI AG
Grosspeteranlage 5
4052 Basel, Switzerland

This is a reprint of the Special Issue, published open access by the journal *Journal of Imaging* (ISSN 2313-433X), freely accessible at: https://www.mdpi.com/journal/jimaging/special_issues/55A1X64G0H.

For citation purposes, cite each article independently as indicated on the article page online and as indicated below:

Lastname, A.A.; Lastname, B.B. Article Title. *Journal Name* **Year**, *Volume Number*, Page Range.

ISBN 978-3-7258-6013-5 (Hbk)
ISBN 978-3-7258-6014-2 (PDF)
https://doi.org/10.3390/books978-3-7258-6014-2

Contents

About the Editor

Daniel Meneveaux

Daniel Meneveaux is a professor at the University of Poitiers (France), affiliated with the XLIM Institute (CNRS UMR 7252). He was awarded a Ph.D. degree in 1998 from the University of Rennes (France), hired as an associate professor in 1999 and a professor in 2010. His fields of interest are lighting simulation, reflectance models, image-based rendering, complex scene management, including topology-based geometric modeling and hierarchical representations.

Preface

The present reprint includes eleven papers focusing on geometry reconstruction from images. The contributions address several well-known challenges in the field, offering insightful advances on a wide range of topics such as underwater constraints, dynamic scenes, neural methods, medical imaging, and multi-view reconstruction. We sincerely thank all the authors for their valuable work, which continues to push the boundaries of accurate geometry reconstruction under constrained conditions.

Daniel Meneveaux
Guest Editor

Journal of *Imaging*

MDPI

Editorial

Editorial on the Special Issue "Geometry Reconstruction from Images (2nd Edition)"

Daniel Meneveaux

XLIM Institute, UMR CNRS 7252, University of Poitiers, 86073 Poitiers, France; daniel.meneveaux@univ-poitiers.fr

In recent decades, research has produced impressive methods for recovering geometric information from real objects [1,2], laying the fundamental foundations for further studies in robotics, industry, medicine, architecture, and visualization approaches. The subject remains a very active field in computer vision, among many other scientific areas. Research advances have long been transferred to industry, but new trends and challenges continuously emerge [3,4], with new sensors, faster computation hardware, and the growing demand for increasingly accurate detail capture. Furthermore, applications for the general public are now appearing, for instance with 3D reconstruction available on mobile phones.

Alongside visualization techniques, machine learning has also increased the quality of reconstruction methods, with approaches such as NeRFs or Gaussian Splatting, which are addressed in this Special Issue. Some specific aspects still require more in-depth fundamental research, such as the management of specular surfaces or accuracy issues in underwater environments.

The 11 articles published in this second edition (https://www.mdpi.com/journal/jimaging/special_issues/55A1X64G0H, accessed on 25 September 2025) tackle several very interesting challenges: reconstructions of objects known for their complexity [5], underwater environments where distortions make depth estimation difficult, interactive systems and dynamic scenes, analyses of existing reconstruction techniques [6], and deep learning approaches.

List of contributions:

The first article, "Adaptive High-Precision 3D Reconstruction of Highly Reflective Mechanical Parts Based on Optimization of Exposure Time and Projection Intensity" by Ci He, Rong Lai, Jin Sun, Kazuhiro Izui, Zili Wang, Xiaojian Liu, and Shuyou Zhang, focuses on reconstructing mechanical parts with highly reflective surfaces. The proposed method relies on an adaptive 3D reconstruction approach that optimizes exposure time and projection intensity, while being further adjusted to the linear dynamic range of the hardware. The resulting image sequence is fused using a combination of a Genetic Algorithm and the Stochastic Adam optimizer to maximize image information entropy. The authors experimentally validate their approach on three sets of typical mechanical components, each with diverse geometric characteristics and varying levels of complexity.

The second article, "Impact of Data Capture Methods on 3D Reconstruction with Gaussian Splatting", by Dimitar Rangelov, Sierd Waanders, Kars Waanders, Maurice van Keulen, and Radoslav Miltchev, investigates how different filming techniques influence the quality of 3D reconstructions for indoor crime scene investigations. The authors examine the impact of factors such as camera orientation, filming speed, data layering, and scanning path on the detail and clarity of 3D reconstructions using Neural Radiance Fields (NeRFs) and Gaussian Splatting. They identify optimal filming methods that help reduce noise and artifacts, and provide valuable guidelines for professionals in forensics, architecture, and cultural heritage preservation to capture realistic, high-quality 3D representations. The study also highlights opportunities for future research, particularly in exploring other algorithms, camera parameters, and real-time adjustment techniques.

In the third article, "Robot-Based Procedure for 3D Reconstruction of Abdominal Organs Using the Iterative Closest Point and Pose Graph Algorithms", by Birthe Göbel, Jonas Huurdeman, Alexander Reiterer and Knut Möller, a procedure is proposed for a robot-based multi-view 3D reconstruction with pose optimization algorithms. In this work, a robotic arm and a stereo laparoscope build the experimental setup. The procedure includes stereo matching for depth measurement and the multiscale color iterative closest point algorithm, along with multiway registration for pose optimization. The procedure is evaluated quantitatively and qualitatively on ex vivo organs. The proposed procedure leads to a plausible 3D model, without hand–eye calibration.

The article "Fitting Geometric Shapes to Fuzzy Point Cloud Data", by Vincent B. Verhoeven, Pasi Raumonen, and Markku Åkerblom, presents procedures and analysis on the reconstruction of geometry-derived data and its associated uncertainty. Instead of treating the data as a discrete point cloud, the authors consider it as a continuous fuzzy point cloud. They introduce a novel approach based on the expected Mahalanobis distance, which is illustrated using laser scanning data of a cylinder. Its performance is compared to that of the conventional least squares method, both with and without random sample consensus (RANSAC). The proposed method achieves a more accurate geometric fit, albeit generally with greater uncertainty, and demonstrates strong potential for geometry reconstruction from laser-scanned data.

The article "Arbitrary Optics for Gaussian Splatting Using Space Warping", written by Jakob Nazarenus, Simin Kou, Fang-Lue Zhang, and Reinhard Koch, addresses the camera models employed in the context of 3D Gaussian Splatting. The authors propose a method to handle arbitrary camera optics, such as highly distorting fisheye lenses. Their approach applies a differentiable warping function to the Gaussian scene representation. They also introduce a learnable skybox for the specific case of outdoor scenes.

The article "A Mathematical Model for Wind Velocity Field Reconstruction and Visualization Taking into Account the Topography Influence", by Guzel Khayretdinova and Christian Gout, proposes a global model for vector field approximation from a given finite set of vectors (corresponding to wind velocity fields or marine currents). The minimization process relies on a Hilbert space energy functional that includes both a data fidelity term and a smoothing term. The continuous problem is then discretized, and topographic effects are incorporated into the wind velocity field.

The article "Multi-Head Attention Refiner for Multi-View 3D Reconstruction", by Kyunghee Lee, Ihjoon Cho, Boseung Yang, and Unsang Park, introduces a post-processing method called the Multi-Head Attention Refiner (MA-R), designed to improve the handling of object edges during the reconstruction process. The method integrates a multi-head attention mechanism into a U-Net-style refiner module. The proposed approach significantly enhances the reconstruction performance of Pix2Vox++ when multiple input images are used.

The article "Three-Dimensional Reconstruction of Indoor Scenes Based on Implicit Neural Representation", by Zhaoji Lin, Yutao Huang, and Li Yao, addresses the problem of indoor scene reconstruction. The authors propose a 3D reconstruction method that combines Neural Radiance Fields (NeRFs) and Signed Distance Function (SDF) implicit representations. The volume density of the NeRF is leveraged to provide geometric information for the SDF field, while the learning of geometric shapes and surfaces is further enhanced through an adaptive normal prior optimization process.

The article "Single-Image-Based 3D Reconstruction of Endoscopic Images", by Bilal Ahmad, Pål Anders Floor, Ivar Farup, and Casper Find Andersen, addresses 3D reconstruction from wireless capsule endoscopes (WCEs) designed for the examination of the human gastrointestinal (GI) tract. The authors propose a single-image reconstruction method using an artificial colon captured with an endoscope that mimics the behavior of a WCE. A Shape-from-Shading (SFS) algorithm reconstructs the 3D shape after geometric and radiometric calibration.

The article "Neural Radiance Field-Inspired Depth Map Refinement for Accurate Multi-View Stereo", by Shintaro Ito, Kanta Miura, Koichi Ito, and Takafumi Aoki, proposes a method to refine depth maps obtained by Multi-View Stereo (MVS) through iterative optimization of Neural Radiance Fields (NeRFs). The proposed approach combines MVS and NeRFs to leverage the strengths of both in depth map estimation and employs NeRFs for depth map refinement. To further improve accuracy, the authors introduce a Huber loss into the NeRF optimization, which reduces estimation errors in the radiance fields by constraining errors larger than a threshold. The method is evaluated against conventional approaches, including COLMAP, NeRF, and DS-NeRF.

The article "Fast Data Generation for Training Deep-Learning 3D Reconstruction Approaches for Camera Arrays", by Théo Barrios, Stéphanie Prévost, and Céline Loscos, focuses on 3D reconstruction from images captured by multi-camera arrays. The authors present a fully virtual data generator for creating large training datasets that can be adapted to any camera array configuration. The generator builds virtual scenes by randomly selecting objects and textures while following user-defined parameters such as disparity range or image properties (e.g., resolution, color space). Its effectiveness is validated by testing the generated datasets with established deep learning methods and depth reconstruction algorithms.

Conflicts of Interest: The author declares no conflicts of interest.

References

1. Zhou, L.; Wu, G.; Zuo, Y.; Chen, X.; Hu, H. A Comprehensive Review of Vision-Based 3D Reconstruction Methods. *Sensors* **2024**, *24*, 2314. [CrossRef] [PubMed]
2. Yang, J.; Sax, A.; Liang, K.J.; Henaff, M.; Tang, H.; Cao, A.; Chai, J.; Meier, F.; Feiszli, M. Fast3R: Towards 3D Reconstruction of 1000+ Images in One Forward Pass. In Proceedings of the IEEE/CVF Conference on Computer Vision and Pattern Recognition (CVPR), Nashville, TN, USA, 10–17 June 2025; pp. 21924–21935.
3. Chen, D.; Li, H.; Ye, W.; Wang, Y.; Xie, W.; Zhai, S.; Wang, N.; Liu, H.; Bao, H.; Zhang, G. PGSR: Planar-Based Gaussian Splatting for Efficient and High-Fidelity Surface Reconstruction. *IEEE Trans. Vis. Comput. Graph.* **2025**, *31*, 6100–6111. [CrossRef] [PubMed]
4. Cui, B.; Tao, W.; Zhao, H. High-Precision 3D Reconstruction for Small-to-Medium-Sized Objects Utilizing Line-Structured Light Scanning: A Review. *Remote Sens.* **2021**, *13*, 4457. [CrossRef]
5. Lee, K.; Cho, I.; Yang, B.; Park, U. Multi-Head Attention Refiner for Multi-View 3D Reconstruction. *J. Imaging* **2024**, *10*, 268. [CrossRef]
6. Rangelov, D.; Waanders, S.; Waanders, K.; Keulen, M.V.; Miltchev, R. Impact of Data Capture Methods on 3D Reconstruction with Gaussian Splatting. *J. Imaging* **2025**, *11*, 65. [CrossRef] [PubMed]

Journal of
Imaging

MDPI

Article

Adaptive High-Precision 3D Reconstruction of Highly Reflective Mechanical Parts Based on Optimization of Exposure Time and Projection Intensity

Ci He [1,2,*], Rong Lai [1], Jin Sun [1], Kazuhiro Izui [3], Zili Wang [2], Xiaojian Liu [2] and Shuyou Zhang [2]

1 School of Mechanical Engineering, Yangzhou University, Yangzhou 225127, China; mz120220865@stu.yzu.edu.cn (R.L.); sunjin@yzu.eu.cn (J.S.)

2 School of Mechanical Engineering, Zhejiang University, Hangzhou 310058, China; ziliwang@zju.edu.cn (Z.W.); liuxj@zju.edu.cn (X.L.); zsy@zju.edu.cn (S.Z.)

3 Department of Precision Engineering, Graduate School of Engineering, Kyoto University, Kyoto 615-8540, Japan; izui.kazuhiro.8c@kyoto-u.ac.jp

* Correspondence: heci_yzu@yzu.edu.cn

Abstract: This article is used to reconstruct mechanical parts with highly reflective surfaces. Three-dimensional reconstruction based on Phase Measuring Profilometry (PMP) is a key technology in non-contact optical measurement and is widely applied in the intelligent inspection of mechanical components. Due to the high reflectivity of metallic parts, direct utilization of the captured high-dynamic-range images often results in significant information loss in the oversaturated areas and excessive noise in the dark regions, leading to geometric defects and reduced accuracy in the reconstructed point clouds. Many image-fusion-based solutions have been proposed to solve these problems. However, unknown geometric structures and reflection characteristics of mechanical parts lead to the lack of effective guidance for the design of important imaging parameters. Therefore, an adaptive high-precision 3D reconstruction method of highly reflective mechanical parts based on optimization of exposure time and projection intensity is proposed in this article. The projection intensity is optimized to adapt the captured images to the linear dynamic range of the hardware. Image sequence under the obtained optimal intensities is fused using an integration of Genetic Algorithm and Stochastic Adam optimizer to maximize the image information entropy. Then, histogram-based analysis is employed to segment regions with similar reflective properties and determine the optimal exposure time. Experimental validation was carried out on three sets of typical mechanical components with diverse geometric characteristics and varying complexity. Compared with both non-saturated single-exposure techniques and conventional image fusion methods employing fixed attenuation steps, the proposed method reduced the average whisker range of reconstruction error by 51.18% and 25.09%, and decreased the median error by 42.48% and 25.42%, respectively. These experimental results verified the effectiveness and precision performance of the proposed method.

Keywords: 3D reconstruction; highly reflective mechanical parts; multi exposure; image fusion; machine vision

1. Introduction

Three-dimensional profilometry based on structured light techniques has become increasingly prevalent in fields such as advanced manufacturing, quality inspection [1–3], intelligent surgery [4], and virtual reality [5], because of its advantages of non-contact operation, full-field measurement, high precision, and efficiency. Phase Measuring Profilometry (PMP) is a widely used frequency encoding method that extracts depth information by analyzing the phase shift resulting from variations in the height of an object's surface. PMP mitigates the common information loss issue in spatial encoding methods and has advantages in reconstruction efficiency and flexibility compared with temporal encoding techniques.

High-quality, high-precision 3D reconstruction relies on high-quality images, which are, however, significantly influenced by surface reflective properties. There are both highly reflective, polished surfaces and rough, diffuse surfaces in mechanical components, so light observed from a metal surface is obtained after both specular and diffuse reflections. It is difficult to obtain information about all surfaces optimally and simultaneously with a single exposure. We usually increase the exposure to capture details on non-polished surfaces, which results in changes to the encoding information of structured light or exceeding the sensor's response range. Reconstruction error or information loss would happen in the saturation area. If the exposure is reduced to recover 3D information on bright, polished areas, the signal-to-noise ratio (SNR) significantly drops in dark regions, leading to large decoding errors and reconstruction accuracy loss. This disparity negatively impacts the efficiency and accuracy of PMP-based 3D profilometry using single-frame projected images.

Various solutions have been proposed to enhance the accuracy of PMP-based 3D profilometry considering the multi-reflection characteristics. Major techniques include polarization filter, adjustment of the intensity of projection patterns, multi-exposure-based methods, phase compensation method, deep learning-based method, improved coding method, and color invariant method [6].

One existing method is the use of a polarization filter to extend polarization information from conventional intensity, frequency, and coherence. Salahieh et al. [7] proposed a multi-polarization fringe projection imaging technique to eliminate saturated or low-contrast fringe regions by optimizing combinations of polarization angles and exposure times. Huang et al. [8] proposed a polarization-coded structured light system designed to enhance both efficient 3D reconstruction and polarimetric target detection. By estimating the degree of linear polarization, targets within the scene were effectively distinguished and reconstructed with high efficiency. Xiang et al. [9] introduced a polarization spatial phase-shifting technique using two orthogonally positioned filtered projectors to cast sinusoidal fringe patterns with distinct phase shifts onto the measured metal surfaces. To ensure accurate alignment of the projected patterns, a fringe registration method based on the epipolar geometry between the projectors was also developed. Zhu et al. [10] proposed a polarization-enhanced fringe pattern method to achieve high dynamic range imaging in a single exposure. The degree of linear polarization is precisely calculated by leveraging the polarization properties of reflected light and a fixed-azimuth linear polarizer. They further extended the approach to build a structured light encoding model in Zhu et al. [11], utilizing the superposition of multiple polarization states. This method enables the generation of polarized structured light containing phase information without the need to rotate the polarizer.

To overcome the image saturation problem in 3D reconstruction, methods for adjusting the intensity of projection patterns have been proposed. Li et al. [12] introduced an adaptive fringe pattern projection method to dynamically adjust the projector's maximum input gray levels based on the local reflectivity of the target surface. In this approach, regions with high reflectivity were illuminated using lower intensity levels to prevent image saturation, while regions with low reflectivity received the highest possible intensity to ensure sufficient intensity modulation for accurate measurement. Li et al. [13] presented an adaptive digital fringe projection technique to calculate the proper intensity for each projector pixel using binary search. The calculation process is simplified compared with previous adaptive techniques because there is no need to obtain the camera response function and homographic mapping between the camera and projector. Chen et al. [14] advanced the fringe projection method to achieve high measurement accuracy. They projected three sets of orthogonal color fringe patterns and a sequence of uniform gray-level patterns with different gray levels onto a measured surface, and captured the deformed patterns by a camera from a different viewpoint. Xu et al. [15] combined adaptive fringe projection with the curve fitting method to compensate for highly reflective surfaces. The optimal light intensity coefficient template of the projection image was calculated to adaptively adjust the projected light intensity based on the pixel mapping and the coefficient template. The 3D reconstruction results were compensated by curve fitting in the horizontal and vertical directions. Zhang et al. [16] projected a small number of uniform grayscale pattern sequences to mark the saturated area and calculate the surface reflection factor and the environment factor of individual pixels. A surface coefficient look-up table was created to calculate the optimal projection intensities of pixels in the saturated regions.

Multi-exposure-based methods improve the quality of the fringe pattern by fusing images acquired at different exposure times. Zhang et al. [17] presented a rapid and fully automatic exposure time determination method by analyzing the texture image acquired by the camera. The reflectivity of the object surface was estimated through a single exposure to determine the global optimal exposure time. Jiang et al. [18] proposed a high dynamic range fringe acquisition to solve the problem caused by a high-reflective surface. They developed a fringe image fusion algorithm to prevent saturation and under-illumination by selecting pixels with the highest fringe modulation intensity from the raw fringe images. Cui et al. [19] introduced a multiple-exposure adaptive selection algorithm. Exposure time nodes are adaptively selected based on the relative irradiance value to cover the highest and lowest gray value of the fringe image. The information entropy theory is introduced in Chen et al. [20] to adaptively optimize the initial exposure value through feature analysis of fringe image entropy. An effective exposure sequence is then generated using the dichotomy method. Zhu et al. [21] proposed an HDR surface 3D reconstruction method based on a shared phase demodulation mechanism and a multi-indicator guided phase fusion strategy. Exposure quality, phase gradient smoothness, and pixel effectiveness were included in a phase sequence fusion model to obtain an optimum phase map for final 3D reconstruction.

Phase compensation methods apply post-processing techniques to images acquired by cameras to recover fringe pattern information that is distorted or lost due to reflective interference. Hu et al. [22] proposed a dynamic phase retrieval algorithm based on unsaturated frame data. By defining a saturation coefficient K to analyze error characteristics and validating the approach through simulations, the method effectively addresses phase distortion issues arising in highly reflective surfaces or scenarios with limited dynamic range. Budianto et al. [23] developed a fringe projection profilometry restoration technique based on geometry-guided iterative regularization. This method employs a Gaussian Mixture Model to detect specular regions, generates an initial fringe structure via geometric sketching, and iteratively refines the result using a dual-tree complex wavelet transform. Ren et al. [24] introduced a specular reflection separation approach based on a global color-line constraint. The method estimates illumination chromaticity by analyzing intersection points of color lines in normalized RGB space, clusters specular-invariant variables, and separates specular components at the pixel level based on their distance to the estimated illumination chromaticity. Chen et al. [25] proposed a high-frequency averaged phase compensation method guided by an optimal frequency strategy. By reducing the number of projected images and incorporating both high-frequency phase compensation and model-driven optimal frequency selection, the method significantly suppresses phase errors caused by gamma nonlinearity.

In a deep learning-based method, Zhang et al. [26] designed a specialized convolutional neural network that takes high dynamic range (HDR) fringe patterns with three-step phase shifting as input, enabling accurate extraction of phase information in both low signal-to-noise ratio (SNR) and HDR scenes. Liu et al. [27] proposed a Skip Pyramid Context Aggregation Network (SP-CAN) to enhance fringe images captured synchronously by a single-exposure camera, while precisely preserving encoded phase details near edges and corners. Shen et al. [28] employed an improved UNet-based deep neural network to establish a "many-to-one" mapping, utilizing π-phase-shifted binary fringes to acquire more saturated fringe information, thereby enabling fast and accurate retrieval of wrapped phase maps for HDR objects. Xi et al. [29] developed an encoder–decoder network guided by reflection priors to restore defective fringe patterns caused by highly reflective surfaces. This method transforms distorted fringes into ideal patterns with uniform grayscale distribution, effectively eliminating reflective artifacts and recovering the missing phase information.

Improved encoding methods have been developed based on advanced projection coding strategies. Song et al. [30] proposed a structured light approach based on fringe edges, which integrates gray code with a positive-negative fringe pattern encoding scheme. An enhanced zero-crossing edge detector is employed to achieve subpixel-level edge localization. Tang et al. [31] introduced a micro-phase measurement profilometry technique that improves both the accuracy and efficiency of shape acquisition while demonstrating strong robustness against global illumination variations. Feng et al. [32] presented a fast 3D measurement method combining dual-camera fringe projection with digital speckle image fusion. By leveraging trifocal tensor constraints to correct phase errors in highly reflective regions and employing a three-step phase-shifting algorithm along with subpixel matching, the method enables efficient and high-precision 3D reconstruction of reflective surfaces in dynamic scenes using only four projected patterns. Zhao et al. [33] developed an adaptive checkerboard high-frequency projection technique, which integrates high-frequency encoded patterns with complementary projection and dynamically adjusts the projection intensity to suppress image saturation caused by specular highlights, thereby significantly enhancing the measurement accuracy and point cloud completeness for highly reflective surfaces.

Color-based specular highlight removal methods are primarily founded on the dichromatic reflection model proposed by Shafer [34], which separates specular and diffuse reflection components by analyzing the color distribution of pixels in RGB images, thereby facilitating the estimation of surface normals for each pixel. Benveniste et al. [35] developed a structured light range scanner based on color invariance, employing binary, ternary, and quaternary encoding schemes to robustly scan both glossy and matte objects under ambient lighting conditions. Xu et al. [36] proposed an adaptive fringe projection framework based on the dichromatic reflection model, which precisely identifies specular regions by decomposing specular and diffuse components. The method suppresses specular reflections by optimizing projection intensity using a power function and restores missing information through pixel inpainting techniques. Feng et al. [37] presented a specular highlight removal method for light field images that combines the dichromatic reflection model (DRM) with exemplar-based patch filling. Specular pixels are first classified using a Gaussian Mixture Model clustering and depth-based segmentation. Non-saturated highlights are removed using a DRM confidence strategy, while saturated highlights are addressed using an exemplar patch matching algorithm that integrates gradient and color difference constraints.

However, limits are exhibited when using the mentioned approaches to measure mechanical components with unknown geometries and highly reflective surfaces. While incorporating a polarization device can effectively suppress high-reflection surfaces, it also reduces the signal-to-noise ratio (SNR) for low-reflection surfaces. Also, precise optical equipment and high environmental requirements are often necessary, which increases hardware costs and makes it difficult to apply to mechanical industrial inspections in complex environments. Multiple exposure techniques can enhance the SNR by integrating images captured at varying exposure times, but they traditionally rely on the empirical knowledge of experimentalists, which may limit the generalizability and reduce reconstruction accuracy. Phase compensation methods are not suitable for high-precision measurements, as they involve complex iterative and growing processes for edge image restoration. Deep learning-based methods offer higher efficiency but require the establishment of extensive datasets for training. Improved encoding methods achieve high reconstruction accuracy, but they necessitate more complex phase unwrapping techniques. Color invariance methods can produce good reconstruction results, but they are influenced by the object's surface color and texture. Further, adjusting the projection intensity can enhance the SNR, but it is subject to the influence of pixel mapping errors and grayscale range limitations between the camera and the projector.

Therefore, adjusting a single imaging parameter cannot effectively adapt to the actual part geometry and surface reflective properties. Existing methods lack effective guidance on the design of these important imaging parameters.

To address this problem, an adaptive high-precision 3D reconstruction of highly reflective mechanical parts based on optimization of exposure time and projection intensity is proposed in this paper. The projection intensity is optimized based on the linear response range of the hardware to generate an image sequence, which is further fused through an optimization process based on Genetic Algorithm and stochastic Adam optimizer. Region segmentation is conducted to adapt the exposure time to surface reflective properties based on histogram analysis.

This paper is organized as follows: Section 2 briefs the basic theory of PMP-based 3D reconstruction. The proposed image fusion based on projection intensity optimization is displayed in Section 3. Histogram-based analysis of the optimal exposure is demonstrated in Section 4. In Section 5, the advantage of the proposed method is shown by three sets of examples. Finally, conclusions and future research are given in Section 6.

2. Basic Theory of PMP-Based 3D Reconstruction

The configuration of PMP-based 3D measurement system implemented a monocular structured light measurement architecture, as shown in Figure 1.

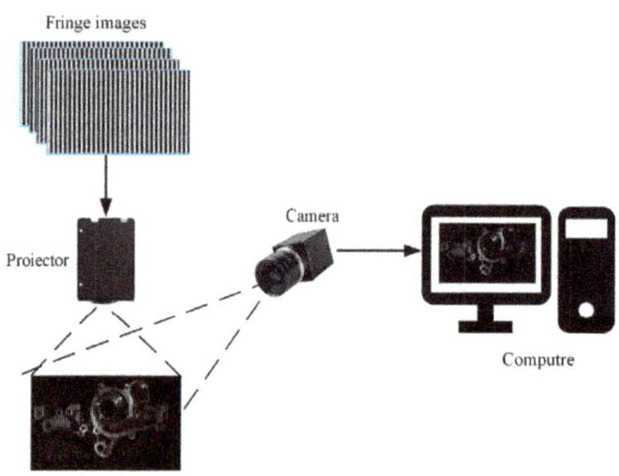

Figure 1. The monocular structured light measurement based on PMP.

Ideal sinusoidal fringe patterns generated by a computer are projected onto the measured object's surface through a Digital Light Processing (DLP) projector. Then the deformed fringe patterns modulated by the object's surface geometry are synchronously captured by a CCD camera.

The computer-generated sinusoidal fringe patterns could be expressed as:

$$I_i(x,y) = I_a(x,y) + I_b(x,y)cos[\varphi(x,y) + \delta_i], \tag{1}$$

where $I_a(x,y)$ represents the average light intensity, $I_b(x,y)$ is the modulated light intensity, $\varphi(x,y)$ is the wrapped phase to be solved, δ_i is the phase shift amount of the i-th image, $i = 1,2,3,\ldots,N$, $\delta_i = 2\pi i/N$, and N denotes the total number of phase-shifting steps. The wrapped phase $\varphi(x,y)$ could be determined as follows:

$$\varphi(x,y) = -arctan\left[\frac{\sum_{i=0}^{N-1} I_i(x,y)sin(\delta_i)}{\sum_{i=0}^{N-1} I_i(x,y)cos(\delta_i)}\right], \tag{2}$$

After obtaining the wrapped phase, the absolute phase containing height information could be retrieved through the heterodyne multi-frequency phase-shifting approach [38]. System calibration is then conducted following Song et al.'s method [39], in which the DLP projector is treated as an inverse camera. The calibrated parameters contribute to geometric reconstruction based on triangulation principles.

3. Image Fusion Based on Projection Intensity Optimization

To accurately recover information in saturated regions, existing studies typically employ a fixed intensity attenuation step size to ensure that the saturated regions are no longer saturated. However, this method is limited by the adaptability and often requires manual intervention when dealing with various surfaces with diverse geometric and material properties.

Therefore, an adaptive method for image fusion based on projection intensity optimization is proposed in this section. The gist of this method is to effectively optimize the projection intensity to ensure that the acquired images in each segmented region remain within the mid-to-high linear response grayscale range, which is typically determined by the camera hardware. As the projection intensity decreases, the upper limit of the best range R_u is utilized to measure the change in the number of pixels in the linear response range: binarize both the initial image $I_0^c\left(I_0^p\right)$ and the projected images $\left\{I_i^c\left(I_i^p\right)\middle| i = 1,2,\ldots,N\right\}$ captured by the camera, in order to extract the difference regions D_i^c with varying grayscale values:

$$\begin{cases} U_i = \sum_{j=1}^{m}\left(D_0^{c(j)} - D_i^{c(j)}\right) \\ D_i^{c(j)} = \begin{cases} 1, & I_i^{c(j)}\left(I_i^p\right) \geq R_u \\ 0, & otherwise \end{cases} \end{cases}, i = 0,1,\ldots,N, \tag{3}$$

where I_0^p represents the initial projection intensity, I_i^p is the projection intensity at iteration i, D_i^c is the difference mask when the projection intensity varies. From Equation (3), it is evident that at iteration i, the difference projection intensity I_i^p are projected onto the workpiece to compute the difference region D_i^c. Since different segmented regions possess distinct optimal projection intensities, they are processed separately and parallelly in subsequent calculations.

To accurately recover information in saturated regions, the projection intensity is typically reduced gradually from 255. When the grayscale range of the difference region aligns with the camera's linear response and high SNR capture interval $[R_l, R_u]$, the projection intensity is considered to achieve its optimal, which should be recognized and encouraged quantitatively in the optimization. Therefore, an objective function is constructed to measure the deviation between the obtained grayscale interval and the optimal range:

$$f_D^{(j)}\left(I_i^p\right) = min\left(\max\left(D_i^{c(j)} \odot U_i\right) - Int_u^c, Int_l^c - \min\left(D_i^{c(j)} \odot U_i\right)\right), \quad (4)$$

where $\odot$ indicates the convolution operator. When the obtained dynamic range of the difference image exceeds the upper limit of the optimal range or falls below the lower limit, the objective function $f_D\left(I_i^p\right)$ returns a large positive value. On the contrary, the objective function would show a negative value if the dynamic range D_i^c falls within the optimal range. Therefore, minimizing this objective function tends to keep the captured image in conformation to the hardware's optimal range.

Then, a simulated annealing algorithm is implemented in this paper to optimize the objective function and achieve the optimal projection intensity. The main steps are as follows:

(1) Calibrate the hardware to obtain the optimal range $[R_l, R_u]$. Set the initial projection intensity I_0^p to the maximum value of 255. Acquire the baseline image $I_0^{c(j)}$ of the j-th cluster with the camera, and perform binarization to get $D_0^{c(j)}$. Initialize $i = 0$.
(2) Let $i = i + 1$. Calculate the i-th projection intensity using an attenuation step S, and conduct a physical experiment to observe the captured image of the j-th cluster $I_i^{c(j)}\left(I_0^p - iS\right)$. Obtain the corresponding difference image $D_i^{c(j)}$ and the value of objective function $f_D^{(j)}\left(I_i^p\right)$ in each cluster based on Equations (3) and (4).
(3) Randomly generate a new projection intensity $\hat{I}_i^p$ in the neighborhood of variable I_i^p. Observe the captured image $f_D\left(\hat{I}_i^p\right)$, and evaluate the objective function in each cluster $f_D^{(j)}\left(\hat{I}_i^p\right)$ to compare with the current one. If the new value is superior, accept the new projection intensity. Otherwise, accept it with a probability P_i, in which T_0 denotes the initial temperature and α is the cooling speed.

$$P_i = exp\left(-\frac{f_D^{(j)}\left(\hat{I}_i^p\right) - f_D^{(j)}\left(I_i^p\right)}{T_0\alpha^t}\right), \quad (5)$$

(4) If the temperature has dropped below a threshold, or the optimal solution has not been updated in multiple consecutive iterations, the iteration process is terminated. Go to step (5). If the iteration has not been terminated and $f_D^{(j)}\left(I_i^p\right) < 0$, store the newly acquired image $f_D^{(j)}\left(I_i^p\right)$, go back to step (2) and substitute the baseline image I_0^c with $f_D^{(j)}\left(I_i^p\right)$.
(5) Consider current I_i^p as the optimal projection intensity. Output all the stored images as an image sequence $\{I_i^s, i = 1, 2, \ldots, L\}$. Terminate the algorithm when the optimization processes of all regions have converged.

Then, an image fusion method is proposed to recover information in saturated regions based on the complementarity of multi-source image sequence. Image information entropy is utilized to quantitatively evaluate the informational richness of fused images, serving as a metric to assess the complexity of grayscale distribution statistics, as shown in Equation (6).

$$H(I_i^s) = -\sum_{j=1}^{n_i} p_{i,j} log(p_{i,j}), i = 1, 2, \ldots, L, \quad (6)$$

where $p_{i,j}$ is the probability of grayscale j in the i-th fused image, and n_i is the number of grayscale levels. Information entropy $H(I_i^s)$ quantifies the disorder and variability in pixel intensity distributions. Oversaturated areas and dark areas show low information entropy because pixels' grayscales are clustered in high-intensity or low-intensity ranges, while higher entropy reflects greater complexity in grayscale distributions and enhanced preservation of fine image details.

The core challenge of image fusion lies in optimizing the retained region from each image in the sequence to ensure that the fused image achieves maximal information entropy. Assume the image sequence is arranged in ascending order of its mean grayscale values. A threshold T_i^f is applied to the segmentation of the i-th grayscale image, in which higher-intensity regions are consistently prioritized for retention to enhance the SNR of the fused output. The segmented region marked by D_i^s is retained in the final fused image. The newly acquired region marked by D_i^s at iteration i is consistently preserved during subsequent image fusion operations.

$$D_i^s(x,y) = \begin{cases} 1, I_i^s(x,y) \geq T_i^f \ and \ D_{i-1}^s(x,y) = 0 \\ 0, otherwise \end{cases}, \tag{7}$$

Therefore, the image fusion process from the obtained image sequence $\{I_i^s, i = 1,2,\ldots,L\}$, can be formulated in:

$$I_L^f = \left(D_1^s \odot I_1^s\right) \cup \left(D_2^s \odot I_2^s\right) \cup \ldots \cup \left(D_L^s \odot I_L^s\right), \tag{8}$$

Then, the L-dimensional segmentation thresholds $\left\{T_i^f, i = 1,2,\ldots,L\right\}$ are obtained through a recursive process of Genetic Algorithm (GA)-based [40] global optimization and local fine-tuning based on stochastic Adaptive Moment Estimation (S-Adam). GA explores optimal solutions by simulating evolutionary processes. GA approaches are able to escape local optima through their stochastic operations, while requiring substantial computational resources and relatively slow convergence rates, especially for multi-dimensional problems. Adam [41] shows superior converging performance in high-dimensional parameter spaces, while remaining susceptible to local optima entrapment and sensitive to the initial value. To overcome these limitations, a hybrid optimization framework is proposed in this paper in which the elite solution from each GA generation serves as the initial parameter set for Adam-based local refinement. In addition, the mini-batch gradient method is used to achieve a balance between computational efficiency and convergence stability.

The chromosome representing segmentation thresholds $\left\{T_i^f, i = 1,2,\ldots,L\right\}$ is encoded using a binary scheme, where each individual consists of L chromosomes. The algorithm begins by generating an initial population, and the fitness function is constructed based on image information entropy $H(I_L^s)$. The top K individuals with the highest fitness values are fine-tuned through the Adam optimizer and subsequently reintroduced into the population. The remaining individuals undergo selection using the roulette wheel method. As the population iteratively evolves, new individuals are generated through crossover and mutation operations. The crossover operation involves exchanging chromosomes between two distinct parent individuals, while the mutation operation randomly alters selected genes.

To mitigate premature convergence to local optima, the variance of population fitness is employed as a convergence metric. The effectiveness of crossover and mutation operations diminishes as evolutionary iterations progress and individual fitness values gradually converge. To address this challenge, an adaptive adjustment mechanism is proposed for modifying the crossover probability p_i^c and mutation probability p_i^m.

$$p_i^c = \begin{cases} p_{i-1}^c - \frac{p_{i-1}^c}{std_i - std_{i-1}}, std_i > std_{i-1} \\ p_{i-1}^c + \frac{p_{i-1}^c}{std_{i-1} - std_i}, std_i < std_{i-1} \end{cases}, i \geq 2, \tag{9}$$

$$p_i^m = \begin{cases} p_{i-1}^m - \frac{p_{i-1}^m}{std_i - std_{i-1}}, std_i > std_{i-1} \\ p_{i-1}^m + \frac{p_{i-1}^m}{std_{i-1} - std_i}, std_i < std_{i-1} \end{cases}, i \geq 2, \tag{10}$$

where std_i is the variance of population fitness at iteration i. Probabilities p_i^c and p_i^m are dynamically tuned according to the convergence state to maintain the exploratory capability.

Adam is then implemented to fine tune the top individuals. An independent learning rate is employed for each parameter, and is adjusted based on the first and second moment estimation of the gradient:

$$\begin{cases} m_t = \beta_1 m_{t-1} + (1 - \beta_1)\hat{g}_t \\ v_t = \beta_2 v_{t-1} + (1 - \beta_2)\hat{g}_t^2 \\ \theta_{t+1} = \theta_t - \dfrac{\eta_1 m_t}{\left(1 - \beta_1^t\right)\sqrt{t\left(\frac{v_t}{1-\beta_2^t} + \varepsilon\right)}} \end{cases}, \tag{11}$$

where m_t represents the first moment of gradient g_t at iteration t, v_t is the second moment, β_1 and β_2 are decaying speed hyperparameters satisfying $\beta_1, \beta_2 \in [0, 1]$, $\beta_1 < \beta_2$, $\theta_t = \left(T_1^f, T_2^f, \ldots, T_L^f\right)$ is the vector of parameters, $\eta_t = \eta_1 / \sqrt{t}$ is a decaying learning rate, ε is an artificial small value to avoid vanishing gradient problem. The gradient g_t here is estimated by the difference of adjacent T_i^f in dimension i. In order to reduce the amount of gradient calculation in each iteration and increase the stability of convergence, the mini-batch method is implemented:

$$\hat{g}_t = \frac{1}{B}\sum_{i=1}^{B} \frac{H(\theta_{t,i} + \Delta I) - H(\theta_{t,i})}{\Delta I}, \tag{12}$$

where B is the batch size, $B < L$, ΔI is a small increase.

After the algorithm converges, the optimal segmentation thresholds are substituted into Equations (7) and (8) to generate the fused image for subsequent calculation of the optimal exposure time.

4. Histogram-Based Analysis of the Optimal Exposure

For mechanical parts with high reflectivity, the exposure time significantly influences fringe pattern quality and consequently impacts 3D reconstruction accuracy when using structured light measurement systems. As shown in Figure 2, the optimal exposure time should be determined adaptively by the surface reflectivity of the measured part and the ambient illumination, which is considered constant in this paper. Basically, shorter exposure times should be employed to prevent overexposure in high reflective regions, while surfaces with lower reflectivity require extended exposure durations to enhance the SNR of captured surfaces. Although this principle is widely recognized, the determination of specific exposure time still predominantly relies on empirical judgment. Such manual approaches lack adaptability under dynamic scenarios of the mechanical parts and ambient lighting, which results in a compromised reconstruction accuracy. Therefore, a histogram-based analysis method is introduced to obtain region-specific optimal exposure times.

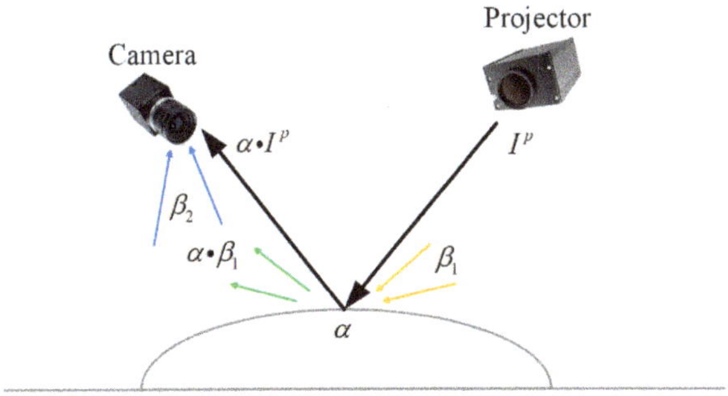

Figure 2. Illustration for PMP-based 3D reconstruction of mechanical parts.

When projecting uniform and high-intensity white light, the intensity distribution of the images captured by the camera can be expressed as:

$$I^c = st\alpha I^P + \alpha\beta_1 + \beta_2, \tag{13}$$

where s and t represent the sensitivity and exposure of the camera, respectively, α is derived from geometric and material properties, I^P is the intensity of projected light, a term $\alpha\beta_1$ denotes ambient light reflected by the measured object, and β_2 is ambient light directly entering the camera. Assume the relationship between image intensity and scene radiance is linear, which means the camera sensitivity s is considered constant. Consider a segmented pixel cluster marked by j in m clusters, Equation (13) can be discretized into a partition expression:

$$I^{c(j)} = st^{(j)}\alpha^{(j)}I^{p(j)} + \alpha^{(j)}\beta_1^{(j)} + \beta_2^{(j)}, j = 1,2,\ldots,m, \tag{14}$$

A radiometric compensation strategy is employed to process high reflective surfaces, in which the projection intensity $I^{P(k)}$ is systematically varied while maintaining the initial exposure time t. A sequence of intensity images I^c_{seq} is obtained by the camera, where pixel saturation in some overexposed regions is progressively alleviated across different intensity levels $\left\{I_1^p, I_2^p, \ldots, I_n^p\right\}$. These multi-intensity segmented pixel regions would be subsequently fused into a composite image using the method detailed in Section 3. Assume n distinct projection intensities are applied to generate a sequence of images $\{I_1^c, I_2^c, \ldots, I_n^c\}$, ensuring a comprehensive coverage of both specular and diffuse reflection characteristics:

$$I_i^c = st\alpha I_i^p + \alpha\beta_1 + \beta_2, i = 1,2,\ldots,n, \tag{15}$$

Therefore, the image fusion process based on spatially and temporally discrete sampling can be formulated based on an integration of Equations (14) and (15):

$$I^f = \sum_{i=1}^{n}\sum_{j=1}^{m}\lambda_{ij}\left(st^{(j)}\alpha^{(j)}I_i^{p(j)} + \alpha^{(j)}\beta_1^{(j)} + \beta_2^{(j)}\right), \tag{16}$$

where λ_{ij} represents the region-adaptive binary weighting function to be obtained in this section, which dynamically prioritizes unsaturated pixel values across temporal and spatial domains. Consider a specific pixel cluster j, if the critical oversaturation intensity is defined as I^{os}, the oversaturation threshold exposure time in region j is given by:

$$t_{os}^{(j)} = \frac{I^{os} - \alpha^{(j)}\beta_1^{(j)} - \beta_2^{(j)}}{s\alpha^{(j)}\sum_{i=1}^{n}\lambda_{ij}\left(I_i^{p(j)}\right)}, \tag{17}$$

To ensure high SNR in the sampled area and avoid saturation of the camera sensor, the critical oversaturation intensity is usually defined as a constant value [42]. Then, the obtained pixel cluster in region j is:

$$I^{c(j)} = \alpha^{(j)} \left(\beta_1^{(j)} - \frac{\beta_1^{(j)} t^{(j)}}{t_{os}^{(j)}} \right) + \frac{I^{os} - \beta_2^{(j)}}{t_{os}^{(j)}} t^{(j)} + \beta_2^{(j)}, \quad (18)$$

Parameters $\beta_1^{(j)}$ and $\beta_2^{(j)}$ for a given region, which are related to ambient illumination, pixel positions, materials, roughness, and other intrinsic properties, remain constant during a continuous measurement session. $t_{os}^{(j)}$ is an unknown constant to be determined. Therefore, as derived from Equation (18), the intensity of a specific region in the acquired image is linearly related to the reflectivity, which further indicates that variations in regional intensity directly reflect changes in surface reflectivity. As shown in Figure 3, pixels with distinct intensity levels can be clustered based on histogram analysis of the foreground's image from images without highly saturated regions, resulting in an identification and segmentation of regions with statistically homogeneous intra-class reflectivity.

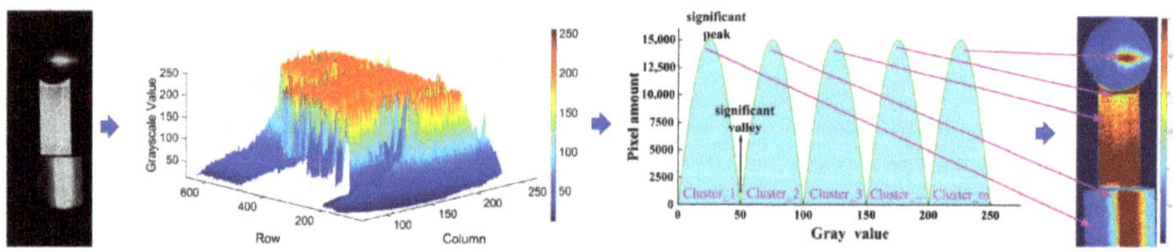

(a)Foreground's image (b)Foreground's image intensity distribution (c)Histogram-based segmentation (d)Segmentation effect

Figure 3. Illustration of histogram-based segmentation.

The method for obtaining the foreground image is derived from the literature [43], and is described in detail as follows. For a grayscale image I, it is composed of a foreground image F, a background image B, and the foreground opacity α_i at each pixel:

$$I_i = \alpha_i F_i + (1 - \alpha_i) B_i, \quad (19)$$

Assuming that the foreground F and background B are approximately constant within a small window around each pixel, this assumption allows the opacity α to be expressed as a linear function of the image I:

$$\alpha_i \approx a I_i + b, \ \forall i \in \omega, \quad (20)$$

where $a = 1/(F - B)$, $b = -B/(F - B)$, and ω is a small image window.

To solve for α, a and b, a cost function is constructed as follows:

$$J(\alpha, a, b) = \sum_{j \in I} \left(\sum_{i \in \omega_j} (\alpha_i - \alpha_j I_i - b_i)^2 + \epsilon a_j^2 \right), \quad (21)$$

where ω_j is a small window around pixel j.

By minimizing $J(\alpha, a, b)$, a quadratic cost function solely with respect to α is obtained:

$$J(\alpha) = \alpha^T L \alpha, \quad (22)$$

where L is the

$$L_{i,j} = \sum_{k|(i,j)\in\omega_k}\left(\delta_{ij} - \frac{1}{|\omega_k|}\left(1 + \frac{(I_i - \mu_k)(I_j - \mu_k)}{\frac{\epsilon}{|\omega_k|} + \sigma_k^2}\right)\right), \tag{23}$$

where μ_k and σ_k^2 are the mean and variance of the intensities in the window ω_k around k, and $|\omega_k|$ is the number of pixels in this window.

Constraints are added, and the resulting sparse linear system is then solved:

$$(L + \lambda D_S)\alpha = \lambda D_S b_S, \tag{24}$$

where D_S is a diagonal, b_S is the constraint vector, and λ is a large constant. The alpha matte α is solved using a sparse linear solver.

Given $I_i = \alpha_i F_i + (1 - \alpha_i)B_i$, where F_i and B_i are unknown, the foreground F and background B can be estimated by minimizing an objective function with smoothness priors:

$$\min_{F,B}\sum_i (\alpha_i F_i + (1 - \alpha_i)B_i - I_i)^2 + \lambda\left(\| \nabla F_i \|^2 + \| \nabla B_i \|^2\right), \tag{25}$$

where ∇F_i and ∇B_i denote the image gradients of the foreground and background.

The clustering-based segmentation is performed using the K-means++ algorithm proposed by Arthur et al. [44], with the following steps:

(1) Randomly and uniformly select one point from the dataset χ as the first initial center c_1.

(2) For each data point $x \in \chi$, compute the distance $D(x)$ to the nearest already selected center:

$$D(x) = \min_{c\in\{c_1,\ldots,c_{i-1}\}} \| x - c \|^2, \tag{26}$$

Select the next center c_i from the dataset with probability proportional to $D(x)^2$:

$$P(x) = \frac{D(x)^2}{\sum_{x\in\chi} D(x)^2}, \tag{27}$$

A new point is then randomly selected as c_i according to this probability distribution.

(3) Assign each data point to the cluster C_i associated with the nearest center c_i:

$$C_i = \{x \in \chi | \| x - c_i \| \leq \| x - c_j \|, \forall j \neq i\}, \tag{28}$$

where C_i denotes the set of points assigned to center c_i.

(4) Recompute the center of each cluster as the mean of all points within the cluster:

$$c_i = \frac{1}{|C_i|}\sum_{x\in C_i} x, \tag{29}$$

(5) Iterate until convergence by repeating steps (3) and (4), either until the set of centers no longer changes or the maximum number of iterations is reached.

(6) To ensure coverage of the exposure time for each class, the data point corresponding to the right boundary of each cluster is selected as the segmentation threshold. This data point should satisfy the following condition:

$$\begin{cases} f(x_{i+1}) - f(x_i) > 0 \\ f(x_i) - f(x_{i-1}) < 0' \end{cases} \tag{30}$$

Through the aforementioned steps, a relatively accurate segmentation threshold can be obtained.

Following the spatial segmentation, the optimal exposure time is determined for each partitioned region. Evidently, regions with distinct reflectivity require different optimal exposure times. To maximize the SNR of captured fringes, the ideal exposure time should correspond to the critical threshold that prevents saturation in all pixels within the region. Since longer exposure durations increase pixel saturation, the goal is to maximize $t_{os}^{(j)}$ while ensuring no saturation occurs:

$$t_{opt}^{(j)} = max\left(t_{os}^{(j)}\right) = max\left(\frac{I^{os} - \alpha^{(i)}\beta_1^{(j)} - \beta_2^{(j)}}{\frac{\delta I^{c(j)}}{\delta t^{(j)}}}\right), \tag{31}$$

Obviously, the optimal exposure time $t_{opt}^{(j)}$ is obtained when the term $\delta I^{c(j)}/\delta t^{(j)}$ reaches its minimum, that is, when the incremental intensity gain per unit exposure time is minimized. Discrete sampling and numerical difference method are implemented to solve this differential term, and obtain the optimal exposure time. At this optimal exposure time $t_{opt}^{(j)}$, acquired image data is effectively clustered in the temporal domain, separating valid intensity responses from saturated artifacts.

The sequence images corresponding to the optimal projection intensity are fused to generate a composite image. By analyzing the histogram of the foreground image, regions with different reflectivities are clustered using the K-means algorithm. The fused fringe patterns, which have been distorted due to modulation by mechanical components under different exposure times, are subsequently used for phase-based 3D reconstruction (PMP). Depth information is then computed to ultimately obtain the point cloud data of the measured workpiece. The main framework of the proposed method in this paper is illustrated in Figure 4.

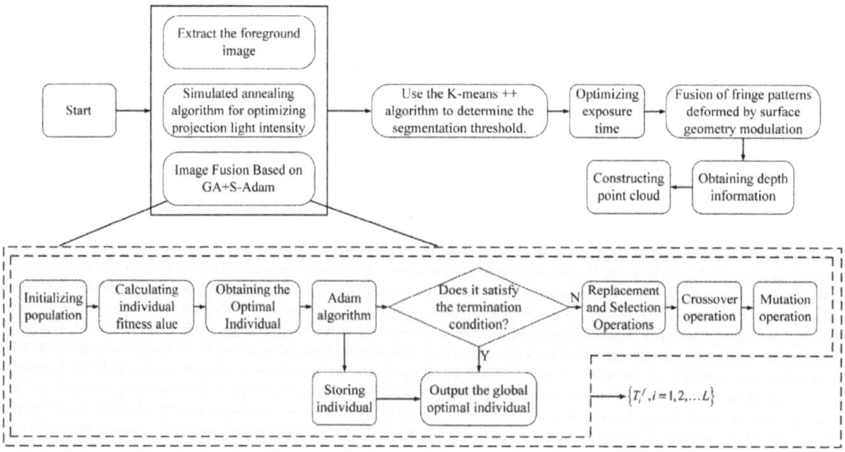

Figure 4. The main framework of the proposed method in this paper.

The steps described in Figure 4 are detailed as follows:

Step 1: As described in Section 3, the optimal projection intensities are obtained using the simulated annealing algorithm. To ensure synchronization, the built-in GPIO interface of the camera is employed to receive external trigger signals from the projector, enabling communication between the camera and the projector.

Step 2: The optimal projection intensity sequence images from Step 1 are fused using the GA+S-Adam algorithm, as detailed in Section 4, to facilitate foreground image acquisition.

Step 3: The fused images from Step 2 are processed using the foreground extraction method proposed in Section 4 to isolate the foreground regions for further analysis.

Step 4: The histogram of the extracted foreground image is clustered using the K-means++ algorithm to determine segmentation thresholds, as illustrated in Section 4.

Step 5: The exposure time is determined based on the method introduced in Section 4, and the camera's exposure settings are subsequently adjusted via computer control.

Step 6: Fringe images, modulated and deformed by mechanical components under varying exposure times, are fused to recover phase information.

Step 7: Finally, the 3D point cloud is reconstructed based on the system calibration parameters and the depth information obtained from decoding the fringe images.

When employing the simulated annealing algorithm to obtain the optimal projection intensity and acquire corresponding images via the camera, the determination of subsequent optimal intensities requires iterative computation. Once two or more images are obtained, image fusion under their respective optimal projection intensities can be performed in parallel. Given the continuous optimization of projection intensity and image fusion, foreground extraction from previously fused images can also be executed concurrently. The extracted foreground images are compared with the final fused image to guide the generation of foreground-specific histograms. These three stages can be processed in a pipelined manner to handle different batches of data, maximizing efficiency and reducing processing time.

5. Case Study

A PMP-based 3D reconstruction system was constructed to validate the effectiveness of the proposed method, whose configuration is shown in Figure 5. This system comprises an MV-CA060-10GC CCD camera with a resolution of 3072 × 2048 pixels, a TJ-23U DLP projector with a resolution of 1280 × 720 pixels, and a high-performance computer. A heterodyne multi-frequency four-step phase-shifting method was employed for phase demodulation. The system software architecture consists of the following main modules: Projection Control Module: Responsible for generating and controlling ideal sinusoidal fringe patterns, which are projected onto the surface of mechanical components through the projector. Image Acquisition Module: Communicates with the camera through custom software to synchronize image acquisition. Data Processing Module: Integrates algorithms for projection intensity optimization, image fusion, foreground extraction, K-means++ clustering, optimal exposure time determination, system calibration, and 3D point cloud construction. Optimization algorithms are encapsulated as independent submodules to enhance maintainability. User Interface Module: Provides a visual interface for operation, supporting parameter configuration, real-time preview, and result output. To facilitate efficient collaboration between the projector, camera, and computing platform, a communication protocol is established. Projector to Host Communication: A USB 3.0 interface is used, with the projector's dedicated SDK enabling real-time transmission and control of the projection patterns. Camera to Host Communication: The camera communicates with the host via the GigE Vision protocol, supporting high-bandwidth image data streaming. To ensure synchronization, the camera's GPIO interface is employed to receive the external trigger signal from the projector, establishing communication between the camera and projector.

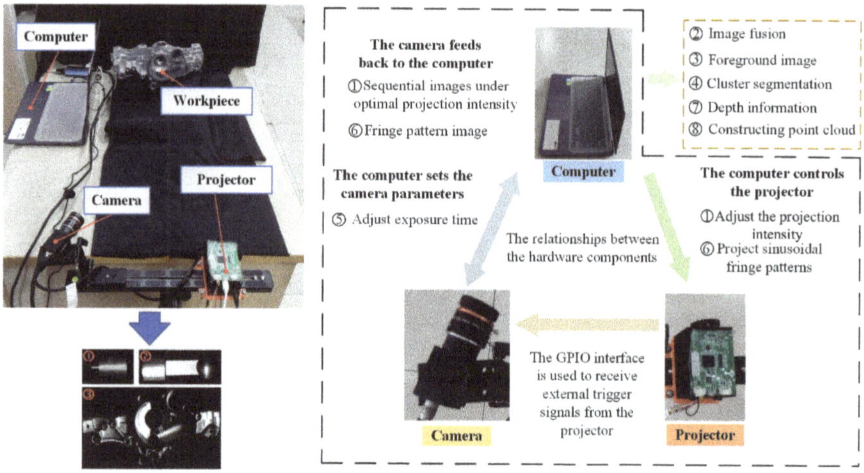

Figure 5. Configuration of the 3D reconstruction system.

Three sets of experiments were conducted. The first group employed a geometrically uniform workpiece with highly reflective surface properties. The second group utilized a component featuring common mechanical geometries, including planar, cylindrical, and spherical surfaces, to validate the robustness of the proposed method across diverse reflective surface topologies. A die-cast automotive component was employed in the third group, in which there were multiple precision-manufactured end surfaces with different depths. Details and results of the three sets of experiments are shown as follows.

(1) Planar metal workpiece

A planar metal workpiece, which is simple in geometry, was selected as the first example. The simulated annealing algorithm was employed to obtain the optimal projection intensities. The initial temperature was 100, the cooling coefficient was 0.95, and the termination temperature was 0.01. After convergence, the optimal projection intensities were 255, 214, 177, 152, 130, 113, 98, and 76, respectively. Image fusion was subsequently conducted based on the image sequence under the optimal projection intensity. The segmentation threshold was determined using a GA-ADAM algorithm, in which the population size was set to 20, the individual length to 8, and the initial crossover and mutation probabilities were set to 0.6 and 0.8, respectively. The hyperparameters for the ADAM algorithm were configured as $\alpha = 1$, $\beta_1 = 0.9$, $\beta_2 = 0.999$. After convergence, a fused image was constructed based on the obtained segmentation threshold, as shown in Figure 6a. Based on the histogram-based analysis method, the pixels were clustered to three segmented regions as shown in Figure 6b, whose optimal exposure time were determined to be 35,264 μs, 5605 μs, and 3664 μs.

For comparison, two sets of conventional methods with fixed attenuation steps were employed. As the most frequently used step in the literature, the projection intensity attenuation step was set to 20 at first with a total of 10 fusion images, resulting in an optimal exposure time of 35,264 μs, 5129 μs and 3664 μs. In the second set, the attenuation step was set to 30 to ensure the same number of fused images with the proposed method, whose optimal exposure time were 35,264 μs, 5391 μs, and 3664 μs. Similarly, fused images were constructed after optimizing the segmentation thresholds using the GA-ADAM algorithm, as shown in Figures 6c and 6d, respectively.

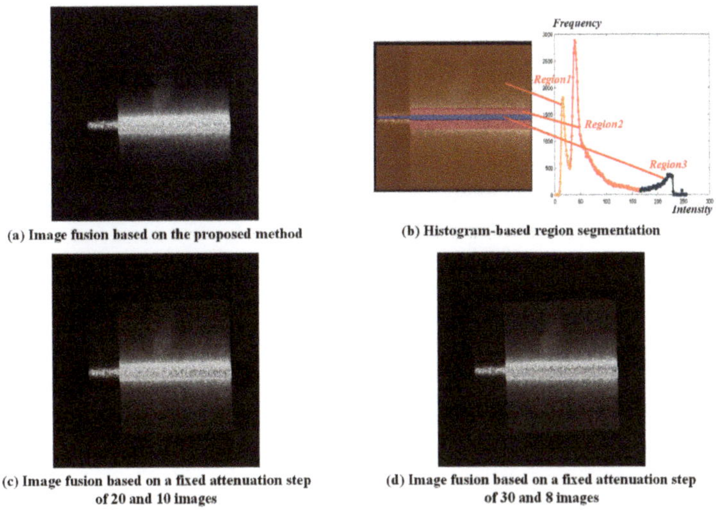

Figure 6. Image fusion based on the proposed methods and two conventional methods.

Three-dimensional reconstruction was then carried out using the conventional PMP method, two sets of fixed attenuation step methods, and the proposed method. A single exposure with the maximum exposure time that just avoids image saturation was employed when using the conventional PMP method, while the obtained optimal exposure time were implemented in the other three scenarios. Three sets of sinusoidal fringes were projected onto the metal part. Each fringe comprised four patterns with different phase shifts, and the fringe periods were set to 90, 99, and 100, respectively. After capturing the projected fringes, the phases were unwrapped using a heterodyne multi-frequency phase-shifting method. The corresponding spatial data points were then localized based on the triangulation principle [45]. The workpiece was thus reconstructed by acquiring images at various exposure times and performing phase fusion. The results of 3D reconstruction based on the conventional PMP method, fixed attenuation steps of 20 and 30, and the proposed method are shown in Figure 7.

Reconstruction accuracy for each spatial point was calculated and statistically analyzed based on least squares planar fitting. A reference plane is fitted to the point cloud using the least squares method, resulting in a plane equation of the form $ax + by + cz = 0$. For a given point $P(x_p, y_p, z_p)$, the distance from the point to the plane is calculated as: $d = |ax_p + by_p + cz_p + d| / \sqrt{a^2 + b^2 + c^2}$. The conventional PMP method which employs a single exposure for the entire region successfully avoided saturation but struggled to enhance the SNR in dark areas, which resulted in a 68.6% more in the number of outliers, 44.8% higher in the whisker range of reconstruction error, and 42.0% higher in the median errors comparing to the proposed method. In contrast, three methods based on image fusion produced reconstruction results with almost no geometric defects and achieved significantly lower reconstruction errors. The whisker range of the reconstruction error boxplot for the proposed method is 29.8% and 37.9% lower than that of the 20-step and 30-step methods, respectively, while the median errors are reduced by 33.7% and 42.3%, respectively. The whisker range is computed as: $max(x_i \mid x_i \leq Q_3 + 1.5 \times IQR) - min(x_i \mid x_i \geq Q_1 - 1.5 \times IQR)$, where Q_1 and Q_3 are the 25th and 75th percentiles, and the interquartile range is defined as $IQR = Q_3 - Q_1$. These results demonstrate the effectiveness of the proposed method and its higher reconstruction accuracy compared with commonly used existing methods.

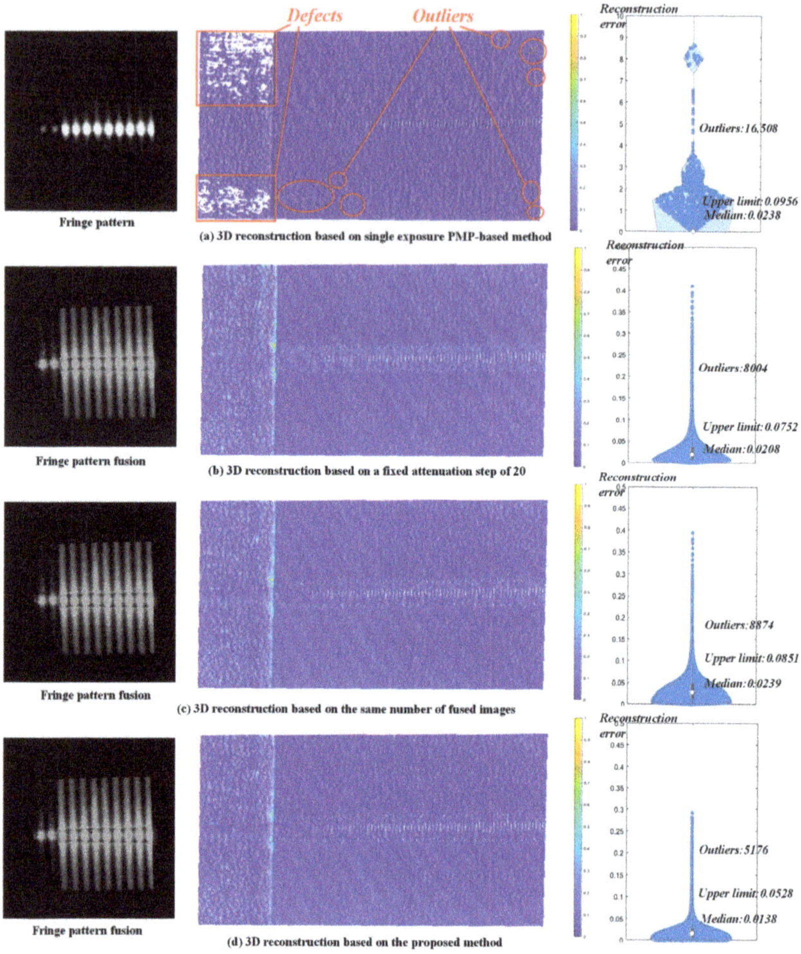

Figure 7. Comparison of 3D reconstruction results in case study 1.

(2) Multi-geometry metal workpiece

Further, a metal component with a slightly more complex geometry, featuring typical mechanical characteristics such as planar surfaces, through-holes, outer cylindrical surfaces, and spherical surfaces, was selected as the second example. Image fusion was conducted based on nine images, which were under the obtained optimal projection intensity: 255, 210, 172, 147, 128, 110, 89, 73, and 58. Three clusters of pixels were segmented based on the proposed method, and the corresponding optimal exposure time were 9159 μs, 18,398 μs, and 32,058 μs. In the conventional PMP-based 3D reconstruction, a single exposure with the maximum exposure time was used to increase the SNR. When using the methods based on fixed attenuation steps, a step size of 25 was employed to ensure that the fixed attenuation step method and the proposed method could use the same number of fused images. The conventional step size of 20 was also used for the comparison experiment. The obtained optimal exposure time were 9159 μs, 14,150 μs, 32,058 μs; 9159 μs, 15,444 μs, 32,058 μs, respectively for steps of 20 and 25. The results of 3D reconstruction are shown in Figure 8.

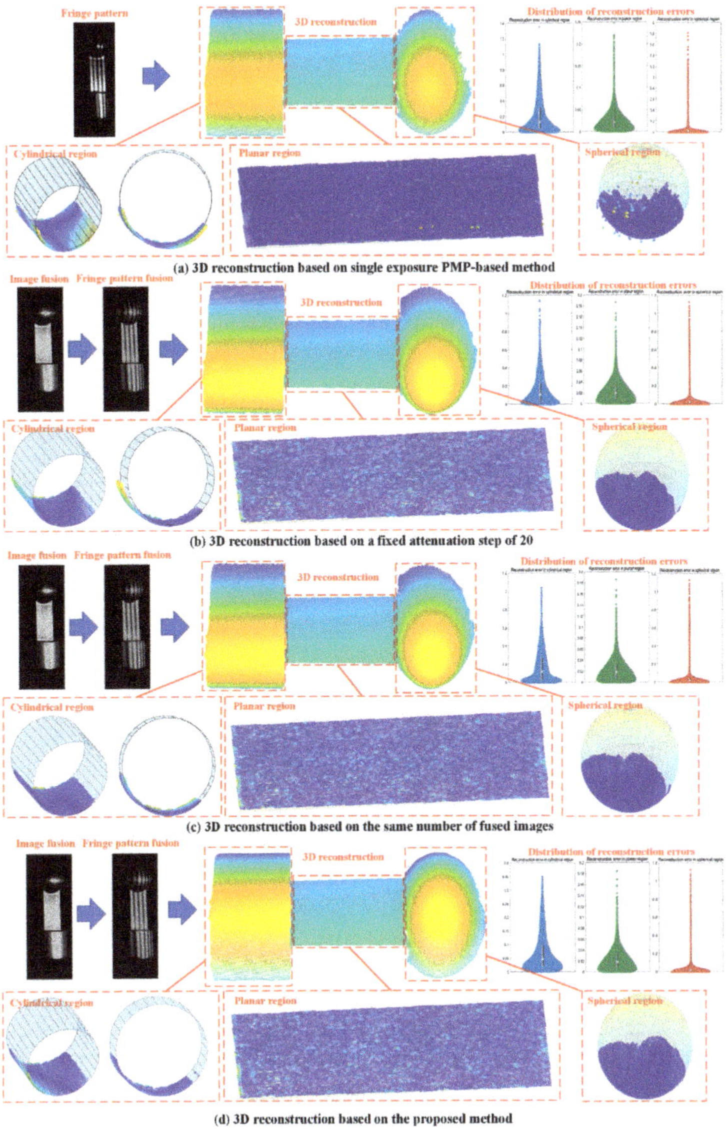

Figure 8. Comparison of 3D reconstruction results in case study 2.

Reconstruction errors were quantified by segmenting local point clouds and performing least-squares fitting to cylindrical, planar, and spherical geometries. When fitting a cylindrical surface, the axis parameters include the direction vector $\vec{D} = (d_x, d_y, d_z)$, a point on the axis $P_0(x_0, y_0, z_0)$, and the radius R. Given a point cloud $\{P_i = (x_i, y_i, z_i) \mid i = 1, 2, \ldots, N\}$, the parameters are estimated by minimizing the objective function: $J = \sum_{i=1}^{N}\left(\sqrt{(x_i - x_0 - t_i d_x)^2 + (y_i - y_0 - t_i d_y)^2 + (z_i - z_0 - t_i d_z)^2} - R\right)^2$, the parameters $x_0, y_0, z_0, d_x, d_y, d_z, R$ are obtained, where $t_i = (x_i - x_0)d_x + (y_i - y_0)d_y + (z_i - z_0)d_z$. For spherical surface fitting, the least squares method is used to estimate the sphere center $C = (x_c, y_c, z_c)$, and radius R. Given a point cloud $\{P_i = (x_i, y_i, z_i) \mid i = 1, 2, \ldots, N\}$, the objective function is defined as: $J = \sum_{i=1}^{N}\left(\sqrt{(x_i - x_c)^2 + (y_i - y_c)^2 + (z_i - z_c)^2} - R\right)^2$, the parameters x_c, y_c, z_c, R are obtained. Comparison of statistics of reconstruction errors in three regions is listed in Table 1. Compared with conventional PMP-based methods, the proposed method reduced the whisker range by 65.2%, 50.9%, and 47.4% for cylindrical, planar, and spherical regions, respectively, while reducing the median error by 63.6%, 13.0%, and 24.7%. When compared with the image fusion method with a fixed attenuation step of 20 which is commonly employed in existing studies, the proposed method showed a reduction in error ranges by 52.7%, 4.3%, and 0.3%, and a reduction in median errors by 33.9%, 8.7%, and 7.3% respectively across the three regions. In comparison with the image fusion method with a fixed attenuation step of 25 to ensure the same number of fused images, the error range is reduced by 57.4%, 1.2%, and 0.3%, and the median error is reduced by 50.6%, 5.4%, and 8.5%. These results validate the superior reconstruction accuracy of the proposed methodology.

Table 1. Comparison of statistics of reconstruction errors in three regions between the single-exposure PMP-based method, image fusion based on a commonly used fixed attenuation step, and the same number of fused images.

Region	Statistical Feature	Single-Exposure PMP-Based Method	Image Fusion with a Fixed Attenuation Step of 20	Image Fusion with the Same Number of Fused Images	The Proposed Method
Cylindrical region	Spread range	0.7387	0.5443	0.6032	0.2572
	Median error	0.1788	0.09855	0.1320	0.06517
Planar region	Spread range	0.1330	0.06818	0.06605	0.06525
	Median error	0.02010	0.01916	0.01848	0.01749
Spherical region	Spread range	0.1551	0.08174	0.08173	0.08152
	Median error	0.02673	0.02170	0.02199	0.02012

A reference point cloud was obtained using a coordinate measuring machine (CMM) to mitigate the influence of machining-induced surface uncertainties in the fitting-based estimation of reconstruction error. Then, the reconstruction errors were benchmarked against the CMM-measured form errors. Because the cylindrical surface was clamped by a three-jaw chuck on the workbench, programmed CMM scanning was only feasible for the upper spherical and planar regions to obtain the reference point cloud. The measurement setup and results are illustrated in Figure 9a. Owing to the finite radius of the CMM's ruby-tipped probe, narrow crevices at the spherical-planar intersections remained unmeasurable, resulting in some holes in the corresponding area. After processing the reference point cloud, the range and median of form error for the planar region are 0.03087 and 0.01265, while the range and median of form error for the spherical region are 0.05183 and 0.01437, respectively. In consideration of the superior accuracy of CMM over that of vision-based measurement methods, these data are considered as ground truth. As shown in Figure 9b, the proposed method achieved an average reduction of 25.5% in the whisker range and 25.9% in the median of reconstruction error for planar regions compared to three existing methods. For spherical regions, the average reductions reached 24.2% in error range and 33.1% in median error.

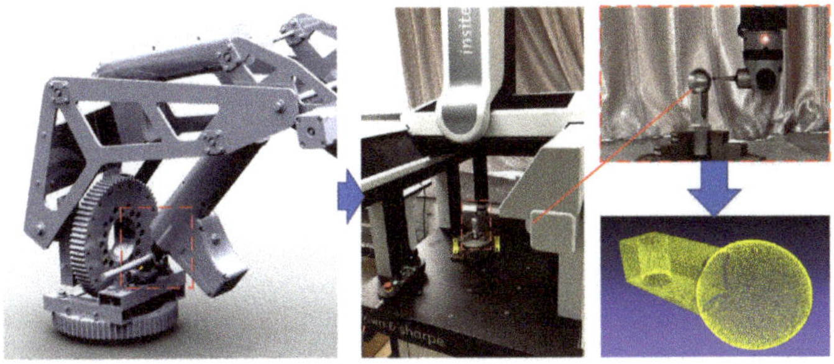

(a) The reference point cloud based on coordinate measuring machine

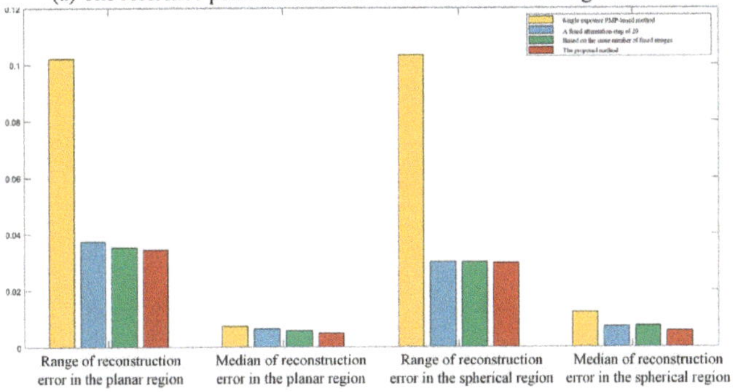

(b) Comparison of reconstruction errors with the reference point cloud

Figure 9. The reference point cloud based on coordinate measuring machine.

(3) High-precision aluminum die-casting component

A precision aluminum die-casting workpiece for new energy vehicles was selected as the third validation case. The die-cast component contains multiple machined hole end faces with different depths, which were machined by precision milling and grinding processes. As these surfaces have passed quality inspection using CMMs with an accuracy of 0.1 μm, they exhibit high-dimensional accuracy, indicating their potential in serving as benchmarks for evaluating the accuracy between vision-based non-contact 3D reconstruction methods.

Using the proposed method, image fusion was conducted based on twelve images, which were under the obtained optimal projection intensity: 255, 216, 184, 155, 131, 110, 92, 76, 63, 51, 45, and 26. The optimal exposure time were 12,823 μs, 17,222 μs, and 31,851 μs. Similarly, the exposure time were maximized in the conventional PMP-based method to increase SNR and reconstruction accuracy. The conventional step size of 20 was used for comparison, in which the obtained optimal exposure time were 12,823 μs, 16,187 μs, and 31,851 μs. The results of 3D reconstruction are presented in Figure 10.

Six precision-machined hole end faces with varying depths were segmented using a random-seed region growing algorithm, and reconstruction errors in six ROIs were estimated through least-squares plane fitting. Compared to the conventional PMP-based method, the proposed method achieved average reductions of 63.6%, 64.1%, 65.6%, and 47.6% in the 25th percentile, median, 75th percentile, and whisker range of reconstruction errors, respectively. The 25th and 75th percentiles are defined as: given a sorted dataset $x_1, x_2, \ldots x_n$, the position of the percentile is calculated by: $i = 1 + (n-1) \times p$, where p is the decimal form of the desired percentile (e.g., $p = 0.25$ for the 25th percentile and $p = 0.75$ for the 75th percentile). Compared to an image fusion method with a fixed step of 20, the proposed method reduced the reconstruction error by 44.6%, 37.1%, 45.8%, and

41.3% on average, respectively. The experimental results empirically validate the superior reconstruction accuracy of the proposed methodology. The comparison of reconstruction error is shown in Figure 11.

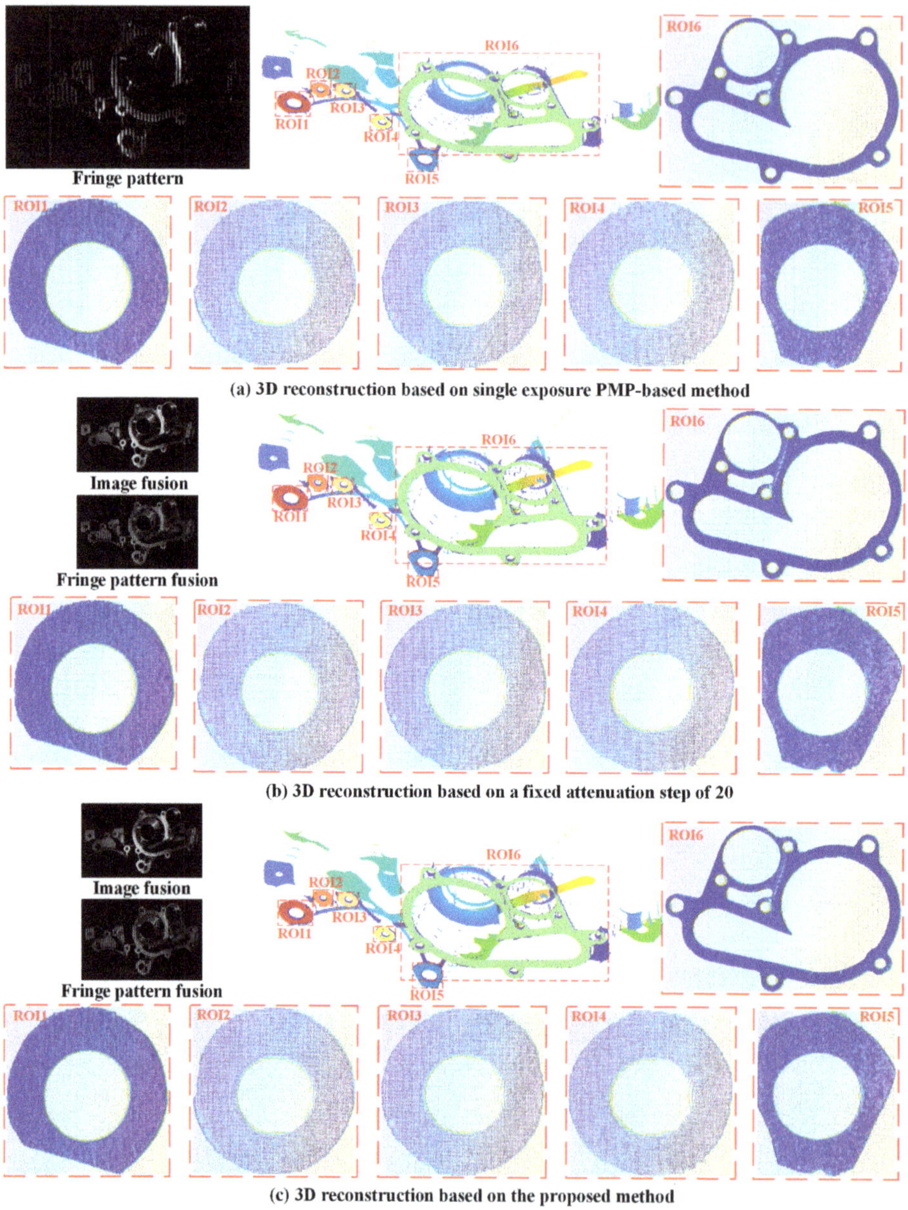

Figure 10. Comparison of 3D reconstruction results in case study 3.

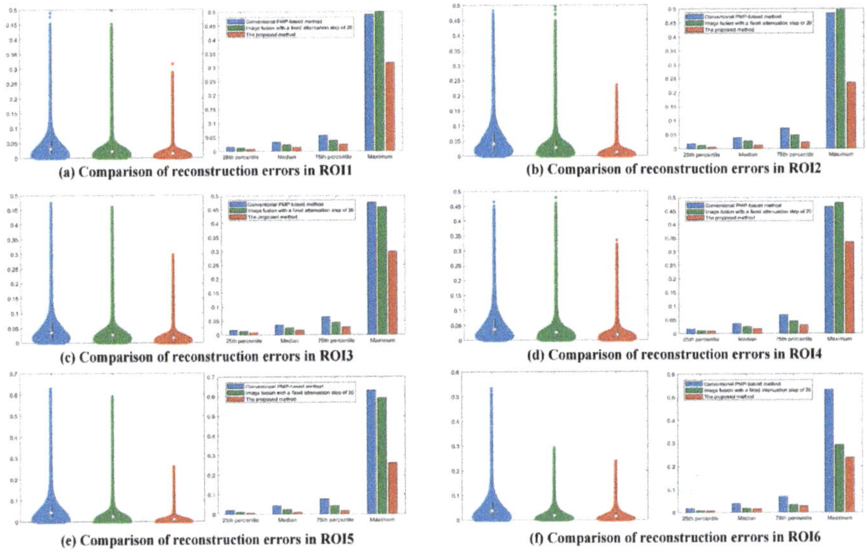

Figure 11. Comparison of reconstruction error in case study 3.

The efficiency of different reconstruction methods for highly reflective metallic components is intuitively illustrated through a statistical analysis of data acquisition and processing cycles. The time-consuming steps primarily include obtaining the optimal projection intensity and corresponding mechanical part images, image fusion, threshold segmentation, and result extraction. It is worth noting that factors such as the algorithm, camera, projector, computing hardware, and the complexity of the workpiece (e.g., geometric complexity and reflective properties) can significantly affect the processing time. A detailed analysis of the aforementioned steps is provided below:

(1) Acquisition of Projection Intensity and Corresponding Mechanical Part Images.

For the three representative cases, the time required to determine a single optimal projection intensity was approximately 2 s, 2 s, and 3 s, respectively. Additionally, the process of determining projection intensity using a fixed-step attenuation strategy, followed by synchronous image acquisition with the camera, took approximately 4 s per cycle.

(2) Image Fusion.

In the three representative cases, the time required for image fusion was approximately 8 s, 9 s, and 12 s, respectively.

(3) Threshold Segmentation.

Threshold segmentation is performed using the K-means++ algorithm. The processing time required for this step was approximately 4 s, 4 s, and 5 s for the three respective cases.

(4) Result Acquisition.

Determining the optimal exposure time takes around 2 s. Capturing a fringe image under a single exposure condition takes approximately 3 s. The phase fusion of fringe images with the same phase shift within the same exposure cycle takes approximately 4 s, 4 s, and 5 s for the three representative cases. Finally, reconstructing the 3D point cloud from the calibration parameters and depth information requires approximately 3 s, 3 s, and 4 s, respectively.

The efficiency of different methods for reconstructing highly reflective metallic components is intuitively demonstrated by statistically analyzing the required processing time. In case 1, the total durations required by the conventional PMP method, the fixed attenuation step method, the proposed method integrated with a consistent attenuation pattern, and the proposed method alone are 6 s, 43 s, 42 s, and 54 s, respectively. In case 2, the corresponding total durations are 6 s, 44 s, 43 s, and 57 s. In case 3, the total durations required by the conventional PMP method, the fixed attenuation step method, and the proposed method are 7 s, 51 s, and 83 s.

From the above analysis, it is evident that the conventional PMP method exhibits the highest efficiency but the lowest reconstruction accuracy. Although the proposed method requires a longer processing time compared to other approaches, it achieves the highest reconstruction accuracy. The fixed-step attenuation method, on the other hand, offers a balance, with processing time and accuracy falling between the conventional PMP method and the proposed method. Therefore, in future applications, a reconstruction method can be selected based on a reasonable trade-off between efficiency and accuracy.

6. Conclusions

As for the lack of effective guidance in the imaging parameters in PMP-based 3D reconstruction, an adaptive high-precision 3D reconstruction of highly reflective mechanical parts based on optimization of projection intensity and exposure time is proposed in this paper. An image sequence is established during the search for optimal projection intensity to achieve the best hardware performance. A GA-SAdam framework is also proposed to maximize the retained details in the image fusion process. The exposure time is adaptively adjusted based on the surface reflective properties.

Three sets of typical mechanical parts were conducted in the case study, which comprises varying geometric shapes and reflective characteristics. Experiment results show that compared with the existing single-exposure method, a commonly-used attenuation step method, and a fixed-step based method with the same number of fused images, the proposed method reduced the average whisker range of reconstruction error by 51.18%, 25.68%, and 24.20%, and decreases the median error by 41.48%, 24.14%, and 26.70%, respectively. The effectiveness and high accuracy performance of the proposed method have been verified.

The currently adopted projector and camera setup may be inadequate for covering larger-scale components. This limitation can be addressed through hardware upgrades to accommodate a wider field of view for both projection and imaging. Enhancing the projector's output power can mitigate the attenuation of projection intensity over large areas, which otherwise degrades fringe quality, particularly on highly reflective surfaces. Real-time inspection demands lightweight algorithms (e.g., compressed deep learning) and hardware optimizations (e.g., FPGAs) to meet latency constraints. Trade-offs between accuracy and speed must be balanced, but hybrid approaches (e.g., adaptive fusion + edge computing) could enable robust deployment in industrial settings.

It is worth noting that the proposed methodology minimized manual intervention through the integration of programmable fringe projectors, customized camera software development, optimization algorithms, and coordinated communication protocols. However, the required number of experiments escalates exponentially with increasing segmented regions and iteration steps. Furthermore, a substantial portion of experimentally acquired images contains underutilized regions, which indicates that correlation analysis of imaging parameters could be focused on to enhance data utilization and searching efficiency. It is worthwhile to investigate a better experimental design to promote real-time performance and practical deployment in industrial scenarios.

Author Contributions: Conceptualization, R.L., C.H. and K.I.; methodology, R.L., C.H. and J.S.; software, R.L., Z.W. and X.L.; validation, C.H., X.L. and Z.W.; formal analysis, X.L. and S.Z.; investigation, R.L. and Z.W.; resources, J.S. and K.I.; data curation, J.S. and S.Z.; writing—original draft preparation, R.L. and C.H.; writing—review and editing, J.S. and K.I.; visualization, Z.W. and X.L.; supervision, K.I. and S.Z.; project administration, J.S. and S.Z.; funding acquisition, C.H. and K.I. All authors have read and agreed to the published version of the manuscript.

Funding: This research was funded by the National Natural Science Foundation of China (Grant Number: 52305285, 52475286), by the State Key Laboratory of Fluid Power and Mechatronic Systems (Grant Number: GZKF202326), and by the Natural Science Foundation of Jiangsu Province (Grant Number: BK20230592).

Institutional Review Board Statement: Not applicable.

Informed Consent Statement: Not applicable.

Data Availability Statement: The data in this study are available upon request from the corresponding author. The data are not publicly available due to privacy.

Conflicts of Interest: The authors declare no conflicts of interest.

References

1. Gorthi, S.S.; Rastogi, P. Fringe projection techniques: Whither we are? *Opt. Lasers Eng.* **2010**, *48*, 133–140. [CrossRef]
2. Chen, R.; Wang, G.; Zhao, J.; Xu, J.; Chen, K. Fringe pattern based plane-to-plane visual servoing for robotic spray path planning. *IEEE/ASME Trans. Mechatron.* **2018**, *23*, 1083–1091. [CrossRef]
3. Li, J.; Chen, Z.; Rao, G.; Xu, J. Structured light-based visual servoing for robotic pipe welding pose optimization. *IEEE Access* **2019**, *7*, 138327–138340. [CrossRef]
4. Gaboutchian, A.V.; Knyaz, V.A.; Korost, D.V. New approach to dental morphometric research based on 3d imaging techniques. *J. Imaging* **2021**, *7*, 184. [CrossRef] [PubMed]
5. Pellegrini, M.D.; Orlandi, L.; Sevegnani, D.; Conci, N. Mobile-based 3d modeling: An in-depth evaluation for the application in indoor scenarios. *J. Imaging* **2021**, *7*, 167. [CrossRef]
6. Zhao, X.; Yu, T.; Liang, D.; He, Z. A review on 3d measurement of highly reflective objects using structured light projection. *Int. J. Adv. Manuf. Technol.* **2024**, *132*, 4205–4222. [CrossRef]
7. Salahieh, B.; Chen, Z.; Rodriguez, J.; Liang, R. Multi-polarization fringe projection imaging for high dynamic range objects. *Opt. Express* **2014**, *22*, 10064–10071. [CrossRef]
8. Huang, X.; Bai, J.; Wang, K.; Liu, Q.; Luo, Y.; Yang, K.; Zhang, X. Target enhanced 3d reconstruction based on polarization-coded structured light. *Opt. Express* **2017**, *25*, 1173–1184. [CrossRef]
9. Xiang, G.; Zhu, H.; Guo, H. Spatial phase-shifting profilometry by use of polarization for measuring 3d shapes of metal objects. *Opt. Express* **2021**, *29*, 20981–20994. [CrossRef]
10. Zhu, Z.; You, D.; Zhou, F.; Wang, S.; Xie, Y. Rapid 3D reconstruction method based on the polarization-enhanced fringe pattern of an HDR object. *Opt. Express* **2021**, *29*, 2162–2171. [CrossRef]
11. Zhu, Z.; You, D.; Zeng, X.; Qiao, S.; Dang, G.; Zhan, Y. 3D reconstruction method based on the multi-polarization superposition coding phase pattern of LRR objects. *Opt. Express* **2023**, *31*, 32350–32361. [CrossRef] [PubMed]
12. Li, D.; Kofman, J. Adaptive fringe-pattern projection for image saturation avoidance in 3d surface-shape measurement. *Opt. Express* **2014**, *22*, 9887–9901. [CrossRef] [PubMed]
13. Li, S.; Da, F.; Rao, L. Adaptive fringe projection technique for high-dynamic range three-dimensional shape measurement using binary search. *Opt. Eng.* **2017**, *56*, 94111. [CrossRef]
14. Chen, C.; Gao, N.; Wang, X.; Zhang, Z. Adaptive pixel-to-pixel projection intensity adjustment for measuring a shiny surface using orthogonal color fringe pattern projection. *Meas. Sci. Technol.* **2018**, *29*, 55203. [CrossRef]
15. Xu, P.; Liu, J.; Wang, J. High dynamic range 3d measurement technique based on adaptive fringe projection and curve fitting. *Appl. Opt.* **2023**, *62*, 3265–3274. [CrossRef]
16. Zhang, M.; Chen, C.; Xie, L.; Zhang, C. Accurate measurement of high-reflective surface based on adaptive fringe projection technique. *Opt. Lasers Eng.* **2024**, *172*, 107820. [CrossRef]

17. Zhang, S.; Yau, S.-T. High dynamic range scanning technique. *Opt. Eng.* **2009**, *48*, 033604. [CrossRef]
18. Jiang, H.; Zhao, H.; Li, X. High dynamic range fringe acquisition: A novel 3-d scanning technique for high-reflective surfaces. *Opt. Lasers Eng.* **2012**, *50*, 1484–1493. [CrossRef]
19. Cui, H.; Li, Z.; Tian, W.; Liao, W.; Cheng, X. Multiple-exposure adaptive selection algorithm for high dynamic range 3d fringe projection measurement. In Proceedings of the Tenth International Symposium on Precision Engineering Measurements and Instrumentation, Guangzhou, China, 13–16 March 2019; SPIE: Bellingham, WA, USA, 2019; Volume 11053, pp. 145–150. [CrossRef]
20. Chen, X.; Du, H.; Zhang, J.; Yang, X.; Xi, J. A self-adaptive multiple exposure image fusion method for highly reflective surface measurements. *Machines* **2022**, *10*, 1004. [CrossRef]
21. Zhu, J.; Yang, F.; Hu, J.; Zhou, P. High dynamic reflection surface 3d reconstruction with sharing phase demodulation mechanism and multi-indicators guided phase domain fusion. *Opt. Express* **2023**, *31*, 25318–25338. [CrossRef]
22. Hu, E.; He, Y.; Chen, Y. Study on a novel phase-recovering algorithm for partial intensity saturation in digital projection grating phase-shifting profilometry. *Optik* **2010**, *121*, 23–28. [CrossRef]
23. Budianto; Lun, D.P.K. Inpainting for fringe projection profilometry based on geometrically guided iterative regularization. *IEEE Trans. Image Process* **2015**, *24*, 5531–5542. [CrossRef]
24. Ren, W.; Tian, J.; Tang, Y. Specular reflection separation with color-lines constraint. *IEEE Trans. Image Process* **2017**, *26*, 2327–2337. [CrossRef] [PubMed]
25. Chen, B.; Wan, Y.; Li, J.; Yang, K.; Luo, L.; Li, H. High-frequency average phase compensation method for gamma nonlinearity based on optimal-frequency strategy. *Phys. Scr.* **2024**, *99*, 115529. [CrossRef]
26. Zhang, L.; Chen, Q.; Zuo, C.; Feng, S. High-speed high dynamic range 3d shape measurement based on deep learning. *Opt. Lasers Eng.* **2020**, *134*, 106245. [CrossRef]
27. Liu, X.; Chen, W.; Madhusudanan, H.; Ge, J.; Ru, C.; Sun, Y. Optical measurement of highly reflective surfaces from a single exposure. *IEEE Trans. Ind. Inform.* **2021**, *17*, 1882–1891. [CrossRef]
28. Shen, M.; He, L.; Zhang, H.; Ma, L.; Li, Y. Deep learning based measurement accuracy improvement of high dynamic range objects in fringe projection profilometry. *Opt. Express* **2024**, *32*, 35689. [CrossRef]
29. Xi, D.; Hou, L.; Wu, F.; Qin, Y. Deep learning-based inpainting of high dynamic range fringe pattern for high-speed 3d measurement of industrial metal parts. *Adv. Eng. Inform.* **2024**, *60*, 102428. [CrossRef]
30. Song, Z.; Chung, R.; Zhang, X. An accurate and robust strip-edge-based structured light means for shiny surface micromeasurement in 3-d. *IEEE Trans. Ind. Electron.* **2013**, *60*, 1023–1032. [CrossRef]
31. Tang, S.; Zhang, X.; Tu, D. Micro-phase measuring profilometry: Its sensitivity analysis and phase unwrapping. *Opt. Lasers Eng.* **2015**, *72*, 47–57. [CrossRef]
32. Feng, S.; Chen, Q.; Zuo, C.; Asundi, A. Fast three-dimensional measurements for dynamic scenes with shiny surfaces. *Opt. Commun.* **2017**, *382*, 18–27. [CrossRef]
33. Zhao, X.; Yu, T.; Kang, L.; Shen, H.; He, Z. Adaptive chessboard-like high-frequency projection method for three-dimensional measurement of shiny surfaces. *Meas. Sci. Technol.* **2024**, *35*, 45025. [CrossRef]
34. Shafer, S.A. Using color to separate reflection components. *Color Res. Appl.* **1985**, *10*, 210–218. [CrossRef]
35. Benveniste, R.; Ünsalan, C. Nary coded structured light-based range scanners using color invariants. *J. Real-Time Image Process.* **2014**, *9*, 359–377. [CrossRef]
36. Xu, F.; Zhang, Y.; Zhang, L. An effective framework for 3d shape measurement of specular surface based on the dichromatic reflection model. *Opt. Commun.* **2020**, *475*, 126210. [CrossRef]
37. Feng, W.; Sun, J.; Liu, Q.; Li, X.; Liu, D.; Zhai, Z. Specular highlight removal of light field image combining dichromatic reflection with exemplar patch filling. *Opt. Lasers Eng.* **2024**, *178*, 108175. [CrossRef]
38. Zuo, C.; Huang, L.; Zhang, M.; Chen, Q.; Asundi, A. Temporal phase unwrapping algorithms for fringe projection profilometry: A comparative review. *Opt. Lasers Eng.* **2016**, *85*, 84–103. [CrossRef]
39. Zhang, S.; Huang, P. Novel method for structured light system calibration. *Opt. Eng.* **2006**, *45*, 083601. [CrossRef]
40. D'Angelo, G.; Palmieri, F. GGA: A modified genetic algorithm with gradient-based local search for solving constrained optimization problems. *Inf. Sci.* **2021**, *547*, 136–162. [CrossRef]
41. Kingma, D.P.; Ba, J. Adam a method for stochastic optimization. *arXiv* **2014**, arXiv:1412.6980. [CrossRef]
42. Feng, S.; Zhang, Y.; Chen, Q.; Zuo, C.; Li, R.; Shen, G. General solution for high dynamic range three-dimensional shape measurement using the fringe projection technique. *Opt. Lasers Eng.* **2014**, *59*, 56–71. [CrossRef]
43. Levin, A.; Lischinski, D.; Weiss, Y. A closed-form solution to natural image matting. *IEEE Trans. Pattern Anal. Mach. Intell.* **2008**, *30*, 228–242. [CrossRef] [PubMed]

44. Arthur, D.; Vassilvitskii, S. K-means++: The advantages of careful seeding. In Proceedings of the ACM-SIAM Symposium on Discrete Algorithms 2007, New Orleans, LA, USA, 7–9 January 2007; pp. 1027–1035. Available online: http://ilpubs.stanford.edu:8090/778/ (accessed on 5 May 2025).
45. Liu, K.; Wang, Y.; Lau, D.; Hao, Q.; Hassebrook, L.G. Dual-frequency pattern scheme for high-speed 3-D shape measurement. *Opt. Express* **2010**, *18*, 5229–5244. [CrossRef] [PubMed]

Journal of
Imaging

MDPI

Article

Impact of Data Capture Methods on 3D Reconstruction with Gaussian Splatting

Dimitar Rangelov [1,2,*], Sierd Waanders [1,3], Kars Waanders [1,3], Maurice van Keulen [2] and Radoslav Miltchev [4]

1 Technologies for Criminal Investigations, Saxion University of Applied Sciences, 7513 AB Enschede, The Netherlands; s.c.waanders@saxion.nl (S.W.); k.t.waanders.01@saxion.nl (K.W.)
2 Faculty of Electrical Engineering, Mathematics and Computer Science, University of Twente, 7522 NB Enschede, The Netherlands; m.vankeulen@utwente.nl
3 Police Academy of The Netherlands, 7334 AC Apeldoorn, The Netherlands
4 Faculty of Industrial Technology, Technical University of Sofia, 1756 Sofia, Bulgaria; rmiltchev@tu-sofia.bg
* Correspondence: d.g.rangelov@utwente.nl

Abstract: This study examines how different filming techniques can enhance the quality of 3D reconstructions with a particular focus on their use in indoor crime scene investigations. Using Neural Radiance Fields (NeRF) and Gaussian Splatting, we explored how factors like camera orientation, filming speed, data layering, and scanning path affect the detail and clarity of 3D reconstructions. Through experiments in a mock crime scene apartment, we identified optimal filming methods that reduce noise and artifacts, delivering clearer and more accurate reconstructions. Filming in landscape mode, at a slower speed, with at least three layers and focused on key objects produced the most effective results. These insights provide valuable guidelines for professionals in forensics, architecture, and cultural heritage preservation, helping them capture realistic high-quality 3D representations. This study also highlights the potential for future research to expand on these findings by exploring other algorithms, camera parameters, and real-time adjustment techniques.

Keywords: 3D reconstruction; neural radiance fields; gaussian splatting; 3D scanner technology; crime scene reconstruction; forensic photogrammetry; forensics

1. Introduction

The introduction should briefly place this study in a broad context and highlight why it is important. Three-dimensional (3D) reconstruction creates detailed digital models of real-world objects or scenes using 2D images, providing precise spatial data for fields like engineering, medicine, and archaeology [1]. One area that could greatly benefit from this technology is crime investigation, where reconstructing crime scenes help investigators understand what occurred. The challenge lies in interpreting the aftermath as evidence which often offers multiple possible scenarios. To develop a complete narrative, the investigators must combine physical traces with additional information such as police reports, witness accounts, forensic data, and their own expertise.

The reconstruction process relies on key principles such as achieving adequate image overlap, minimizing motion blur, and ensuring consistent lighting, all of which are critical for producing high-quality models [2]. While extensive research has been conducted on 3D reconstruction algorithms, the literature on image acquisition methods remains limited, particularly for constrained indoor environments like crime scenes. Existing studies emphasize the significance of camera movement and stability for enhancing reconstruction fidelity, especially in settings with limited space or intricate spatial layouts [3,4]. Moreover, approaches tailored for indoor reconstructions, such as photogrammetry-based workflows and structured light systems, have demonstrated varying degrees of success [5–8].

Traditional photogrammetry remains a widely used method in 3D reconstruction, particularly for forensic and architectural applications. It relies on structure-from-motion (SfM) [9] and multi-view stereo (MVS) [10] techniques to generate 3D models from 2D images. While photogrammetry provides accurate reconstructions with measurable geometric fidelity, it often requires an extensive manual effort for camera calibration, feature matching, and mesh generation. Additionally, photogrammetric methods struggle in scenes with low-texture surfaces, reflective materials, or non-uniform lighting conditions, leading to incomplete reconstructions and increased noise. These limitations pose significant challenges in forensic applications, where scene complexity and environmental variability must be accounted for.

Recent advancements in deep-learning-based reconstruction, such as Neural Radiance Fields (NeRF) [11] and Gaussian Splatting (GS) [12], present alternatives that address some of these challenges. Unlike photogrammetry, NeRF synthesizes new viewpoints by learning a continuous volumetric representation of the scene, allowing for superior handling of complex lighting and occlusions. However, NeRF is computationally demanding and requires long training times, making it less practical for time-sensitive forensic reconstructions. Gaussian Splatting, on the other hand, offers a more efficient approach by representing 3D structures as a set of discrete Gaussian functions, enabling fast and high-fidelity rendering while reducing manual processing efforts. While both methods improve visual quality and automation, their suitability for forensic applications depends on balancing accuracy, processing efficiency, and real-world usability.

In this research, the focus is on optimizing the data acquisition process for 3D reconstruction in indoor crime scenes by evaluating filming parameters such as camera orientation, operator speed, and path selection. These parameters were chosen due to their fundamental impact on scene coverage, motion blur reduction, and data alignment, critical factors for achieving high-fidelity reconstructions in constrained forensic environments. While factors like lighting conditions, exposure settings, and camera calibration influence reconstruction quality, they are either hardware-dependent or have been extensively explored in prior research, whereas filming methodologies remain an underexplored yet crucial aspect of 3D reconstruction optimization. Examining how the method of filming a crime scene impacts these reconstructions' quality provides valuable insights not only for forensic applications but also for other fields. For example, architectural visualization can benefit from accurate indoor 3D reconstructions for design and renovation projects [13–15]. In cultural heritage preservation, detailed reconstructions of historical sites and artefacts can aid in documentation and restoration efforts [16,17]. Similarly, virtual reality (VR) and augmented reality (AR) applications in gaming and simulation training can achieve higher realism and immersion with optimized 3D reconstruction techniques [18].

To achieve improvements in 3D reconstruction, Neural Radiance Fields (NeRF) [11] is utilized. NeRF is an innovative approach that employs deep learning to generate highly detailed and accurate 3D models from 2D images. It achieves this by modeling the complex interplay between spatial geometry and light behavior in the scene, capturing fine details such as texture and reflectance. NeRF's ability to synthesize realistic views from novel perspectives makes it particularly suitable for applications demanding precision and realism. By utilizing volumetric rendering, NeRF constructs continuous 3D representations, which are further enhanced when combined with complementary techniques such as Gaussian Splatting, to produce models with superior visual fidelity and minimized artifacts. Gaussian Splatting further enhances this process by refining and smoothing the spatial data points, leading to better-quality reconstructions. Additionally, 3D Gaussian Splatting (3D-

GS) [12] is a promising technique in computer graphics for 3D rendering of scenes, gaining traction due to its ability to efficiently render high-quality images while maintaining a compact scene representation. Unlike NeRF, which relies on neural networks conditioned on viewpoint and position, 3D-GS employs Gaussian functions that can be rasterized directly into images, facilitating faster rendering speeds and improved visual fidelity. Similar Gaussian-based methods have been successfully applied in broader fields, such as data assimilation for environmental modeling and field reconstruction, where they help interpolate sparse observations and reconstruct dynamic spatial fields [19,20]. These methods demonstrate the versatility of Gaussian-based approaches in various applications, reinforcing their potential in 3D reconstruction.

NeRF is not the first step in this new evolution of 3D reconstructions but is, rather, the building block of a family of algorithms, which includes SNeRF [21], Tetra-NeRF [22], NeRFacto [23], Instant-NGP [24], SPIDR [25], MERF [26], and so on. In fact, each one of them solves particular problems and allows for the increment of overall capabilities in different ways.

Preliminary research highlighted the importance of the camera, lens, and their settings as critical factors. The impact of each parameter on 3D reconstruction was thoroughly examined [27,28]. Additionally, the method used to film an environment was found to have a significant influence on the quality of the 3D reconstruction. Building on these findings, specific camera movement techniques were employed to ensure comprehensive scene coverage and high-quality data acquisition.

This paper concentrates solely on the method of filming indoor environments. The experimental setup consists of a living room, kitchen, and hallway. Four comparison experiments were conducted, varying one parameter at a time, to identify the optimal scannig method for high-quality 3D reconstructions in indoor crime scene scenarios.

We investigate various factors that influence the quality of 3D reconstructions, focusing primarily on camera orientation, operator walking speed, layering techniques, and scanning paths. The orientation of the camera during capture plays a significant role. Landscape mode offers a broader horizontal field of view, ideal for wide scenes, while portrait mode enhancing vertical detail, which is better suited for tall subjects. Capturing the same environment in both modes enables an analysis of differences in data quality, level of detail, and the accuracy of 3D reconstructions.

Another key factor is the walking speed of the camera operator, as rapid movement can cause autofocus issues, resulting in blurry frames that negatively impact the reconstruction quality. By testing various walking speeds and comparing the resulting video lengths, we assess how motion clarity influences the final 3D model. Furthermore, the use of various layering techniques is examined to enhance data capture from multiple heights, along with the significance of scanning paths to achieve comprehensive environmental coverage. Together, these tests aim to optimize the methodology for achieving high-quality 3D reconstructions.

Accurate 3D reconstructions in crime scene investigations may provide crucial insights into the sequence of events and assist investigators in analyzing evidence more efficiently. The traditional way of representation, like sketching or photographing the scene, neglects important details. On other hand, 3D reconstruction provides an overall interactive way to view the scene. This technology aids in visualizing the crime scene, understanding the spatial relationships between various pieces of evidence, and displaying discoveries in court.

In this paper, the main contributions are provided to the field of 3D reconstruction with a special emphasis on its application within crime scene investigation. The contributions of research can be divided into theoretical and practical advancements.

Theoretical Contributions

- Novel framework for filming method optimization: This study introduces a structured framework for systematically analyzing and optimizing key parameters, camera orientation, filming speed, camera layers, and filming path, for 3D reconstructions. While based on existing methods, this framework uniquely addresses the specific challenges of indoor forensic environments, such as limited space, variable lighting, and object clutter, which have not been comprehensively addressed in prior research.
- Extension of theoretical framework: This research extends the theoretical framework related to 3D reconstruction. For example, a theoretical framework for cameras to be optimized could be any application that it has not seen before. There is big potential of using advanced techniques in reconstruction for architecture, archaeology, and digital media fields.

Practical Contributions

- Crime Scene Investigation Guidelines: The guidelines in the research demonstrate how forensic investigators can achieve an accurate 3D reconstruction of a crime scene. The primary focus of this paper was on the methodology for filming a crime scene. The key takeaway is the importance of capturing the scene from multiple angles by circling around specific objects in different zones.
- Application of advanced 3D reconstruction techniques: This study examines the use of Gaussian Splatting within existing 3D reconstruction workflows, focusing on its effectiveness in handling complex scenes. These include environments with intricate spatial layouts, dense object arrangements, and varied textures, which are particularly challenging in forensic applications. While no modifications to the underlying algorithms were made, this research contributes by detailing how these techniques can be systematically applied and optimized to address the unique challenges of reconstructing detailed, accurate models of such environments.
- 3D Modelling Processes Optimization: The results of the research give very practical recommendations for the optimization of 3D reconstruction processes within numerous fields. These recommendations can significantly improve the quality of 3D reconstructions used in architectural visualization, cultural heritage preservation, and VR/AR applications for enhanced realism and immersion.

This paper introduces significant advancements in the methodology of 3D reconstructions for crime scene investigations by combining state-of-the-art techniques, such as Neural Radiance Fields and Gaussian Splatting, with novel optimization strategies tailored for forensic applications. The research offers innovative insights into how filming parameters, such as camera orientation, walking speed, layering, and scanning paths, can be systematically optimized to achieve higher-quality reconstructions. These contributions not only enhance existing methodologies but also provide practical guidelines for real-world forensic scenarios, thereby expanding the scope of 3D reconstruction technology across various professional domains.

This paper is organized as follows: Section 2 describes the methodology used in this study, including the methods employed during image capture and how they meet the demands of crime scene investigations. Section 3 presents the results of our experiments and comparative analysis of different methods. Section 4 provides a detailed discussion of these findings, comparing them with the existing literature and exploring their implications for various applications. Finally, Section 5 concludes this paper, summarizing our key contributions and suggesting directions for future research.

2. Materials and Methods

2.1. Experimental Setup

The experimental setup consists of a living room, kitchen, and hallway with several key features, such as tables, chairs, tableware, closets, ceiling, and lamps. These elements were chosen to provide a diverse set of objects with different geometric and textural information for the 3D-GS. The floor plan of the room is displayed in Figure 1, which illustrates the arrangement of the environment used in the experiments.

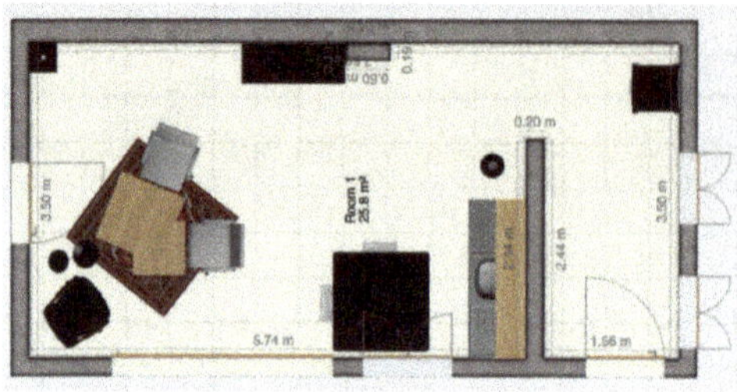

Figure 1. Experimental setup.

The primary equipment that was used was a SONY a7c, Sony Corporation, Tokyo, Japan [29] camera with a 24.2 MP full-frame Exmor R CMOS BS, Sony Corporation, Tokyo, Japan [29]. This camera was chosen for its full-frame sensor, which results in high image quality. The lens combined with the camera is a Sigma, Sigma Corporation, Kanagawa, Japan 14 mm f/1.4 DG DN [30] Art lens, which is a wide-angle lens. With this lens, capturing more of a room in single capture is possible. To create an even more consistent capturing method, a DJI RS 4,DJI, Shenzhen, China [31] gimbal was added to the camera setup. The gimbal will result in a smoother capture of the environment, with less vibration and thus, fewer unclear frames. Controlled lighting conditions were maintained across all captures.

To improve the quality of 3D reconstructions, particularly in noise reduction, and detail accuracy, Postshot [32] was utilized, an end-to-end software for Radiance Fields to generate the 3D-GS. The goal was to optimize the capturing method of rooms such as mentioned above, while generating accurate 3D models.

In this implementation, Gaussian Splatting was employed to enhance scene geometry representation using a set of 3D Gaussian functions. These functions were rasterized directly into 2D images, which allowed for efficient rendering while preserving high visual fidelity. The Postshot software automates key aspects of this process but provides options for manual parameter adjustments to fine-tune results. For this study, we utilized a Splat ADC profile, which optimizes spatial data distribution by regularizing density values, thereby reducing artifacts and enhancing fine details.

Key parameters included a downsample resolution of 1600 pixels for the input images and a splat density adjustment layer set to 1. These settings were calibrated through iterative testing to balance computational efficiency and reconstruction quality. The training process was configured to stop at 54 kSteps, based on convergence criteria observed during early experimentation.

The integration with NeRF involved aligning Gaussian splats with the neural radiance field's density and color predictions. Camera poses were computed from the captured images using Postshot's internal algorithms, ensuring accurate alignment of spatial data with radiance field outputs. By combining NeRF's volumetric rendering and Gaussian Splatting's spatial refinement, the method achieved improved clarity and reduced noise in the final 3D models.

A key distinction between NeRF and Gaussian Splatting lies in how they represent and process 3D scene information as shown in Figure 2. NeRF constructs a scene using a volumetric neural representation, where the space is densely populated with points that encode density and color information. These points do not exist as discrete entities but are instead sampled through a neural function that maps spatial coordinates to radiance values. Rendering a scene requires ray-marching through this continuous representation, accumulating color and transparency at each sampled point along a given viewing direction. While this method produces highly detailed and photorealistic reconstructions, it is computationally expensive and struggles with real-time performance.

In contrast, Gaussian Splatting represents the scene as a discrete set of 3D Gaussian functions ('splats') [33], each carrying spatial position, opacity, and shape properties. Unlike NeRF's volumetric points, which require extensive neural computation to render, splats are directly projected onto the image plane, allowing for rasterization-based rendering that is significantly faster and more efficient. Furthermore, the overlapping and blending nature of Gaussian splats helps to fill in missed spots and smooth out inconsistencies in areas with sparse input data, reducing gaps that might otherwise appear in the reconstruction.

For this research, Gaussian Splatting was chosen over NeRF's point-based volumetric representation due to its ability to produce faster and cleaner reconstructions. By using splats instead of a densely sampled neural point cloud, our method achieves clearer object boundaries, fewer artifacts, and improved texture fidelity, which is essential for forensic investigations. This choice ensures that reconstructions remain both high-quality and practical for real-world applications where clarity and efficiency are critical.

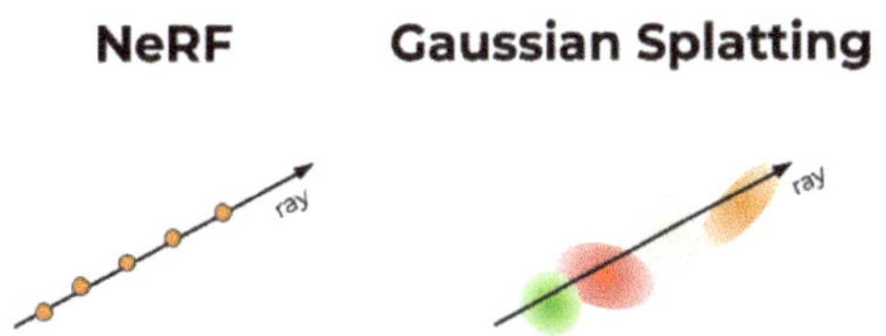

Figure 2. A conceptual difference between NeRF and GS, **Left**: Query a continuous MLP along the ray, **Right**: Blend a discrete set of Gaussians relevant to the given ray [34].

We conducted four comparison experiments, each varying one parameter of capturing at a time, using the same optimal camera setting and setup during all comparisons. The video footage were processed into 3D reconstructions, which were evaluated based on noise and detail accuracy. These criteria were chosen for their importance in forensic applications where both clarity and realism are essential for analyzing evidence.

This approach allowed us to identify the optimal scan method for high-quality 3D reconstructions in indoor crime scene scenarios, providing valuable insights for future forensic investigations.

2.2. Camera Settings

Various camera settings significantly influence the quality of a 3D reconstruction. This study does not focus on camera settings. Therefore, the camera is set to automatic mode. One parameter that is not set in automatic mode is the aperture. A higher aperture results in a more detailed background [35]. The aperture is set to f/5.6 during the capturing, which is the maximum aperture of the Sony a7c.

2.3. Data Collection

2.3.1. Capturing Techniques

The camera was handheld with a gimbal throughout the capture process, allowing for flexible, controlled, and stable movements as needed. The selection of capturing techniques was guided by their ability to maximize scene coverage, minimize occlusions, and enhance reconstruction fidelity in forensic applications. In indoor crime scenes, capturing a comprehensive and accurate dataset is crucial, as missing or distorted details can compromise forensic analysis. The chosen techniques, truck, pedestal, boom, and arc were selected due to their effectiveness in maintaining smooth motion, consistent framing, and minimizing abrupt perspective shifts that could degrade reconstruction quality. These techniques offer a structured approach to data acquisition, ensuring that all key objects and spatial relationships within the crime scene are accurately represented [28]. Each technique was strategically used at different points along the paths based on the spatial characteristics of the scene and the specific details we aimed to capture. Other potential methods, such as static multi-angle photography or robotic scanning, were considered but deemed less practical due to their increased setup time and limited adaptability to dynamic environments. By integrating these controlled capturing movements, this study aims to provide an optimized and reproducible approach to filming for high-quality 3D reconstructions.

These techniques were not used for comparison but combined into the different capturing methods to complement each other during the capturing process. Each technique was applied based on the specific spatial requirements of the scene.

2.3.2. Scanning Methods Comparison

Orientation Impact on 3D Reconstruction Quality

This test examines the impact of landscape and portrait modes on 3D reconstruction. Landscape mode (horizontal orientation) captures wide scenes, providing a broader field of view and more horizontal overlap between frames. This will ensure comprehensive scene coverage. Portrait mode (vertical orientation) is ideal for tall objects, offering better vertical detail with more overlap on the vertical axis. By capturing the same environment in both modes, the differences in data quality, level of detail, and accuracy of the 3D reconstructions were assessed.

Effect of Walking Speed on 3D Reconstruction Quality

During filming, the camera operator navigates through the environment to collect data. If the camera operator moves too quickly, some video frames may become blurry because the autofocus cannot keep up. This test aims to examine the impact of blurry frames on 3D reconstruction. The difference in movement speed was assessed by comparing video lengths. Specifically, the duration of the video recorded at a slower movement speed is expected to be twice as long as the video recorded at a faster movement speed.

Layering Technique for Enhanced 3D Reconstruction

Based on preliminary research, the importance of capturing an environment at various heights has been identified. For methods 1 to 10, we utilized a 3-layer technique, with cameras positioned at 0.3 m, 1 m, and 1.7 m. This approach allows for more comprehensive data capture, leading to a better overall picture of the environment and improved 3D reconstruction. However, capturing the environment with more layers does increase the time required. The test aims to examine the difference in 3D reconstruction quality when capturing the environment with 1, 3, or 5. The same object/zone was filmed using identical settings and setup, but the number of layers was varied. The 5 passes that have been taken are shown in Figure 3. For 1 pass we used camera pose 3, for 3 passes we used 1, 3, and 4, for 5 passes we used all 5 of them.

Figure 3. Layering technique, the numbers indicate the position of the layers.

Scanning Paths

The previously mentioned parameters play a crucial role in data collection. Another significant factor is the specific path taken through an environment. Each path differs in some way from each other. The results from testing the different filming paths were a base for the next paths. This continued until there was a suitable path for the use case. All the paths have been filmed with the same camera with the same settings. The different scanning paths are captured with the optimal film technique which are concluded from tests 1, 2, and 3.

Below the scanning paths are displayed and shortly explained:

Method 1: Following the contours of the whole environment

The camera operator follows the contours of the environment while pointing the camera inwards. The path is shown in Figure 4a. The camera operator starts in the upper left corner.

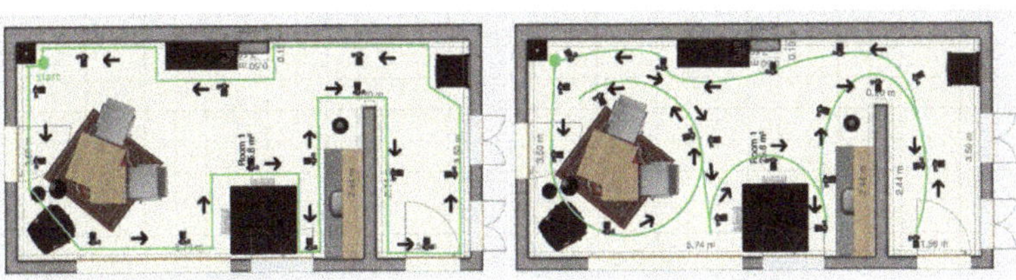

(**a**) (**b**)

Figure 4. (**a**) Method 1: Following the contours of the whole environment, (**b**) Method 2: Circle around the main objects in the rooms in one path.

Method 2: Circle around the main objects in the rooms in one path

The camera operator follows circles partly around the main objects in the room in one connected path. The objects are a dining table, kitchen table, kitchen, closet, and hallway. This method is expected to create more detail of the individual objects because the focus was more on that. The path is shown in Figure 4b. The camera operator starts in the upper left corner.

Method 3: Circle around the main objects in the room and separate hallway

This method is similar to method 2, the main difference is that the hallway is filmed separately from the kitchen/living room. At the end of the path in the living room, there is an extra part added to obtain more data from the room from a distance. The path in the hallway follows the same principle as in method 1. The two videos are combined in the 3D reconstruction algorithm separately to create two reconstructions. The paths are shown in Figure 5a.

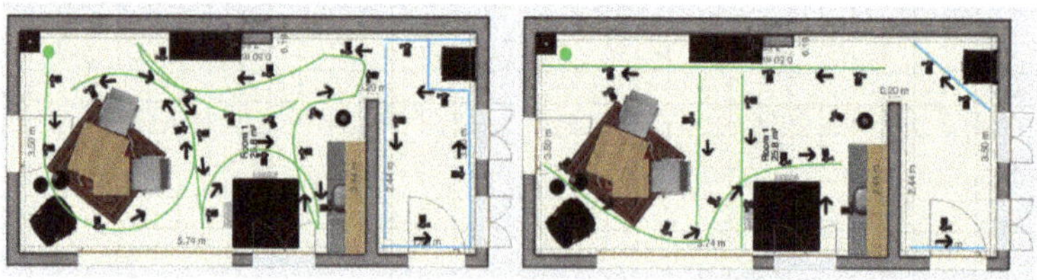

(a) (b)

Figure 5. (**a**) Method 3: Circle around the main objects in the room and separate hallway, (**b**) Method 4: Four passes in the living room and separate hallway.

Method 4: Four passes in the living room and separate hallway

The camera operator makes two separate passes while facing the camera inward on the horizontal axis. On the vertical axis in the living room, the camera was facing outwards. In the hallway, the cameraman should make two horizontal passes on either side while facing the camera inwards. The paths are shown in Figure 5b.

Method 5: Individual objects

Every main object/zone will be filmed separately to create more detail. The different paths are shown in the image to the left. The path is shown in Figure 6a. The room is separated in the following objects/zones: leather seat, side table, dining table, kitchen table, closet, kitchen, corner closet, and hallway. With this method, the videos will be put in the 3D reconstruction algorithm separately.

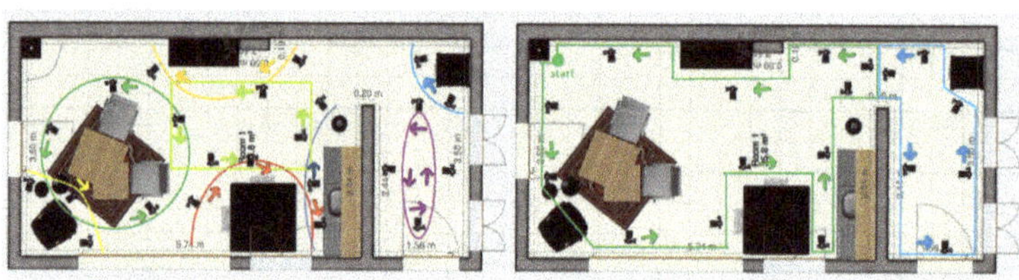

(a) (b)

Figure 6. (**a**) Method 5: Individual objects, (**b**) Method 6: Alternative to Method 1.

Method 6: Alternative to Method 1

This method is the same as method 1 but now the living room and the hallway are separated. The two videos will be put in the 3D reconstruction algorithm separately. The path is shown in Figure 6b.

Method 7: Three-zone scans

The room is separated into three zones; each zone was captured separately. The path is shown in Figure 7a. Zone one (blue) captures the living area, zone two (red) captures the kitchen area, and the third zone (purple) captures the hallway. These three zones/videos will each be put in the 3D reconstruction algorithm separately.

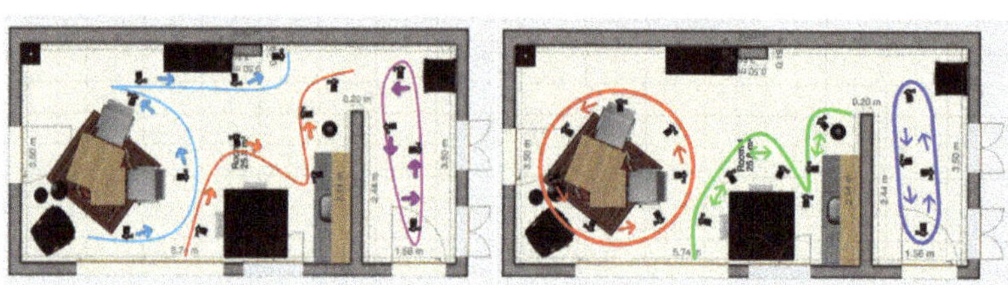

(a) (b)

Figure 7. (**a**) Method 7: Three-zone scans, (**b**) Method 8: Three-zone scans v2.

Method 8: Three-zone scans v2

It was found in the previous testing round that the best results were obtained when the room was divided into multiple zones. In Method 8, as shown in Figure 7b, the setup involves two loop closures: one at the coffee table and one in the hallway.

Method 9: Two zone scans

Method 9 involves capturing two videos and is presented in Figure 8a. One circles around the coffee table, while the other follows a continuous path around the dining table, kitchen, and hallway. This approach is expected to improve capture of the hallway. However, it may overlook some details, specifically the windows and the closet in the hallway corner. To address this, it is important to focus on capturing these features when rounding the corner.

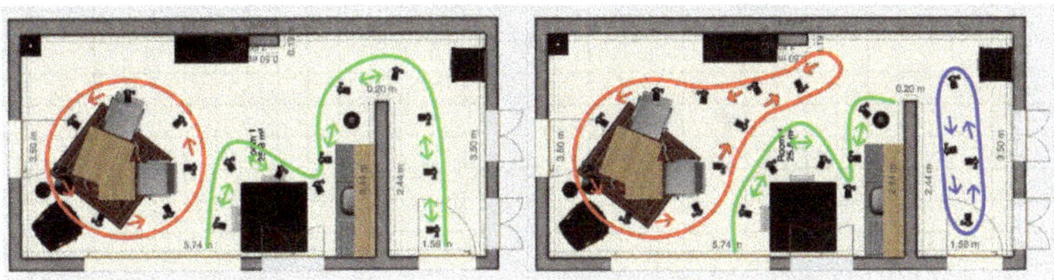

(a) (b)

Figure 8. (**a**) Method 9: Two zone scans, (**b**) Method 10: Alternative of Method 8.

Method 10: Alternative of Method 8

This method is similar to method 8; the only difference is the extra loop around the closet. This path is shown in Figure 8b. This method is expected to have a better reconstruction of the black cabinet.

2.4. Comparative Analysis Criteria

In this study, we captured a continuous video sequence using a handheld camera, which was then processed into a 3D reconstruction model. The 3D reconstruction is assessed using certain viewpoints in the 3D reconstruction. These viewpoints were chosen for their ability to represent different textures, lighting conditions, and surfaces common in indoor crime scenes.

There are three focus areas: the living room with a coffee table, the kitchen with a dining table, and the hallway. These areas offer varying textures and lighting challenges, making them ideal for comparison. The 3D reconstructions were evaluated based on three criteria: (1) Noise (artifacts in the model), and (2) Details (fidelity of the reconstruction), (3) Splats. In Table 1, these criteria are rated on a 1–5 scale, allowing us to systematically compare different scanning methods. This framework adds a novel approach for assessing 3D reconstructions in forensic applications.

Table 1. Criteria for comparative analysis.

Criteria	1	2	3	4	5
Noise	There is too much noise present, and nothing can be seen	There is too much noise present, but the room is visible	There is some noise present; however, the outline of the room is still visible	Almost no noise is present, and the room is quite clear in visibility	There is no noise

Table 1. *Cont.*

Criteria	1	2	3	4	5
Details (focus on blue bottle, walls, whiteboard, TV, and TCI coffee mug)	The reconstruction appears pixelated, yet it is discernible that an object should be present in that location	The reconstruction is pixelated, but it is still possible to discern the object type (e.g., table, chair, paper)	Identification of the object types is easily achievable	Capable of accurately identifying the object and providing brand information	Extremely detailed; there is no discernible difference between the model and the video
Splats (count)	Below 250,000 splats, very sparse data leading to low reconstruction fidelity	Between 250,000 and 300,000 splats, sparse data with visible gaps	Between 300,000 and 400,000 splats, moderate data density but some inconsistencies in texture mapping	Between 400,000 and 450,000 splats, high data density with uniform coverage and minimal artifacts	Above 450,000 splats, exceptionally dense data providing superior reconstruction and texture quality

3. Results

3.1. Orientation Impact on 3D Reconstruction Quality

The purpose of this test is to evaluate the difference in the quality of 3D reconstructions between environments captured in portrait mode and landscape mode. During the scanning process, Method 6 is employed. The difference between portrait and landscape modes is slight but noticeable. As illustrated in Figure 9, the quality is generally comparable. However, in portrait mode, the wall with leaves exhibits a duplicated black closet, and the wall continues beyond this duplicated closet. This anomaly is absent in landscape mode.

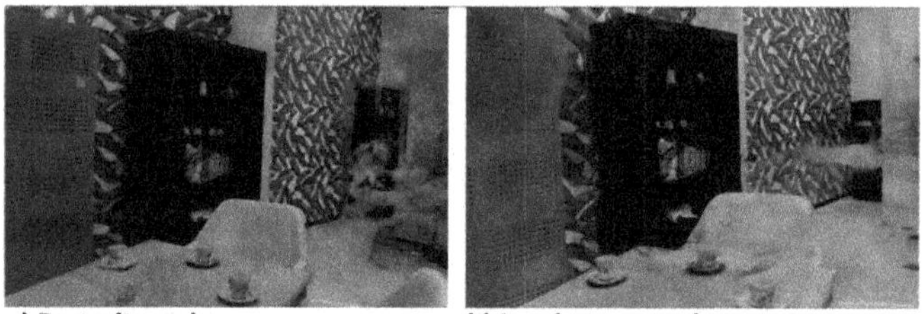

Figure 9. A comparison of 3D reconstructions in (**a**) Portrait mode, and (**b**) Landscape mode. The portrait mode reconstruction shows a duplicated black closet artifact, extending beyond the actual scene. This anomaly is absent in landscape mode, which provides better spatial accuracy by ensuring more horizontal overlap for feature matching.

Additional scans comparing the two orientations consistently showed that landscape mode yields slightly better results. The primary conclusion is that filming in landscape mode is superior to portrait mode. We hypothesize that portrait mode is slightly inferior because most data overlaps on the vertical axis rather than the horizontal axis. Given that the camera operator moves along the horizontal axis, horizontal overlap is preferable to ensure features are recognized and matched more frequently.

3.2. Effect of Walking Speed on 3D Reconstruction Quality

The duration of the video recording determines the velocity of the camera operator. The trajectory followed by the camera operator remains consistent across both videos. However, the speed at which the operator moves varies. This study evaluated two walking speeds: slow speed at 0.085 m/s and normal speed at 0.164 m/s. The speed velocity was measured by Garmin Fenix 6 Pro Solar [36] and later verified by calculation. The duration of the video recorded at normal speed (3:43 min) is approximately twice as long as that of the faster video (1:56 min).

The differences between the fast and slow recordings are clearly evident. At a slower speed, more details, particularly smaller objects, become visible. Additionally, there is significantly less noise in the 3D reconstructions produced from the slower video. This experiment demonstrates that slower filming speeds enhance the detail of the reconstruction. The underlying reason is that a longer video provides more frames for the algorithm to select from. Furthermore, slower movements allow the camera more time to focus on the environment, thereby improving the quality of the 3D reconstruction. Another factor is that slower movements result in greater frame overlap, which increases the number of features that can be matched between frames. All this is visible in Figure 10.

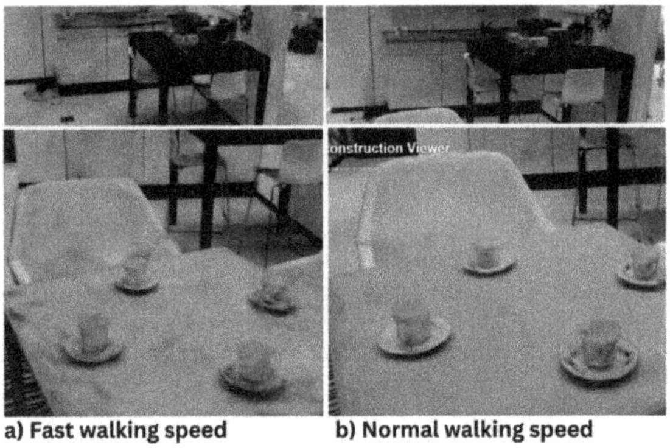

Figure 10. (**a**) Fast walking speed, (**b**) Normal walking speed.

In contrast, faster walking speeds, such as 0.164 m/s or above, significantly reduce the recording time and data size, making the process more efficient in terms of storage and computational load. However, this comes at the cost of increased motion blur and reduced frame overlap, which can compromise the accuracy of feature matching. For scenarios where reconstruction speed is prioritized over precision, faster speeds may still be acceptable, but they are not ideal for environments requiring high-detail reconstructions.

In summary, a slower filming speed increases the amount of data, enhances the quality of the data, and facilitates the matching of more features between frames, all of which contribute to higher-quality 3D reconstructions. It is recommended to use a walking speed of approximately 0.09 m/s to 0.12 m/s to balance clarity and efficiency.

3.3. Layering Technique for Enhanced 3D Reconstruction

The choice of using 1, 3, and 5 layers for the layering technique was guided by preliminary testing and practical constraints. A single-layer scan provides a baseline for comparison, representing the simplest and fastest capturing method. The three-layer configuration was chosen based on prior studies suggesting that capturing data at multiple heights enhances reconstruction fidelity without significantly increasing processing time. The five-layer configuration was included to test whether additional layers provide further improvements. Other configurations (e.g., 2, 4, or 6 layers) were initially considered, but early experiments indicated diminishing returns beyond three layers, as additional passes introduced excessive data without significantly enhancing reconstruction quality. Moreover, higher layer counts increased processing time and data complexity, making them less practical for real-world forensic applications.

Preliminary research underscores the significance of capturing an environment at varying heights to maximize data acquisition. This test aims to analyze the impact of using 1, 3, or 5 layers on the quality of 3D reconstruction. To evaluate the quality of each reconstruction, three reference images are utilized: the top of the table, the bottom of the table, and the ceiling. For an optimal 3D reconstruction, all these components should be clearly visible and of high quality.

Top of the Table:

- All three reconstructions (1, 3, and 5 layers) provided clear visibility of the tabletop.
- The scan with a single layer showed the sharpest 3D reconstruction with minimal noise.
- The scans with 3 and 5 layers, while acceptable, were slightly less sharp compared to the single-layer scan.

Bottom of the Table:

- The single-layer scan failed to capture the bottom of the table and the sides, including the legs of the chairs.
- Both the 3-layers and 5-layers scans successfully captured the bottom of the table and the chair legs.
- The 3-layers scan performed better in visualizing the sides and bottom of the table compared to the 5-layers scan.

Ceiling:

- The ceiling in the single-layer scan appeared pitch black due to the absence of data, which the AI algorithm filled with black.
- The difference between the 3-layers and 5-layers scans was minimal, with the 3-layers scan being slightly sharper.

A single-layer scan is effective for capturing specific details, such as the surface of a table, but falls short when it comes to achieving a complete 3D reconstruction of an entire environment. The difference between the 3-layers and 5-layers scans was minimal, with the 3-layer scan often outperforming the 5-layer scan. This may be attributed to the 5-layers scan providing excessive data, complicating the algorithm's ability to accurately align the frames. For this case, a 3-layers scan is sufficient. All of the above is visible in Figure 11.

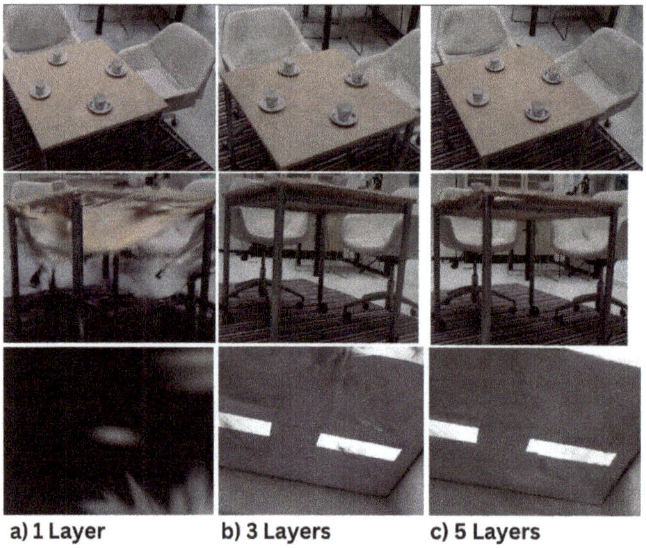

Figure 11. Layering Technique for Enhanced 3D Reconstruction (**a**) 1 Layer, (**b**) 3 Layers, (**c**) 5 Layers.

3.4. Most Optimal Scan Method

3.4.1. Summary of Methods

All the methods mentioned above have been tested. The chosen method within an environment significantly impacts the quality of the 3D reconstruction. Each scan was evaluated based on the amount of noise present, the level of detail achieved, and the number of splats (data density). The results varied across the different scans. The poorest quality was observed in Method 3, where the 3D reconstructed environment was barely recognizable, with a splats count of only 250,265 (rating 2). In contrast, Methods 5, 8, and 10 yielded excellent results, characterized by high detail, minimal noise, and high splats counts of 432,640, 489,000, and 474,357 (ratings 4 and 5), respectively. These three reconstructions will be discussed in detail in the next section.

The analysis primarily concludes that capturing the hallway, living room, and kitchen in a single video is ineffective. The algorithm receives an excessive amount of data, resulting in a poorly reconstructed model. Dividing the environment into multiple scans enhances the reconstruction quality, making it more detailed. The splats metric further supports this conclusion, as it quantifies the density of data points (splats) in the reconstruction. A low splats count, such as in Method 9 (212,211, rating 1), often indicates insufficient data density, leading to sparse or incomplete reconstructions. These models may exhibit noticeable gaps, lower texture quality, and reduced accuracy, making them unsuitable for applications requiring high fidelity, such as forensic investigations.

Conversely, a high splats count, as observed in Methods 5, 8, and 10 (432,640, 489,000, and 474,357 splats, ratings 4 and 5), reflects a dense and uniform distribution of data points. This results in smoother surfaces, reduced noise, and improved texture mapping quality, all of which contribute to more realistic and detailed 3D reconstructions. However, an excessively high splats count can also increase computational requirements, making the processing more resource intensive. This trade-off highlights the need to balance splats density with practical workflow considerations.

For the room used in these tests, segmenting it into a maximum of three segments ensures an optimal balance. This approach achieves sufficiently high splats counts for accurate reconstructions without overwhelming the reconstruction algorithm or requiring excessive computational resources. By leveraging the splats metric as an additional evaluation criterion, it becomes evident that methods focusing on segmentation and optimized data density produce the best results for detailed and reliable 3D reconstructions.

Reflecting on Method 5, scanning around a specific object (loop closures) proved beneficial. For instance, the scan around the coffee table produced a high-quality reconstruction. For scans 8 and 10, the environment was divided into three paths, which resulted in the best reconstructions. However, reconstructing the hallway remains difficult due to its height, narrowness, and large monotone surfaces, which complicate the reconstruction process. The analysis of all scanning methods is visible in Table 2.

Table 2. Comparison of Scan Methods.

Scan Methods	Noise	Details	Splats
1	3	4	4
2	2	3	3
3	2	2	2
4	4	3	5
5	4	4	4
6	3	3	3
7	2	4	2

Table 2. *Cont.*

Scan Methods	Noise	Details	Splats
8	4	4	5
9	2	3	1
10	4	5	5

3.4.2. Method 5

Method 5 yielded promising results. In this scan, each object or area was individually scanned, resulting in nine separate 3D reconstructions. Most of these individual reconstructions were high quality. Among the nine reconstructions, images were captured of the kitchen, the kitchen table, and the coffee table, as illustrated below. Figure 12 showcases the reconstruction of the kitchen table, where the table, coffee cups, and chairs are clearly visible with minimal noise in the 3D data. The objects on both the kitchen table and counter are well-defined and easily identifiable as shown in Figure 13.

Figure 12. Method 5—Scanning kitchen and tabletop.

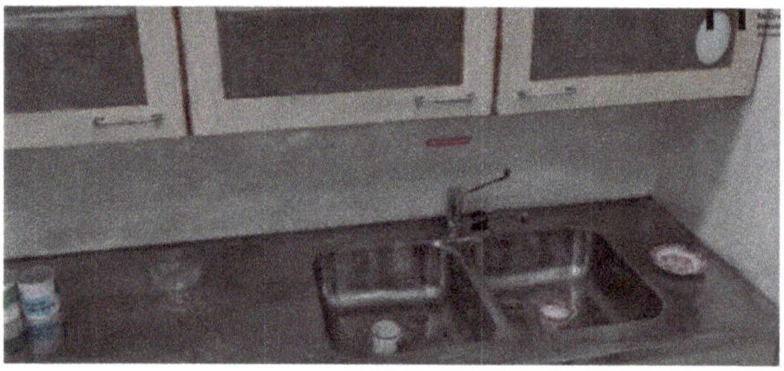

Figure 13. Method 5—Kitchen counter.

Overall, scanning specific objects or zones led to more detailed and accurate 3D reconstructions, confirming that isolating individual areas improves reconstruction quality. However, one limitation is that current 3D reconstruction technology still requires manual work to merge separate scans. This means that each scan must be cropped, resized, and aligned to assemble a complete 3D reconstruction of an entire room. While larger zones can be connected, merging nine separate reconstructions would be time-consuming and challenging. Given the current limitations in splat file editing applications, this process is cumbersome and impractical for now. In summary, while focusing on individual zones enhances the quality of 3D reconstructions, this method is not yet feasible for whole-room reconstructions until more advanced algorithms or tools are developed to automate merging and clean up multiple 3D scans.

3.4.3. Methods 8 and 10

Methods 8 and 10 are identical, with the exception that in Method 10, the area around the coffee table is extended to include the side of the black cabinet, resulting in a complete reconstruction of the cabinet. Figures 14 and 15 show screenshots of the cabinet, kitchen, and coffee table, all of which are captured detailed. Some items even have readable text. However, there is still some noise in the reconstructions, although this can be removed with further processing.

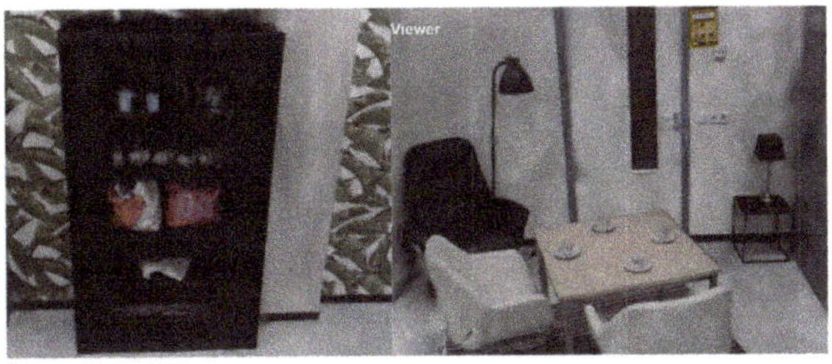

Figure 14. Results from Method 8 and Method 10.

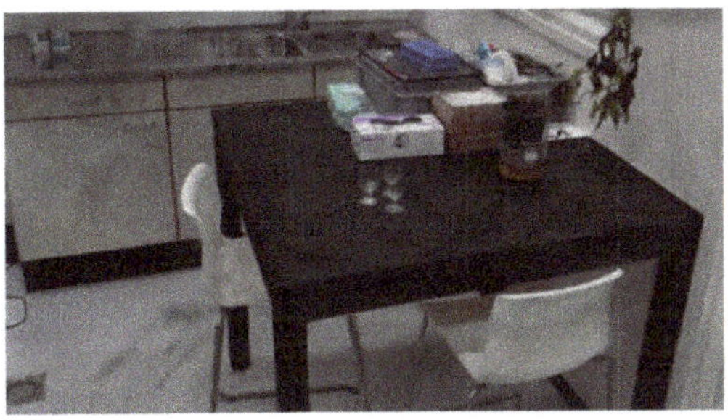

Figure 15. Tabletop results from Method 8 and Method 10.

In summary, Method 5 produced very detailed 3D reconstructions of individual objects and areas. However, due to the lack of effective techniques for stitching 3D reconstructions, this method is not suitable. It takes too much time to attach nine smaller 3D reconstructions to each other. Method 10 emerged as the best scan path. The general living/kitchen area was reconstructed well while maintaining an efficient workflow. However, despite its effectiveness in more complex areas, limitations persist, particularly in narrow or featureless spaces like hallways. The inability to effectively stitch reconstructions of individual objects into a cohesive whole reflects current technological constraints in handling multiple separate scans.

4. Discussion

This section evaluates the implications of the findings on capturing techniques for 3D reconstruction, emphasizing their practical applications, limitations, and alignment with the existing literature. By comparing the observed results with previous studies, this section identifies how specific capturing methods influence reconstruction quality, noise reduction, and detail accuracy. Additionally, it highlights the potential for refining current workflows, addresses the limitations imposed by technological constraints, and proposes avenues for future research to enhance the efficiency and adaptability of 3D reconstruction processes.

4.1. Comparison with the Literature

The importance of slow, deliberate camera movements in improving 3D reconstruction quality is well-documented. Zhang et al. [37] highlighted that optimized camera trajectories, such as those implemented in ROSEFusion, enhance feature matching and reduce noise artifacts even under dynamic conditions. Consistent with their findings, our study demonstrated that slower camera movements produced higher-quality reconstructions, with improved detail and reduced noise. This is attributed to increased frame overlap and better focus, aligning with the broader consensus in the literature.

Camera orientation also plays a pivotal role in the quality of 3D reconstructions. While portrait and landscape modes offer unique advantages, landscape orientation is generally preferred for its broader field of view and greater horizontal overlap. This observation is supported by principles in photogrammetry, which emphasize the importance of maximizing overlap for better feature recognition [38]. Our results reinforce this understanding, with landscape mode consistently yielding superior reconstructions compared to portrait mode, especially for horizontally expansive environments.

Reconstructing narrow or featureless spaces remains a persistent challenge. Lu et al. [39] identified that such environments hinder feature detection, impacting reconstruction accuracy. This was evident in our study, where hallways posed significant difficulties for reconstruction algorithms due to their monotone surfaces and lack of distinctive features. However, our segmentation-based methods, particularly Method 10, showed promise in mitigating these issues by dividing environments into manageable zones. This approach echoes similar strategies proposed in prior research but extends them by incorporating Gaussian Splatting techniques to further enhance reconstruction fidelity.

4.2. Potential Applications

The findings of this research have significant implications across multiple fields requiring high-quality 3D reconstructions. In crime scene investigations, accurate 3D models enhance evidence analysis and presentation by capturing spatial relationships and event sequences. Beyond forensics, these optimized methods support architecture, archaeology, and cultural heritage preservation, enabling precise, non-invasive documentation and restoration efforts while enhancing public engagement through virtual experiences.

In VR and AR, high-quality 3D models improve realism and immersion, benefiting training simulations, gaming, and education. By applying the recommended capturing techniques, developers can achieve clearer, more detailed models, broadening the applicability of 3D reconstruction technology across diverse professional domain

4.3. Limitations and Future Research

4.3.1. Limitations

This study had several limitations. The experimental setup was constrained by the available space and objects, which may not be fully representative of typical crime scenes. Additionally, the inability to fully automate the merging of individual object scans into a single coherent model limited this study's application to real-world crime scene reconstruction. This study was also limited by the specific technology and software used for 3D Gaussian Splattering (3D-GS), which may not have been the most advanced available at the time. The manual intervention required in aligning multiple scans remains a challenge, emphasizing the need for more sophisticated algorithms and tools for future studies.

Another limitation to consider is domain shift, which occurs when the developed methods are applied to environments significantly different from the controlled indoor settings used in this study. For example, outdoor crime scenes or disaster sites introduce unique challenges, such as dynamic lighting, larger spatial dimensions, and less predictable textures. These shifts could affect the performance of Gaussian Splatting and NeRF algorithms, necessitating domain-specific adaptations or improvements to maintain reconstruction quality. This will be a topic for future research. Initial experiments have already started, and the domain shift is considered.

While this study identified optimal filming techniques for indoor crime scene reconstructions, their generalizability to other environments, such as outdoor scenes or highly reflective surfaces, remains an open question. Outdoor crime scenes introduce additional challenges, including variable lighting, weather conditions, and larger spatial areas, which may impact the effectiveness of the scanning paths and layering techniques used in this study. Similarly, scenes with highly complex or featureless surfaces, such as reflective floors or monochromatic walls, could require adjustments to filming strategies to ensure sufficient feature matching. Future work should investigate how these techniques perform under different environmental conditions and whether adaptive filming strategies, such as exposure compensation or alternative scanning paths, are necessary to maintain reconstruction quality across diverse crime scene scenarios.

4.3.2. Future Research

Future work should focus on overcoming the limitations of merging multiple scans into a cohesive whole, perhaps by advancing the software algorithms used in 3D reconstruction. Further exploration into using more advanced stabilizers or integrating markers within the scene could enhance the quality of the reconstructed models. Additionally, investigating how different lighting conditions or environments affect the 3D Gaussian Splattering process could broaden the applicability of this method in real-world crime scene reconstructions. There is also room for improvement in the reconstruction of narrow, featureless spaces like hallways, where more specialized techniques may be necessary.

Expanding the scope of future work, we also plan to explore the application of these methods in diverse indoor and outdoor environments. While this study focused on simulated indoor scenarios for ease of replication, real-world environments such as abandoned buildings, train stations, or underground tunnels and outdoor areas present unique challenges, including variable lighting and complex spatial layouts [40]. For instance, outdoor environments may introduce domain shift due to differences in lighting conditions, object textures, and spatial dimensions, potentially affecting the performance of Gaussian Splatting and NeRF algorithms. Addressing these challenges through adaptive techniques, such as fine-tuning algorithms for specific domains or integrating more generalizable approaches, would significantly enhance the robustness of these methods. As an example, we are currently experimenting with 3D reconstructions of arson scenarios, which require capturing detailed and accurate representations of fire-damaged structures and debris. Initial work with cases of arson in both indoor and outdoor locations, as shown in Figure 16, demonstrates the need to test the adaptability and scalability of the proposed methods under diverse conditions. These efforts aim to ensure that the methods remain effective across varying scenarios, thereby expanding their utility in real-world forensic applications.

Future research could focus on integrating real-time feedback mechanisms during the filming process to enhance data acquisition. Such systems could identify areas with insufficient data coverage, where low frame density might lead to poor textures, artifacts, or cloud-like distortions in the final reconstruction. By providing immediate visual cues or alerts, operators could adapt their filming strategy on the spot, ensuring better coverage of critical areas before processing begins. This would help minimize gaps and inconsistencies, reducing the need for re-scanning and improving the overall quality of 3D reconstructions. Implementing this approach could lead to more efficient workflows, as adjustments could be made in real-time rather than relying solely on post-processing corrections

Figure 16. 3D reconstruction of arson.

5. Conclusions

This study explored various parameters influencing the quality of 3D reconstructions in forensic contexts, focusing on capturing methods using handheld cameras in indoor environments. By systematically adjusting factors such as camera orientation, filming speed, layering techniques, and scanning paths, the research identified the most effective strategies for achieving detailed and accurate reconstructions of crime scenes.

The findings highlight the advantages of using a landscape orientation over a portrait mode, as it offers superior horizontal overlap, which is essential for aligning features effectively. This minimizes anomalies and ensures a more consistent reconstruction quality. Furthermore, the results demonstrated that slower camera movements significantly enhance the quality of reconstructions. The additional frames captured during slower movements allow for improved data quality and better feature matching, leading to reconstructions with higher detail and reduced noise.

In terms of layering, this study found that capturing the environment at three different heights provides an optimal balance between detail and processing efficiency. While additional layers occasionally introduced noise, the three-layer approach captured sufficient vertical data to reconstruct both the tops and bottoms of objects effectively. Scanning individual objects or areas, as seen in Method 5, produced high-quality reconstructions but was impractical due to the challenges of merging separate scans into a cohesive whole. Method 10, which employed segmented scanning with an additional loop for specific areas, emerged as the most practical approach, balancing detail and workflow efficiency. However, difficulties remained in reconstructing narrow, featureless spaces like hallways, underscoring limitations in current algorithms.

These findings provide valuable insights for forensic investigations, offering guidance on optimizing 3D reconstruction techniques with current technologies. While promising, this study also highlights the need for advancements in software to automate the merging of segmented scans into unified models. Future research could explore marker-based scanning techniques and more sophisticated reconstruction algorithms to address these challenges, enhancing the practical applicability of the proposed methods in real-world forensic scenarios.

By improving the quality and efficiency of 3D reconstructions, this study contributes to forensic science, architecture, and other fields requiring accurate spatial modeling, paving the way for more reliable and immersive applications of 3D technology.

Author Contributions: Conceptualization, K.W., S.W., R.M., M.v.K. and D.R.; methodology, D.R.; software, K.W., S.W. and D.R.; validation, K.W., S.W. and D.R.; formal analysis, D.R.; investigation, K.W., S.W. and D.R.; resources, D.R. and R.M.; data curation, D.R.; writing—original draft preparation, D.R.; writing—review and editing D.R., M.v.K. and R.M.; visualization, K.W. and S.W.; supervision,

M.v.K. and R.M.; project administration, D.R.; funding acquisition, D.R. All authors have read and agreed to the published version of the manuscript.

Funding: This research received financial support from the University of Twente and research group Technologies for Criminal Investigations, part of the Saxion University for Applied Sciences and the Police Academy of The Netherlands. Also, this research is part of the project; AI-aided robotic system for crime scene investigation and reconstruction; funded by the Police and Science grant of the Police Academy of The Netherlands.

Institutional Review Board Statement: Not applicable.

Informed Consent Statement: Not applicable.

Data Availability Statement: Data are contained within this article.

Acknowledgments: Special gratitude to the University of Twente in The Netherlands, the Technology for Criminal Investigations research group, part of the Saxion University of Applied Sciences and the Police Academy in The Netherlands, the Technical University of Sofia in Bulgaria and all researchers in the CrimeBots research line part of research group TCI. Also, thanks for the dedicated opportunity and funding from the Police Academy with the Police and Science grant.

Conflicts of Interest: The authors declare no conflicts of interest.

References

1. Mostafa, Y.M.; Al-Berry, M.N.; Shedeed, H.A.; Tolba, M.F. Data Driven 3D Reconstruction from 2D Images: A Review. In *Lecture Notes on Data Engineering and Communications Technologies, Proceedings of the 8th International Conference on Advanced Intelligent Systems and Informatics 2022, Cairo, Egypt, 20–22 November 2022*; Springer: Cham, Switzerland, 2023; Volume 152, pp. 812–823. [CrossRef]
2. Moons, T.; Van Gool, L.; Vergauwen, M. 3D Reconstruction from Multiple Images Part 1: Principles. *Found. Trends® Comput. Graph. Vis.* **2008**, *4*, 287–404. [CrossRef]
3. Kang, Z.; Yang, J.; Yang, Z.; Cheng, S. A Review of Techniques for 3D Reconstruction of Indoor Environments. *ISPRS Int. J. Geo-Inf.* **2020**, *9*, 330. [CrossRef]
4. Sharma, O.; Arora, N.; Sagar, H. Image Acquisition for High Quality Architectural Reconstruction. In Proceedings of the 45th Graphics Interface Conference, Kingston, ON, Canada, 28–31 May 2019.
5. Apollonio, F.I.; Fantini, F.; Garagnani, S.; Gaiani, M. A Photogrammetry-Based Workflow for the Accurate 3D Construction and Visualization of Museums Assets. *Remote. Sens.* **2021**, *13*, 486. [CrossRef]
6. Zhang, C.; Maga, A.M. An Open-Source Photogrammetry Workflow for Reconstructing 3D Models. *Integr. Org. Biol.* **2023**, *5*, obad024. [CrossRef] [PubMed]
7. Feng, Y.; Wu, R.; Liu, X.; Chen, L. Three-Dimensional Reconstruction Based on Multiple Views of Structured Light Projectors and Point Cloud Registration Noise Removal for Fusion. *Sensors* **2023**, *23*, 8675. [CrossRef]
8. Fan, J.; Wang, X.; Zhou, C.; Zhang, P.; Jing, F.; Hou, Z. Structured Light Vision 3-D Reconstruction System for Different Media Considering Refraction: Design, Modeling, and Calibration. *IEEE/ASME Trans. Mechatron.* **2023**, *29*, 1997–2008. [CrossRef]
9. Eltner, A.; Sofia, G. Structure from motion photogrammetric technique. *Dev. Earth Surf. Process.* **2020**, *23*, 1–24. [CrossRef]
10. Wang, X.; Wang, C.; Liu, B.; Zhou, X.; Zhang, L.; Zheng, J.; Bai, X. Multi-view stereo in the Deep Learning Era: A comprehensive review. *Displays* **2021**, *70*, 102102. [CrossRef]
11. Mildenhall, B.; Srinivasan, P.P.; Tancik, M.; Barron, J.T.; Ramamoorthi, R.; Ng, R. NeRF: Representing Scenes as Neural Radiance Fields for View Synthesis. In *Lecture Notes in Computer Science (Including Subseries Lecture Notes in Artificial Intelligence and Lecture Notes in Bioinformatics), Proceedings of the 16th European Conference, Glasgow, UK, 23–28 August 2020*; Springer: Berlin/Heidelberg, Germany, 2020; Volume 12346 LNCS, pp. 405–421. [CrossRef]
12. Kerbl, B.; Kopanas, G.; Leimkuehler, T.; Drettakis, G. 3D Gaussian Splatting for Real-Time Radiance Field Rendering. *ACM Trans. Graph.* **2023**, *42*, 14. [CrossRef]
13. David, A.; Joy, E.; Kumar, S.; Bezaleel, S.J. Integrating Virtual Reality with 3D Modeling for Interactive Architectural Visualization and Photorealistic Simulation: A Direction for Future Smart Construction Design Using a Game Engine. In *Lecture Notes in Networks and Systems, Proceedings of the Second International Conference on Image Processing and Capsule Networks: ICIPCN 2021, Bangkok, Thailand, 27–28 May 2021*; Springer: Cham, Switzerland, 2022; Volume 300 LNNS, pp. 180–192. [CrossRef]
14. Remondino, F.; El-Hakim, S.; Girardi, S.; Rizzi, A.; Benedetti, S.; Gonzo, L. 3D Virtual Reconstruction and Visualization of Complex Architectures—The '3D-ARCH' Project. Available online: www.stefanobenedetti.com (accessed on 15 July 2024).

15. Bevilacqua, M.G.; Russo, M.; Giordano, A.; Spallone, R. 3D Reconstruction, Digital Twinning, and Virtual Reality: Architectural Heritage Applications. In Proceedings of the 2022 IEEE Conference on Virtual Reality and 3D User Interfaces Abstracts and Workshops (VRW), Christchurch, New Zealand, 12–16 March 2022; pp. 92–96. [CrossRef]
16. Gomes, L.; Bellon, O.R.P.; Silva, L. 3D reconstruction methods for digital preservation of cultural heritage: A survey. *Pattern Recognit. Lett.* **2014**, *50*, 3–14. [CrossRef]
17. Cefalu, A.; Abdel-Wahab, M.; Peter, M.; Wenzel, K.; Fritsch, D. Image based 3D Reconstruction in Cultural Heritage Preservation. In Proceedings of the 10th International Conference on Informatics in Control, Automation and Robotics, Reykjavík, Iceland, 29–31 July 2013. [CrossRef]
18. Rodriguez-Garcia, B.; Guillen-Sanz, H.; Checa, D.; Bustillo, A. A systematic review of virtual 3D reconstructions of Cultural Heritage in immersive Virtual Reality. *Multimed. Tools Appl.* **2024**, *83*, 89743–89793. [CrossRef]
19. Zhong, C.; Cheng, S.; Kasoar, M.; Arcucci, R. Reduced-order digital twin and latent data assimilation for global wildfire prediction. *Nat. Hazards Earth Syst. Sci.* **2023**, *23*, 1755–1768. [CrossRef]
20. Cheng, S.; Liu, C.; Guo, Y.; Arcucci, R. Efficient deep data assimilation with sparse observations and time-varying sensors. *J. Comput. Phys.* **2023**, *496*, 112581. [CrossRef]
21. Nguyen-Phuoc, T.; Liu, F.; Xiao, L. SneRF: Stylized Neural Implicit Representations for 3D Scenes. *ACM Trans. Graph.* **2022**, *41*, 11. [CrossRef]
22. Kulhanek, J.; Sattler, T. Tetra-NeRF: Representing Neural Radiance Fields Using Tetrahedra. 2023. Available online: https://github.com/jkulhanek/tetra-nerf (accessed on 15 July 2024).
23. Tancik, M.; Weber, E.; Ng, E.; Li, R.; Yi, B.; Wang, T.; Kristoffersen, A.; Austin, J.; Salahi, K.; Ahuja, A.; et al. Nerfstudio: A Modular Framework for Neural Radiance Field Development. In Proceedings of the SIGGRAPH 2023, Los Angeles, CA, USA, 6–10 August 2023. [CrossRef]
24. Müller, T.; Evans, A.; Schied, C.; Keller, A. Instant Neural Graphics Primitives with a Multiresolution Hash Encoding. *ACM Trans. Graph.* **2022**, *41*, 102. [CrossRef]
25. Liang, R.; Zhang, J.; Li, H.; Yang, C.; Guan, Y.; Vijaykumar, N. SPIDR: SDF-Based Neural Point Fields for Illumination and Deformation. *arXiv* **2022**, arXiv:2210.08398. Available online: https://arxiv.org/abs/2210.08398v3 (accessed on 15 July 2024).
26. Reiser, C.; Szeliski, R.; Verbin, D.; Srinivasan, P.; Mildenhall, B.; Geiger, A.; Barron, J.; Hedman, P. MERF: Memory-Efficient Radiance Fields for Real-time View Synthesis in Unbounded Scenes. *ACM Trans. Graph.* **2023**, *42*, 1–12. [CrossRef]
27. Rangelov, D.; Knotter, J.; Miltchev, R. 3D Reconstruction in Crime Scenes Investigation: Impacts, Benefits, and Limitations. In *Lecture Notes in Networks and Systems, Intelligent Systems and Applications, IntelliSys 2024, Amsterdam, The Netherlands, 5–6 September 2024*; Springer: Cham, Switzerland, 2024; Volume 1065 LNNS, pp. 46–64. [CrossRef]
28. Rangelov, D.; Waanders, S.; Waanders, K.; van Keulen, M.; Miltchev, R. Impact of Camera Settings on 3D Reconstruction Quality: Insights from NeRF and Gaussian Splatting. *Sensors* **2024**, *24*, 7594. [CrossRef]
29. Sony. Sony Alpha 7C Full-Frame Mirrorless Camera—Black | ILCE7C. Available online: https://electronics.sony.com/imaging/interchangeable-lens-cameras/all-interchangeable-lens-cameras/p/ilce7c-b?srsltid=AfmBOoo5N6vG9O3tR3d9p7ZKy9YqWMPZSzdnQnfZjfl4XP9WE2vRx1bz (accessed on 22 December 2024).
30. Sigma. 14mm F1.4 DG DN | Art | Lenses | SIGMA Corporation. Available online: https://www.sigma-global.com/en/lenses/a023_14_14/ (accessed on 22 December 2024).
31. DJI. DJI RS 4—Gripping Storytelling—DJI. Available online: https://www.dji.com/bg/rs-4 (accessed on 22 December 2024).
32. Jawset. Jawset Postshot. Available online: https://www.jawset.com/ (accessed on 22 December 2024).
33. Tosi, F.; Zhang, Y.; Gong, Z.; Sandström, E.; Mattoccia, S.; Oswald, M.R.; Poggi, M. How NeRFs and 3D Gaussian Splatting are Reshaping SLAM: A Survey. *arXiv* **2024**, arXiv:2402.13255. [CrossRef]
34. Yurkova, K. A Comprehensive Overview of Gaussian Splatting | Towards Data Science. Available online: https://medium.com/towards-data-science/a-comprehensive-overview-of-gaussian-splatting-e7d570081362 (accessed on 3 February 2025).
35. Canon. What Is Aperture Photography? | Canon U.S.A., Inc. Available online: https://www.usa.canon.com/learning/training-articles/training-articles-list/what-is-aperture (accessed on 22 December 2024).
36. Garmin fēnix® 6 Pro Solar | Multisport Solar Watch. Available online: https://www.garmin.com/en-US/p/702902 (accessed on 12 January 2025).
37. Zhang, J.; Zhu, C.; Zheng, L.; Xu, K. ROSEFusion: Random Optimization for Online Dense Reconstruction under Fast Camera Motion. *ACM Trans. Graph.* **2021**, *40*, 56. [CrossRef]
38. Lefcourt, D. Portrait vs Landscape Orientation in Photography (Which Is Better?)—Lefcourt Photography. Available online: https://www.lefcourtphotography.com/portrait-vs-landscape-orientation-in-photography-which-is-better (accessed on 22 December 2024).

39. Lu, F.; Zhou, B.; Zhang, Y.; Zhao, Q. Real-time 3D scene reconstruction with dynamically moving object using a single depth camera. *Vis. Comput.* **2018**, *34*, 753–763. [CrossRef]
40. Fang, X.; Easwaran, A.; Genest, B.; Suganthan, P.N. Your Data Is Not Perfect: Towards Cross-Domain Out-of-Distribution Detection in Class-Imbalanced Data. *arXiv* **2024**, arXiv:2412.06284. Available online: https://arxiv.org/abs/2412.06284v1 (accessed on 3 February 2025). [CrossRef]

Journal of *Imaging*

MDPI

Article

Robot-Based Procedure for 3D Reconstruction of Abdominal Organs Using the Iterative Closest Point and Pose Graph Algorithms

Birthe Göbel [1,2,*], Jonas Huurdeman [2,3], Alexander Reiterer [1,4] and Knut Möller [5]

1 Department of Sustainable Systems Engineering, NATECH, University of Freiburg, Emmy-Noether-Straße 2, 79110 Freiburg, Germany; alexander.reiterer@mail.inatech.uni-freiburg.de
2 KARL STORZ SE & Co. KG, Dr.-Karl-Storz-Str. 34, 78532 Tuttlingen, Germany; jonas.huurdeman@karlstorz.com
3 Institute for Bioinformatics and Medical Informatics (IBMI), University of Tübingen, Geschwister-Scholl-Platz, 72074 Tübingen, Germany
4 Fraunhofer Institute for Physical Measurement Techniques IPM, Georges-Köhler-Allee 301, 79110 Freiburg im Breisgau, Germany
5 Institute of Technical Medicine, ITeM, Furtwangen University (HFU), Jakob-Kienzle-Straße 17, 78054 Villingen-Schwenningen, Germany; knut.moeller@hs-furtwangen.de
* Correspondence: birthe.goebel@mail.inatech.uni-freiburg.de

Abstract: Image-based 3D reconstruction enables robot-assisted interventions and image-guided navigation, which are emerging technologies in laparoscopy. When a robotic arm guides a laparoscope for image acquisition, hand–eye calibration is required to know the transformation between the camera and the robot flange. The calibration procedure is complex and must be conducted after each intervention (when the laparoscope is dismounted for cleaning). In the field, the surgeons and their assistants cannot be expected to do so. Thus, our approach is a procedure for a robot-based multi-view 3D reconstruction without hand–eye calibration, but with pose optimization algorithms instead. In this work, a robotic arm and a stereo laparoscope build the experimental setup. The procedure includes the stereo matching algorithm Semi Global Matching from OpenCV for depth measurement and the multiscale color iterative closest point algorithm from Open3D (v0.19), along with the multiway registration algorithm using a pose graph from Open3D (v0.19) for pose optimization. The procedure is evaluated quantitatively and qualitatively on ex vivo organs. The results are a low root mean squared error (1.1–3.37 mm) and dense point clouds. The proposed procedure leads to a plausible 3D model, and there is no need for complex hand–eye calibration, as this step can be compensated for by pose optimization algorithms.

Keywords: Image-based 3D reconstruction; laparoscopy; multi-view reconstruction; robot-assisted intervention; stereo vision

1. Introduction

Multi-view image-based 3D reconstruction describes the stitching of multiple depth maps of an object into a single 3D model. In laparoscopy, this could include the stitching of abdominal organs such as the liver or gallbladder. The advantage of multi-view reconstruction is that multiple perspectives of an organ can be observed, leading to a complete 3D model (rather than only parts of an organ), which is necessary for applications such as image-guided navigation and robot-assisted interventions. It requires two separate tasks: depth estimation and camera pose estimation. This article describes a procedure to achieve a complete 3D model of abdominal organs, with a focus on the second task, pose estimation.

A robot delivers an initial guess for the camera pose, which is then optimized by two pose optimization algorithms. For the experimental setup, a stereo laparoscope and a robotic arm for camera guidance are utilized.

1.1. Related Work

In the literature, there are a few articles about multi-view reconstruction based on stereo laparoscopes. Reference [1] used stereo frames for depth estimation and combined optical tracking and an iterative closest point (ICP) algorithm for camera pose estimation. The accuracy was not measured. Similar to this, Reference [2] took calibrated stereo images and compared three different pose estimation algorithms, including simultaneous localization and mapping (SLAM), visual odometry (VO), and structure from motion (SfM), to reconstruct a phantom surgical cavity. Their findings indicated that SLAM is the most promising algorithm due to its fastest computation time and high accuracy and precision (a little less than SfM). For accuracy measurement, the root mean squared error (RMSE) was applied and resulted in 13.97 mm.

In contrast to these two studies, Reference [3] designed a stereo endoscope with two different states of stereo bases, which can be applied by pushing a button. By using a feature algorithm, they calculated a homography matrix to stitch together two monocular images taken from a bigger stereo base. Here, the accuracy was not measured. Reference [4] applied SLAM for camera pose estimation and a stereo matching algorithm for 3D reconstruction. In the end, they wanted to overlay augmented information during minimally invasive surgery (MIS). The measured RMSE for a simulated abdominal scene amounted to 2.37 mm.

Reference [5] proposed a method for 3D reconstruction based on a SLAM algorithm, in which feature patch extraction replaced feature point extraction, making the algorithm more robust. The method was validated on the laparoscopic in vivo Hamlyn dataset, and the RMSE varied between 5.2 mm, 2.3 mm, and 1.3 mm depending on the image resolution—the lower, the worse the RMSE. Reference [6] described the development of an autonomous scanning and mosaicking system with the help of the da Vinci ® surgical robot (Intuitive Surgical, Sunnyvale, CA, USA), an endomicroscope, and a stereo laparoscope. The camera motion estimation was performed by a tracking marker on the endomicroscope and the acquired images. There was no information offered about the 3D reconstruction accuracy.

Reference [7] used stereoscopy for 3D reconstruction and a 2D feature tracking algorithm for motion estimation to register the reconstructed points to a preoperative volumetric scan, reporting a registration error for intraoperative data of 1.6493 mm (error between points on the registered stereo reconstruction and the corresponding points on the preoperative volume). Reference [8] took stereo laparoscopic images and a SLAM algorithm called EMDQ-SLAM for 3D reconstruction, as well as camera motion and tissue deformation estimation. The accuracy was not computed.

1.2. Camera Pose Estimation by Uncalibrated Robot Position Combined with Pose Optimization Algorithms

Most of the aforementioned papers only focus on handheld laparoscopes, which is fine, as this is the state-of-the-art for most surgeons. However, the amount of robot-assisted surgery (RAS) increases yearly [9–12]. Research even gets a step further in the direction of autonomous robotic surgery as in [13–15]. In Reference [14], 3D reconstruction of the surgical scene is the basis for an autonomous suturing experiment. Thus, 3D reconstruction is an enabling technology for autonomy in RAS. Moreover, a robot-guided camera brings advantages to our 3D reconstruction approach, such as the precise execution of movements, the standstill during image acquisition, and the known robot position. Accordingly, this work presents a procedure to create a 3D model by utilizing the stereo laparoscope TipCam Rubina 1S 30° (KARL STORZ SE & Co. KG, Tuttlingen, Germany) guided by the UR5 CB3 robotic arm (Universal Robots A/S, Odense, Denmark).

When a camera is mounted on a robot's flange, hand–eye calibration is usually executed to know the exact transformation between the camera and the robot frame [16]. There are different approaches to achieving this, and often such a process requires targets such as chessboard patterns or QR code-like patterns, making it complicated and error-prone [17]. Especially, surgeons and their assistant staff cannot be expected to execute it. Ref. [18] developed a hand–eye calibration routine without such printed targets but instead used the instruments during the intervention as targets. It resulted in a 3 mm accuracy of the 3D instrument positioning error, which might not be enough for a submillimeter-accurate 3D reconstruction. As discussed in [19], small measurement errors in the robot flange position lead to amplified errors at the laparoscope tip. Both the complicated hand–eye calibration process and the risk of errors motivate us to examine if hand–eye calibration can be left out and be substituted by plain image-processing or point cloud-processing software. Thus, the camera pose estimation does not rely solely on the robot but on the robot position combined with an ICP and a pose graph optimization algorithm.

2. Methods

The challenge of multi-view compared to single-shot 3D reconstruction is that not only the depth is estimated, but also these separate depth maps must be stitched together. Thus, two tasks can be separated:

1. Depth estimation for point cloud generation;
2. Camera pose estimation and optimization for point cloud stitching (the focus of this work).

There are different image-based methods to realize the first task, e.g., Structure-from-Motion (SfM), Shape-from-Shading (SfS), Time-of-Flight (ToF) cameras, and stereoscopy, as presented in [20]. In this work, stereoscopy is chosen for depth estimation (see Figure 1, showing the laparoscope tip with two image sensors). The reason is that compared to mono camera systems, the second camera delivers additional information, which enables 3D vision and accurate depth estimation as shown in [21]. Moreover, stereo camera systems are state-of-the-art systems for laparoscopic interventions [20].

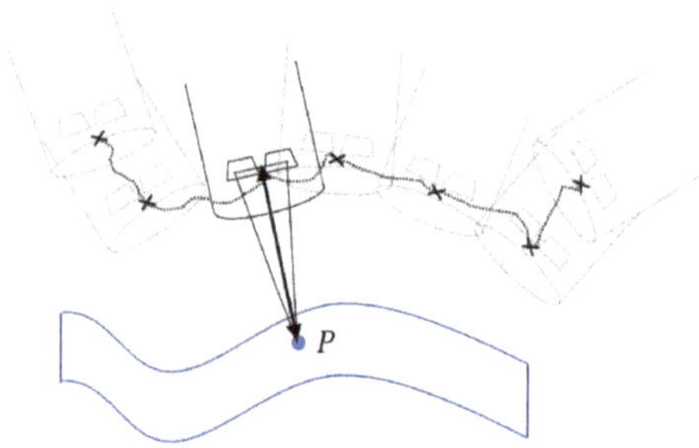

Figure 1. Schematic overview of the 3D reconstruction method stereoscopy. It shows the surface to be reconstructed, with point P in blue, the laparoscope tip with two image sensors generating the estimated depth (black arrow), and the estimated camera positions (x marks and dotted line).

The second task is relevant for aligning each single point cloud with its neighboring point cloud, leading to a complete merged point cloud in the end. This requires knowledge about all the passed camera positions (see Figure 1, black-dotted line). Ideally, the camera pose is known by a sensor, e.g. inertial measurement unit (imu), but standard laparoscopes do not include such and an imu suffers from drift, which leads to inaccurate poses. The alternative is a pose estimation algorithm based on image/point cloud processing or robotic guidance (where the camera position is known by the robot).

In this work, the laparoscope is guided by a robotic arm, which brings the advantage of knowing the robot flange position, and based on that, the camera position is calculated by applying a transformation matrix. For submillimeter accuracy, hand–eye calibration of the exact camera position relative to the robot flange would be necessary. As stated in the introduction, an alternative procedure is examined to avoid hand–eye calibration by instead utilizing two image-/point cloud-based algorithms for camera pose optimization, namely an ICP and a pose graph optimization algorithm. To show the effectiveness of this approach, three different stitching approaches based on the same stereo frames and the same robot positions are created and compared. The difference between each stitching is the method of pose estimation:

1. Rob only (only the uncalibrated robot positions are considered for stitching);
2. Rob + ICP (in addition to the robot position, the ICP algorithm is also applied for pose optimization);
3. Rob + ICP + pose graph (our final approach, which utilizes the uncalibrated robot position as an initial position guess alongside the ICP and pose graph algorithm for pose optimization).

The following chapter presents the experimental setup, the software architecture (including a detailed description of the ICP and pose graph optimization algorithm), and the evaluation of the results.

2.1. Experimental Setup

The experimental setup includes the robotic arm UR5 CB3 (Universal Robots A/S, Odense, Denmark) for camera guidance, the stereo laparoscope TipCam Rubina 1S 30° (KARL STORZ SE & Co. KG, Tuttlingen, Germany) (see Figure 2) for image acquisition, and the EinScan H2 laser scanner (Shining 3D Tech Co., Ltd., Hangzhou, China) for ground truth acquisition. The stereo camera is angled at 30° and has an 80° FOV field-of-view (FOV). Each left and right image frame offers a Full HD resolution of 1920 × 1080 pixels, and image processing is partly run on an RTX A4000 graphics card (NVIDIA Corporation, Santa Clara, CA, USA). Ex vivo pig organs were used for the qualitative and quantitative evaluation. In this work, the fulcrum point—also known as the trocar point, which is the entry point of the laparoscope through the abdominal wall—is not considered. This constraint complicates the robotic movement for image acquisition because it reduces the six degrees of freedom (DOF) to four DOF. Thus, the robot is positioned so that the camera viewing direction is perpendicular to the object surface, and images are taken in a grid-shaped movement pattern. In total, 63 stereo frames are taken with a 70 mm distance between the laparoscope tip and the object and around 10 mm distance between neighboring image frames.

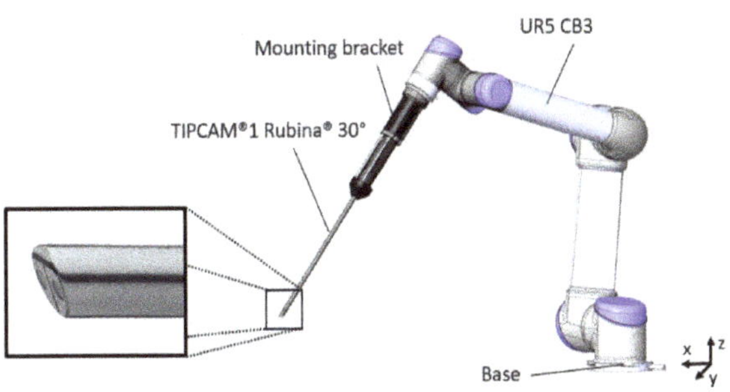

Figure 2. Schematic overview of the experimental setup. The TipCam Rubina 1S 30° is held by the UR5 CB3 robotic arm (Universal Robots A/S, Odense, Denmark). The video laparoscope is equipped with a stereo camera system with chip-on-the-tip technology, which is angled at 30° and has an 80° FOV.

2.2. Architecture of the Robot-Based 3D Reconstruction Procedure

The procedure includes four parallel running processes: Robot movement, image capturing, point cloud creation and transformation (from camera coordinate system to world coordinate system), and point cloud processing (see Figure 3). The latter process includes the sub-processes point cloud appending, ICP computing and pose graph updating, and visualization. After point cloud processing, two postprocessing steps—pose graph optimization and mesh creation—are applied before user interaction begins. As soon as the robot reaches the first position, the first stereo frame is taken to be processed and visualized. As the robot moves on, more frames are taken, processed, and visualized. After the last frame is acquired and processed, postprocessing begins. This multiprocessing approach offers the advantage that the growing 3D model is already visualized during the robotic movement.

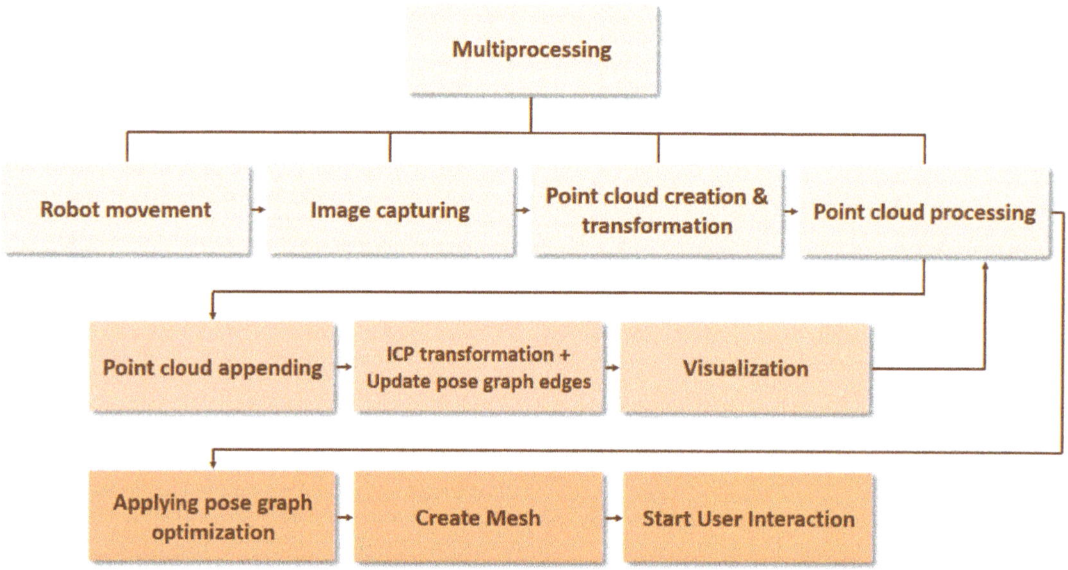

Figure 3. Visualization of the architecture of the robot-based 3D reconstruction procedure.

2.2.1. Robot Movement

The robotic arm has a repeatability of ±0.1 mm and is controlled by Python via the real-time data exchange (RTDE) interface provided by Universal Robots A/S (Universal Robots A/S, Odense, Denmark). The command used for robotic movement is called movel (end_pose, a = 0.1, v = 0.05), which requires the target position, acceleration, and velocity as input parameters. The target position is a 6 DOF vector containing x-, y-, and z-coordinates and rx, ry, and rz as parts of the rotation vector. By default, this command works with respect to the base coordinate system. The movement in this experiment follows a grid-shaped pattern, which only changes in the x- and y-direction and is fixed in the z-direction (compare with the world coordinate system in Figure 2).

2.2.2. Image Capturing

The images are captured as soon as the robot has reached its first resp. next target position and has come to a stop. Images are only taken if the robot stands still. For this experiment, the frame grabber USB Capture HDMI 4K Plus (Nanjing Magewell Electronics Co., Ltd., Nanjing, China) is used. The stereo camera has been calibrated to undistort each channel and align left to right.

2.2.3. Point Cloud Creation and Transformation

Based on the calibrated stereo images, disparity maps are generated by the Semi Global Matching (SGM) algorithm with the command cv2.cuda.createStereoSGM() by OpenCV version 4.8.0, which can then be further converted into point clouds by the OpenCV command cv2.reprojectImageTo3D().

Depending on the robot trajectory, the object is seen from different perspectives, and for each perspective, a point cloud is generated. With the known robot position, these point clouds are stitched together, leading to a complete point cloud of the scanned object. The transformation from the robot flange to the camera is essential for stitching because, based on this information, each perspective's camera position is estimated. The transformation from the camera tip position to the robot flange with respect to the robot base coordinate system involves a 180° turn around the y-axis ($\phi = 180°$), a $-30°$ turn around the x-axis ($\theta = -30°$), and a + 142° turn around the z-axis ($\psi = 142°$), which creates the rotation matrix as in (1) and (2) and a translation in the z-direction of $z = 580$ mm as in (3). These values arise from the CAD model (see Figure 2) and the OpenCV image frame convention, which defines the coordinate origin to lie in the upper left corner of the image, the x-axis pointing to the right and the y-axis pointing to the bottom. Due to the omitted hand–eye calibration, the transformation from camera to robot flange is only based on the known geometry of the mounting. This leads to uncertainties in the robot-based camera position estimation and errors in the stitched point cloud.

$$R = R_{y,\phi} \times R_{z,\theta} \times R_{x,\psi} \tag{1}$$

$$R = \begin{pmatrix} \cos\phi\cos\theta & -\cos\phi\sin\theta\cos\psi + \sin\phi\sin\psi & \cos\phi\sin\theta\sin\psi + \sin\phi\cos\psi \\ \sin\theta & \cos\theta\cos\psi & -\cos\theta\sin\psi \\ -\sin\phi\cos\theta & \sin\phi\sin\theta\cos\psi + \cos\phi\sin\psi & -\sin\phi\sin\theta\sin\psi + \cos\phi\cos\psi \end{pmatrix} \tag{2}$$

$$t = \begin{pmatrix} 0 \\ 0 \\ z \end{pmatrix} \tag{3}$$

2.2.4. Point Cloud Processing

As said before, the missing hand–eye calibration causes positional errors in the 3D model. Consequently, point cloud processing includes an ICP algorithm with the purpose of an optimized surface alignment between neighboring point clouds. The algorithm used in this work is the multiscale colored ICP algorithm from Open3D (v0.19) (cuda compatible) and is based on the colored ICP registration algorithm from Reference [22], including both geometry and color information by computing a joint photometric and geometric optimization objective (o3d.t.pipelines.registration.multi_scale_icp()). The goal is to find a transformation T that aligns two colored point clouds by taking the colors (photometric term E_C) and the geometry (geometric term E_G) as well as a weight $\sigma \in [0, 1]$ into account, as in (4) [22].

$$E(\mathrm{T}) = (1-\sigma)E_C(T) + \sigma E_G(\mathrm{T}) \tag{4}$$

A further pose optimization is performed by the multiway registration algorithm via pose graph by Open3D (v0.19) with the command o3d.pipelines.registration.PoseGraph(), which is based on References [23,24]. It is applied when multiple point clouds must be aligned in a global space, and it tackles the challenge of pruning incorrect transformations by modeling the validity of loop closure pieces with low overlap. The pose graph algorithm defines nodes and edges, from which in our case the nodes contain the position information coming from the robot, and the edges contain the transformation information computed by the ICP. The edges act as constraints between the robot poses (nodes). The goal of the algorithm is to find a configuration of the nodes that is consistent with the information in the edges. After each pairwise ICP transformation, the pose graph edges are updated with this information. The final step of the pose graph algorithm is executed in the postprocessing step because it is necessary that all positions of the robot trajectory have been reached to connect not only neighboring and overlapping point clouds but also nodes of non-overlapping areas. After ICP transformation and edge updating, the growing point cloud is visualized.

2.2.5. Postprocessing

When all images are acquired, the visualized point cloud is closed, and postprocessing starts. At this stage, pose graph optimization is executed by the command o3d.pipelines.registration.global_optimization(), which requires the previously created map containing nodes and edges, as well as the optimization method based on the Levenberg–Marquardt algorithm, which is known for a faster convergence compared to the Gauss–Newton method. The global optimization is performed in two steps: the first one considers all edges, and the second step prunes uncertain edges. After this, the point cloud is finally transformed into a mesh by the Open3D (v0.19) command o3d.geometry.TriangleMesh.create_from_point_cloud_poisson() based on Reference [25], which generates a watertight closed surface. It includes the color information, which is stored in each point cloud and is transferred from there to every vertex (point building a triangle) in the mesh. The colors between vertices are interpolated, which leads to a smooth color distribution and reduces light reflections and non-uniform illuminations from the captured images. Finally, the mesh is displayed to the user and ready for use.

2.3. *Qualitative and Quantitative Evaluation of 3D Reconstruction*

The qualitative evaluation focuses on a dense point cloud with fewer holes, color fastness, and a plausible and steady topology. The quantitative evaluation focuses on the root mean squared error (RMSE) of the reconstruction error, which is the distance between 3D points on the reconstructed surface $\hat{y}_i$ and the corresponding points on the real surface y_i in mm, as in (5), with N representing the number of reconstructed points. The reconstruction error is measured by aligning the reconstructed surface to the ground truth surface, first manually and then by applying the ICP for fine-tuning.

$$\mathrm{RMSE} = \sqrt{\frac{1}{N}\sum_{i=1}^{N}(\hat{y}_i - y_i)^2} \quad (5)$$

3. Results

The final experimental results are visualized as point clouds and as mesh (closed surface) by the 3D mesh processing software MeshLab (Instituto di Scienza e Tecnologie dell'Informazione, San Giuliano Terme, Italy). The qualitative results show a dense point cloud (see Figure 4 (right) and Figure 5) with plausible topology and colors when compared with the photography (see Figure 4 (left)).

The first task, "depth estimation for point cloud generation", is performed by the stereo matching algorithm SGM from OpenCV, and a part of the results can be seen in Figure 6. There, six example images (always the left image) are presented showing excerpts of the heart, liver, and gallbladder, and for each image, the corresponding depth map is shown. The colors in the depth maps represent the depth with nan values in white, closer objects in more yellowish tones, and objects further away in more blueish colors. The depth maps are mostly dense except for areas that are hidden from one of the two cameras or areas with reflections.

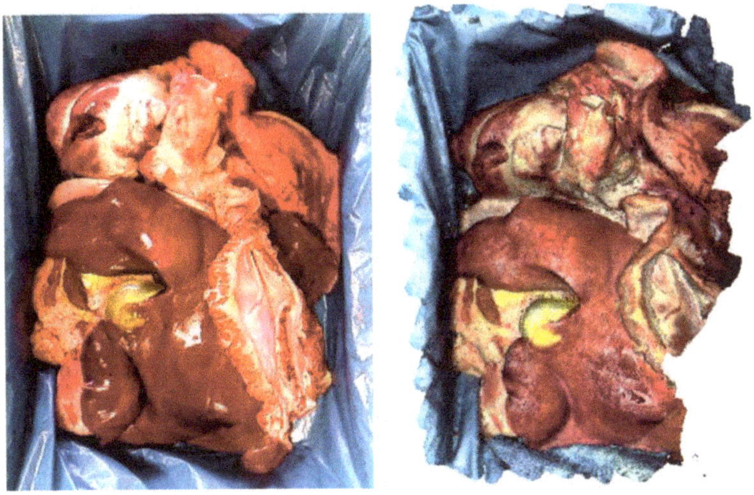

Figure 4. Photography of pig organs (**left**) and screenshot of the reconstructed point cloud created by our approach (**right**).

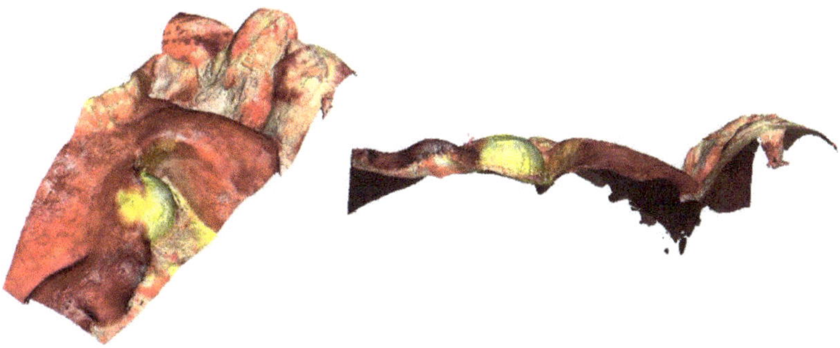

Figure 5. Screenshot of a section of the reconstructed point cloud in front view (**left**) and side view (**right**).

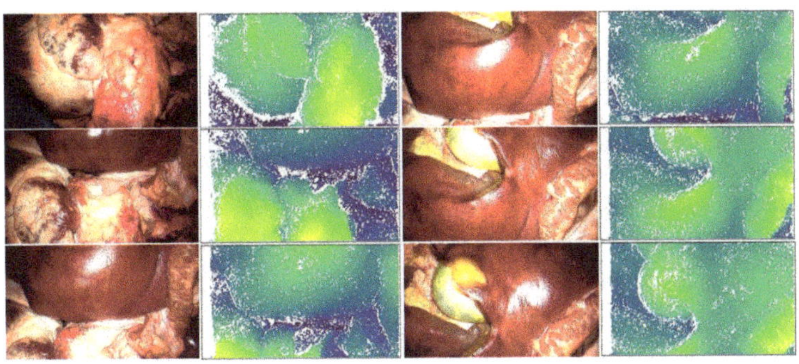

Figure 6. Six example images from the examined dataset (always left images) (**left**) and the corresponding depth maps with nan values in white, closer objects in more yellowish tones, and objects further away in more blueish colors (**right**).

The second task, "camera pose estimation and optimization for point cloud stitching", is performed by first transforming each point cloud based on the known robot position and second by applying the ICP and pose graph algorithm. To show their effectiveness, three different approaches are distinguished: Rob only, Rob + ICP, and Rob + ICP + pose graphs. The corresponding resulting point clouds are presented in Figure 7. The Rob only approach results in a stitched point cloud with edges between neighboring point clouds, which do not fit very well (see an excerpt of the gallbladder in Figure 7 (top right)). The additional use of the ICP improves the transitions between neighboring point clouds but lacks the consideration of an overall position optimization (see Figure 7 (middle)). The reconstructed gallbladder is not as round as it is. Thus, the pose graph algorithm ensures further position optimization, which leads to an improved stitched point cloud (see Figure 7 (bottom)). Here, the reconstructed gallbladder is round, the transitions between the neighboring point clouds are smooth, and no edges can be observed.

Figure 7. Screenshots of the pig organ point clouds (**left**) and an excerpt with a focus on the gallbladder (**right**) created by three different approaches: camera position estimation only by robot kinematics (Rob only) (**top**), by robot kinematics + ICP (Rob + ICP) (**middle**), by robot kinematics + ICP + pose graphs (Rob +ICP + pose graph) (**bottom**).

Compared to the point clouds seen in Figures 4 and 5, the point clouds in Figure 7 miss any filtering regarding the reflections. Moreover, the brightness distribution within each single point cloud corresponds with the original colors of the image frames. There, the corners of the image are darker than its center. The reflections are reduced by filtering out those pixels with values over 0.97 (when the range for R, G, and B is from zero to one). As mentioned in Section 2, the mesh function of Open3D (v0.19) interpolates colors between vertices, leading to a smooth color distribution.

The quantitative results are generated with the software CloudCompare 2.14.alpha (www.cloudcompare.org). The reconstruction error is calculated by overlaying the reconstructed point cloud onto the ground truth point cloud, on which the eight markers A0 to A7 on the ground truth and R0 to R7 on the reconstructed point cloud are selected for alignment (see Figure 8). Afterward, the ICP is applied for RMSE computation. The RMSE value is 3.37 mm for the whole point cloud seen in Figure 8 and Table 1 (Rob + ICP + pose graph), while the RMSE value is 1.1 mm for the excerpt seen in Figure 5. In comparison to that, the RMSE value for Rob only is 6 mm, and for Rob + ICP without pose graph, it is 4.37 mm, which shows the improvement caused by the pose graph algorithm.

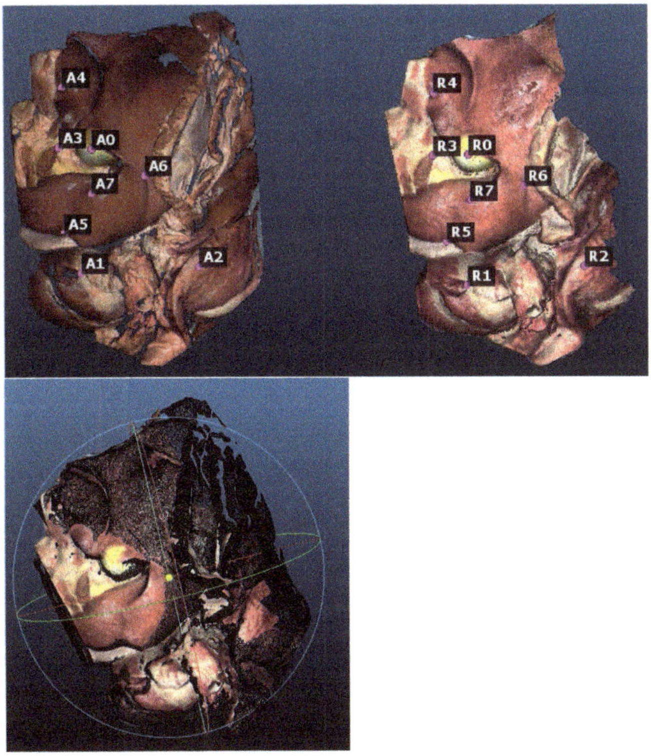

Figure 8. Screenshot of the ground truth point cloud (**left**) and the reconstructed point cloud by our approach (**middle**) with markers A0–A7 and R0–R7. The markers are used for point cloud alignment to compute the reconstruction error as RMSE. Screenshot of the overlaid ground truth in black and the reconstructed point cloud in colors to compute the reconstruction error (**right**).

Table 1. Overview of the quantitative evaluation by computing the RMSE of the reconstruction error depending on the pose estimation method: Rob only, Rob + ICP, and Rob + ICP + pose graph.

Pose Estimation Method	RMSE in mm
Rob only	6.0
Rob + ICP	4.37
Rob + ICP + pose graph	3.37

4. Discussion

This work examined the impact of pose optimization algorithms on the appearance and accuracy of stitched point clouds. The main achievement and contribution of this work is that the pose estimation error, caused by an uncalibrated robot-guided camera system, can be corrected successfully by the Open3D (v0.19) pose optimization algorithm multiscale colored ICP and the multiway registration algorithm via pose graph. This can be observed in Section 3 (Figure 5). The results show that the combination of robotic camera guidance with ICP and pose graph algorithms (Rob + ICP + pose graph) leads to an accurate and realistic-looking 3D model of ex vivo organs. The appearance is similar to the ground truth point cloud and the photograph. The accuracy of 1.1–3.37 mm is within the range of other approaches dealing with ex vivo organs. Ref. [21] is a review focusing on the reconstruction error with the finding that the error for ex vivo organs lies between 1.1 and 10.8 mm (RMSE), where some references only focused on a single shot and some on multi-view reconstruction. Thus, our approach can compete with approaches with a similar setup.

In this work, only ex vivo organs are used for validation. Open-source in vivo data, including stereo images and robotic poses, do not exist, and an animal test is not yet possible, which leads us to validate our approaches on ex vivo organs. The difference between in vivo and ex vivo data is mainly the occurrence of instruments, blood, smoke, and movements (breathing and intestinal peristalsis) in vivo. Instruments, blood, and smoke will not be a problem in our case because the plan is to offer the 3D model as an assistive tool for the surgeon during the intervention or to use it for autonomous camera guidance, which means image acquisition must occur right before the surgeon starts the intervention. The two things that happen in vivo and affect our 3D reconstruction are movements emerging from breathing and intestinal peristalsis. Breathing takes place at a certain frequency, which must be detected to compensate for it. If detected, images could be taken directly after the exhalation cycle (empty lungs) or inhalation cycle (full lungs). That would lead to two 3D models—fully inhaled and fully exhaled. This must be examined in our next experiment and could be simulated with ex vivo organs by implementing a motor that generates a breathing-like movement. A potential solution to avoid breathing movements is to stop the breathing of the patient during image acquisition. If permitted by the medical staff, the time for the whole organ scan must be reduced significantly to around 10–20 s. The occurrence of intestinal peristalsis is not predictable and leads to errors in the stitching procedure (excessive movement and neighboring point clouds not matching). The scan must be repeated.

In summary, for a final validation of our approach, an animal lab is required for in vivo data. But for the current project state, the validation on ex vivo data is sufficient and more ethically defensible.

According to Ref. [21], a standard validation scenario should be set up to compare our approaches with those of other researchers. Because of this, the collected image data of the ex vivo organs including the corresponding robot positions and the ground truth point cloud of this experiment can be obtained from the authors. Ex vivo organs can already give an indication of whether an approach works successfully in a real interventional scenario, but in vivo data will additionally be required to critically validate an approach. Thus, our approach must be validated in an animal lab to address challenges such as movements (heartbeat, breathing, and intestinal peristalsis).

The fact that the fulcrum point is not considered in this work is a critical limitation. The fulcrum point complicates adopting the ideal perspectives (perpendicular) from the camera onto the tissue because it decreases the laparoscopic movement from six to only four DOF [26]. To achieve coverage of all areas of the tissue, the laparoscope's motion must be a combination of rotation around its longitudinal axis and pivoting around the fulcrum point. In this case, the translation of the camera position between neighboring frames consists of rotation and translation related to various axes. In comparison to that, the omission of the fulcrum point allows a linear camera motion following only translational changes in the x- and y-direction.

Moreover, reflections disturb the quality of the resulting point cloud, and filtering is necessary. Also, the illuminance of the laparoscope's light decreases with increasing distance (following the inverse square law for illuminance) [27], which leads to a difference in brightness within each frame. The result is darker colors in the image corners and brighter colors in the image center, which also degrades the quality of the resulting point cloud (mosaic-like appearance). However, not only the inverse square law for illuminance causes the brightness difference, but also the image acquisition from different perspectives and different working distances (distance between laparoscope tip and object). This leads to different color appearances of the objects. This must be fixed by filtering.

Currently, the whole procedure takes three minutes depending on the number of images. Most of the time is consumed by the initialization of the robot and the script (~30 s), the robot's movement (~400 ms between frames), and the ICP and pose graph optimization (~510 ms per frame pair). The next improvement for latency reduction is the installation of a PCI express frame grabber by Magewell (Nanjing Magewell Electronics Co., Ltd., Nanjing, China), which should reduce the current time (~250 ms) to grab a stereo frame.

5. Conclusions

This work presents a procedure to achieve a complete 3D model of the abdomen and the internal organs. It requires a robotic arm and a stereo laparoscope for image acquisition and pose estimation. The depth estimation is based on a stereo matching algorithm and the camera pose estimation is a combination of the robot position and the optimization algorithms ICP and multiway registration via pose graph—both provided by Open3D (v0.19). For validation, ex vivo organs, including the heart, liver, gallbladder, and lung, were collected along with a laser scanner for ground truth acquisition. The qualitative results show dense point clouds, and the quantitative results show a high accuracy of 1.1–3.37 mm (RMSE of the reconstruction error) depending on the removal of outliers and the region of interest (the larger the reconstructed area, the higher the reconstruction error).

The next steps will be the automatic trajectory planning and execution based on a rough quick scan directly after laparoscope insertion (only a few millimeters inside the abdomen), the automatic detection of holes in the 3D model as well as a method to fill the holes, the improvement for real-time computation, and a proposal on how the surgeon can apply the 3D reconstruction as an interactive 3D model to automatically guide the laparoscope to a certain region of interest.

Author Contributions: Conceptualization, B.G., A.R. and K.M.; methodology, B.G., J.H.; validation, B.G., J.H.; formal analysis, B.G.; investigation, B.G., J.H.; data curation, B.G., J.H.; writing—original draft preparation, B.G.; writing—review and editing, B.G.; visualization, B.G., J.H.; supervision, A.R. and K.M.; project administration, B.G., A.R. and K.M. All authors have read and agreed to the published version of the manuscript.

Funding: This research received no external funding.

Data Availability Statement: The data that supports the findings of this work are available from the corresponding author.

Conflicts of Interest: Birthe Göbel is a part-time employee at KARL STORZ SE & Co. KG. The remaining authors declare that the research was conducted without any commercial or financial relationships that could be construed as a potential conflict of interest.

References

1. Reichard, D.; Bodenstedt, S.; Suwelack, S.; Mayer, B.; Preukschas, A.; Wagner, M.; Kenngott, H.; Müller-Stich, B.; Dillmann, R.; Speidel, S. Intraoperative On-the-Fly Organ- Mosaicking for Laparoscopic Surgery. *J. Med. Imaging* **2015**, *2*, 045001. [CrossRef] [PubMed]
2. Su, Y.-H.; Lindgren, K.; Huang, K.; Hannaford, B. A Comparison of Surgical Cavity 3D Reconstruction Methods. In Proceedings of the 2020 IEEE/SICE International Symposium on System Integration (SII), Honolulu, HI, USA, 12–15 January 2020; pp. 329–336.
3. Kim, D.T.; Cheng, C.-H.; Liu, D.-G.; Liu, J.; Huang, W.S.W. Designing a New Endoscope for Panoramic-View with Focus-Area 3D-Vision in Minimally Invasive Surgery. *J. Med. Biol. Eng.* **2020**, *40*, 204–219. [CrossRef]
4. Chen, L.; Tang, W.; John, N.W. Real-Time Geometry-Aware Augmented Reality in Minimally Invasive Surgery. *Healthc. Technol. Lett.* **2017**, *4*, 163–167. [CrossRef] [PubMed]
5. Wei, G.; Shi, W.; Feng, G.; Ao, Y.; Miao, Y.; He, W.; Chen, T.; Wang, Y.; Ji, B.; Jiang, Z. An Automatic and Robust Visual SLAM Method for Intra-Abdominal Environment Reconstruction. *J. Adv. Comput. Intell. Intell. Inform.* **2023**, *27*, 1216–1229. [CrossRef]
6. Zhang, L.; Ye, M.; Giataganas, P.; Hughes, M.; Yang, G.-Z. Autonomous Scanning for Endomicroscopic Mosaicing and 3D Fusion. In Proceedings of the 2017 IEEE International Conference on Robotics and Automation (ICRA), Singapore, 29 May–3 June 2017; pp. 3587–3593.
7. Speers, A.D.; Ma, B.; Jarnagin, W.R.; Himidan, S.; Simpson, A.L.; Wildes, R.P. Fast and Accurate Vision-Based Stereo Reconstruction and Motion Estimation for Image-Guided Liver Surgery. *Healthc. Technol. Lett.* **2018**, *5*, 208–214. [CrossRef]
8. Zhou, H.; Jayender, J. EMDQ-SLAM: Real-Time High-Resolution Reconstruction of Soft Tissue Surface from Stereo Laparoscopy Videos. In Proceedings of the Medical Image Computing and Computer Assisted Intervention–MICCAI 2021: 24th International Conference, Strasbourg, France, 27 September–1 October 2021; Proceedings, Part IV 24. Springer International Publishing: Berlin/Heidelberg, Germany, 2021; Volume 12904, pp. 331–340.
9. Handa, A.; Gaidhane, A.; Choudhari, S.G. Role of Robotic-Assisted Surgery in Public Health: Its Advantages and Challenges. *Cureus* **2024**, *16*, e62958. [CrossRef]
10. Bramhe, S.; Pathak, S.S. Robotic Surgery: A Narrative Review. *Cureus* **2022**, *14*, e29179. [CrossRef]
11. Boyina, K.K.; Dasukil, S. Robotic Surgery-Safety and Effectiveness, in Comparison with Traditional Surgery, Present Context and Recent FDA Safety Warning. *Indian J. Surg. Oncol.* **2020**, *11*, 613–614. [CrossRef]
12. Sui, X.; Zhang, Y.; Zhao, X.; Tao, B. Binocular-Based Dense 3D Reconstruction for Robotic Assisted Minimally Invasive Laparoscopic Surgery. *Int. J. Intell. Robot. Appl.* **2024**, *8*, 866–877. [CrossRef]
13. Attanasio, A.; Scaglioni, B.; Momi, E.D.; Fiorini, P.; Valdastri, P. Autonomy in Surgical Robotics. *Annu. Rev. Control Robot. Auton. Syst.* **2021**, *4*, 651–679. [CrossRef]
14. Saeidi, H.; Opfermann, J.D.; Kam, M.; Wei, S.; Leonard, S.; Hsieh, M.H.; Kang, J.U.; Krieger, A. Autonomous Robotic Laparoscopic Surgery for Intestinal Anastomosis. *Sci. Robot.* **2022**, *7*, eabj2908. [CrossRef] [PubMed]
15. Han, J.; Davids, J.; Ashrafian, H.; Darzi, A.; Elson, D.S.; Sodergren, M. A Systematic Review of Robotic Surgery: From Supervised Paradigms to Fully Autonomous Robotic Approaches. *Int. J. Med. Robot. Comput. Assist. Surg.* **2022**, *18*, e2358. [CrossRef] [PubMed]
16. Tsai, R.Y.; Lenz, R.K. A New Technique for Fully Autonomous and Efficient 3D Robotics Hand/Eye Calibration. *IEEE Trans. Robot. Autom.* **1989**, *5*, 345–358. [CrossRef]
17. Enebuse, I.; Foo, M.; Ibrahim, B.S.K.K.; Ahmed, H.; Supmak, F.; Eyobu, O.S. A Comparative Review of Hand-Eye Calibration Techniques for Vision Guided Robots. *IEEE Access* **2021**, *9*, 113143–113155. [CrossRef]
18. Zhong, F.; Wang, Z.; Chen, W.; He, K.; Wang, Y.; Liu, Y.-H. Hand-Eye Calibration of Surgical Instrument for Robotic Surgery Using Interactive Manipulation. *IEEE Robot. Autom. Lett.* **2020**, *5*, 1540–1547. [CrossRef]
19. Gruijthuijsen, C.; Dong, L.; Morel, G.; Vander Poorten, E. Leveraging the Fulcrum Point in Robotic Minimally Invasive Surgery. *IEEE Robot. Autom. Lett.* **2018**, *3*, 2071–2078. [CrossRef]
20. Maier-Hein, L.; Mountney, P.; Bartoli, A.; Elhawary, H.; Elson, D.; Groch, A.; Kolb, A.; Rodrigues, M.; Sorger, J.; Speidel, S.; et al. Optical Techniques for 3D Surface Reconstruction in Computer-Assisted Laparoscopic Surgery. *Med. Image Anal.* **2013**, *17*, 974–996. [CrossRef]
21. Göbel, B.; Reiterer, A.; Möller, K. Image-Based 3D Reconstruction in Laparoscopy: A Review Focusing on the Quantitative Evaluation by Applying the Reconstruction Error. *J. Imaging* **2024**, *10*, 180. [CrossRef]
22. Park, J.; Zhou, Q.-Y.; Koltun, V. Colored Point Cloud Registration Revisited. In Proceedings of the 2017 IEEE International Conference on Computer Vision (ICCV), Venice, Italy, 22–29 October 2017; pp. 143–152.
23. Choi, S.; Zhou, Q.-Y.; Koltun, V. Robust Reconstruction of Indoor Scenes. In Proceedings of the IEEE Conference on Computer Vision and Pattern Recognition, Boston, MA, USA, 7–12 June 2015; pp. 5556–5565.
24. Grisetti, G.; Kümmerle, R.; Stachniss, C.; Burgard, W. A Tutorial on Graph-Based SLAM. *IEEE Intell. Transp. Syst. Mag.* **2010**, *2*, 31–43. [CrossRef]
25. Kazhdan, M.; Hoppe, H. Screened Poisson Surface Reconstruction. *ACM Trans. Graph.* **2013**, *32*, 29. [CrossRef]

26. Gallagher, A.G.; McClure, N.; McGuigan, J.; Ritchie, K.; Sheehy, N.P. An Ergonomic Analysis of the Fulcrum Effect in the Acquisition of Endoscopic Skills. *Endoscopy* **1998**, *30*, 617–620. [CrossRef]
27. Wang, L.; Wu, B.; Wang, X.; Zhu, Q.; Xu, K. Endoscopic Image Luminance Enhancement Based on the Inverse Square Law for Illuminance and Retinex. *Int. J. Med. Robot. Comput. Assist. Surg.* **2022**, *18*, e2396. [CrossRef] [PubMed]

Journal of
Imaging

MDPI

Article

Fitting Geometric Shapes to Fuzzy Point Cloud Data

Vincent B. Verhoeven *, Pasi Raumonen * and Markku Åkerblom

Faculty of Information Technology and Communication Sciences, Mathematics Research Centre, Tampere University, Korkeakoulunkatu 1, 33720 Tampere, Finland; markku.akerblom@tuni.fi
* Correspondence: vincentius.verhoeven@tuni.fi (V.B.V.); pasi.raumonen@tuni.fi (P.R.)

Abstract: This article describes procedures and thoughts regarding the reconstruction of geometry-given data and its uncertainty. The data are considered as a continuous fuzzy point cloud, instead of a discrete point cloud. Shape fitting is commonly performed by minimizing the discrete Euclidean distance; however, we propose the novel approach of using the expected Mahalanobis distance. The primary benefit is that it takes both the different magnitude and orientation of uncertainty for each data point into account. We illustrate the approach with laser scanning data of a cylinder and compare its performance with that of the conventional least squares method with and without random sample consensus (RANSAC). Our proposed method fits the geometry more accurately, albeit generally with greater uncertainty, and shows promise for geometry reconstruction with laser-scanned data.

Keywords: uncertainty quantification; geometry reconstruction; laser scanning; point cloud

1. Introduction

In a typical shape-fitting problem with data covering the object's surface (a so-called point cloud), the distance between the points and the shape is used as the metric to quantify how well a given shape fits the data. For example, in the least squares approach the sum of the squared distances is used as the objective function to be minimized in order to find the optimal shape parameters for the fitting problem. However, this approach does not consider the inherent uncertainties in the measured point cloud data explicitly.

There are many sources of point location uncertainty, and moreover, the uncertainty can vary a lot between the data points. For example, in light detection and ranging (LiDAR) instruments, such as fixed position terrestrial laser scanning (TLS) instruments, the point location uncertainty is affected by the range and resulting beam size and the incidence angle between the laser beam and the measured object's surface.

In this paper, we present a generalization of the shape-fitting problem with point cloud data which considers the assumed/known uncertainty of each data point. In this generalization, we model the uncertainty of each data point explicitly with three-dimensional (3D) distributions such as Gaussians, which may be of varying magnitude and orientation to account for the differences in uncertainty between the data points. Thus, our data set for the fitting problem is not a 3D point cloud but a so-called *fuzzy point cloud*; a cloud of 3D distributions, and each of these distributions can be different in size and shape. This more accurately describes the underlying data and should therefore enable a more accurate fit of the geometry and the ability to determine its uncertainty. The outlined approach comes from the field of uncertainty quantification and is an improvement over other Bayesian methods that assume homoscedasticity [1], assume the distance to the geometry to follow a Gaussian distribution [1,2], or assume a distribution of the geometry parameters [3]. Additionally, the method requires no uniform sampling or full coverage of the geometry.

The distance between the points and the shape to be reconstructed is replaced with the expected distance between the distributions and the shape. The distance metric does not need to be the Euclidean distance and is instead substituted by the Mahalanobis distance to incorporate anisotropic uncertainty, i.e., [4]. To the best knowledge of the authors, the application of the Mahalanobis distance for geometry fitting is rare. Our approach deviates from an existing study in the approach to reduce bias and by evaluating the entire geometry for every distribution rather than the Mahalanobis nearest point [5]. Finally, the objective function for the shape-fitting problem is the average of the expected distances over the fuzzy point cloud.

Geometry reconstruction of laser-scanned objects has been applied in both the built and natural environment in a large body of research [6–9]. As such, we demonstrate the method in detail by fitting a cylinder to simulated LiDAR data. The choice for this shape is motivated by its commonality in both the natural and built environment, for instance, to approximate part of a tree or a pipe [10,11].

We start this paper by describing the fuzzy data in Section 2, followed by our geometry fitting approach in Section 3. The aforementioned approach is illustrated on cylinder fitting with laser scanning data in Section 4, followed by a discussion in Section 5, and finally, the conclusion in Section 6.

2. Fuzzy Data

2.1. Fuzzy Point Clouds

The data consist of a fuzzy point cloud FC which is a generalization of a point cloud: Instead of having a finite set of points X in $\mathbb{R}^m$ (typically $\mathbb{R}^3$), we have a finite set of n distributions $\mathcal{N}_i$ defined over $\mathbb{R}^m$, i.e. $FC = \{\mathcal{N}_i \,|\, i = 1,2,\ldots,n\}$. Notice that each distribution can be different in shape and variance. Some can be finitely supported and others have infinite support, but the support is m-dimensional. To model laser scanning in the real world, the distributions consist of trivariate normal distributions given by $\mathcal{N}_i = N_i(\hat{\mu}_i, \Sigma_i)$, where $\hat{\mu}_i = [x_i, y_i, z_i] \in X \subset \mathbb{R}^3$ is the ith measured point and Σ_i its covariance matrix, modeling the uncertainty of the measured location. Thus, the advantage and difference in using fuzzy point clouds as opposed to point clouds is that we can rigorously consider the uncertainty of each point, which locally can vary significantly depending on the geometrical shape. An example fuzzy point cloud is shown in Figure 1 for a cylinder section measured by two laser scanners.

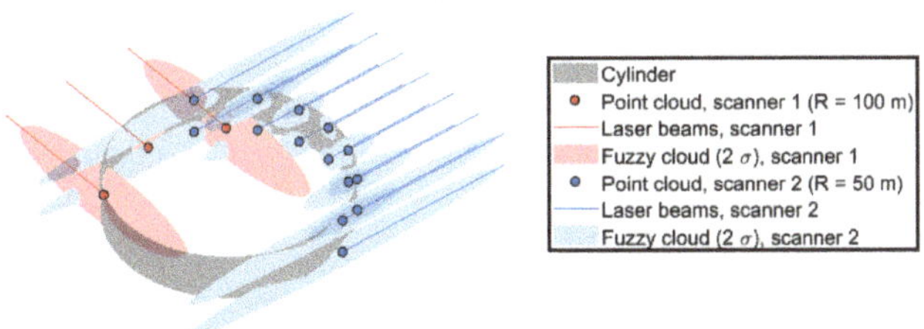

Figure 1. Example fuzzy point cloud generated for a cylinder slice scanned by two laser scanners at a range of 50 m and 100 m, respectively, 90 degrees apart.

2.2. Determination of Fuzziness

One important and immediate question regarding fuzzy point clouds is the ability to define the distributions for a given problem. For example, if the data come from laser scanners, then the uncertainty of the measurement not only depends on the instrument's specifications but also on the distance (range) to the object. When the range of the object is much greater than its dimensions, it may be assumed independent of the object's exact geometry and considered a constant. More importantly, however, is the incidence angle between the laser beam and the measured object's surface, which is often not known, at least not accurately. We can perhaps estimate quite accurately the minimum uncertainty for each point, e.g., in the case of laser scanning by assuming zero incidence angle, or if the object being measured is large enough so that small changes in the laser beam locations lead only to insignificant changes in the incidence angle so that the angle can be estimated well. Otherwise, we could have a situation in which uncertainty is very sensitive to the shape whose determination is the objective of the fitting problem. This circular relationship is discussed in more detail in the next section.

3. Geometry Fitting

This section describes the approach used to fit the optimal geometry to a fuzzy point cloud. It is written in a general way, such that it is independent of the geometric shape being fitted and is not specific to laser scanning data.

3.1. Objective Function

There are many possibilities for formulating meaningful objective functions for our shape-fitting problem. First, the basic idea is to minimize some distance between the data and the shape by varying the shape parameters. We have multiple choices for the distance $d : \mathbb{R}^m \times \mathbb{R}^m \to \mathbb{R}$, such as the often used *Euclidean distance*, but also others, such as the *Mahalanobis distance*.

Because our data consist of continuous distributions and not discrete points, we need to define how the distance is defined between a distribution and a shape. With points, this is simply the minimum distance from the point to the shape. With distributions, we can similarly define the minimum distance from any point within the distribution to the shape and then integrate these distances over the support of the distribution to produce a single distance. A natural choice is the expected distance, where this point-wise minimum distance is weighted by the distribution's probability density value.

Let g denote the set of geometry parameters describing the shape $G \subset \mathbb{R}^m$, where a point lying on its surface is denoted by $\hat{p} \in G$. The minimum distance from a point $\hat{x} \in \mathbb{R}^m$ to the shape G is then denoted by $d_G(\hat{x}) = \min_{\hat{p} \in G} d(\hat{p} \mid \hat{x})$. Then, the expected distance from a distribution $\mathcal{N}$ with probability density $f(\hat{x})$ to the shape G is given by

$$E[d_G(\mathcal{N})] = \int_{\text{supp}(\mathcal{N})} f(\hat{x}) d_G(\hat{x}) d\hat{x}. \quad (1)$$

A natural alternative to the expected distance $E[d_G(\mathcal{N}_i)]$ with distributions $\mathcal{N}_i$ would be the maximum likelihood

$$L(G|\mathcal{N}_i) = \max_{\hat{p} \in G} \mathcal{N}_i(\hat{p}). \quad (2)$$

The fit is then quantified by how likely each geometry is for a given distribution. The objective function would be the mean of the likelihoods over the data. However, we use the expected distance instead of the maximum likelihood despite it perhaps being computationally easier to determine, because the objective function based on the likelihood suffers the major problem that the likelihood values can easily become negligible, and thus, their contribution to the objective function and indeed its gradient become negligible. This would result in said distribution being ignored by the optimizer, hence our choice to use the expected distance. Similarly, the log-likelihood suffers from extreme gradients for low likelihood values.

The objective function $O(shape \mid data) = O(G \mid FC)$ is the mean of the expected distances over the fuzzy point cloud FC. The optimal geometry is thus the one that minimizes the following sum:

$$\arg\min_G O(G \mid FC) = \arg\min_G \frac{1}{n} \sum_{i=1}^{n} E[d_G(\mathcal{N}_i)]. \tag{3}$$

For non-trivial problems, this minimization problem is highly complex and thus solved iteratively, starting from an initial geometry estimate G_0, from which the final geometry estimate G follows. To make the objective function values comparable irrespective of the number of points or uncertainty magnitude, the objective function is divided by the value given by the initial estimate. The geometry fitting problem can now be formulated as the minimization problem

$$\arg\min_G O(G \mid FC) = \arg\min_G \frac{\sum_{i=1}^{n} E[d_G(\mathcal{N}_i)]}{\sum_{i=1}^{n} E[d_{G_0}(\mathcal{N}_i)]}. \tag{4}$$

3.2. Mahalanobis Distance

We alluded already that one possible and often used distance metric for shape fitting is the Euclidean distance; however, it has some potential problems: First, it is scale-dependent. More natural would be to have the distance metric depend on the magnitude of the uncertainty, i.e., how far a point is from the shape in terms of standard deviations of the distributions, instead of it depending on arbitrary units such as meters or inches. Second, the Euclidean distance is symmetric or isotropic, meaning it treats every direction equally. We can, however, generally not assume the uncertainty to be isotropic due to anisotropy in measurements, and thus, it makes sense for the distance function to also include this possible anisotropy. A more suitable distance function is the Mahalanobis distance. Given two points $\hat{x}, \hat{y} \in \mathbb{R}^m$, the Mahalanobis distance M_i for the distribution $\mathcal{N}_i$ is defined as

$$M_i(\hat{x}, \hat{y}) = \sqrt{(\hat{x} - \hat{y})^T P_i (\hat{x} - \hat{y})}, \tag{5}$$

where P_i is the $m \times m$ precision matrix (i.e., the inverse of the positive-definite covariance matrix Σ_i) of distribution $\mathcal{N}_i$.

The Mahalanobis distance is unitless, scale-invariant, and considers the directional dependence (anisotropy) of the distribution's covariance structure. We note that the definition is a unitless equivalent of the Euclidean distance when the distribution's covariance matrix equals the identity matrix.

3.3. Uncertainty Independent of Geometry

The expected values $\hat{\mu}_i \in X$ are the measured point locations, and thus our input data for shape fitting; however, the issue is raised that the covariance of the distributions depends not only on the given point locations but also on the unknown geometry, i.e., $\Sigma_i = f(\hat{\mu}_i, G)$. Contrary to the Euclidean distance, the Mahalanobis distance depends on the covariance, so we have a cyclical problem $G = h(X, G)$, where the function h denotes the minimization of the objective function.

An intuitive way to break this cycle is to make the distributions depend on a previous estimate of the geometry, especially as a minimization of the objective function is an iterative process. However, making the distributions dependent on previous iterations leads to several issues in minimizing the objective function that are outside of the scope of this paper. Instead, we first provide an initial estimate of the geometry using the discrete point cloud, i.e., through least squares, which we use to create the fuzzy point cloud used for fitting. As the distributions may be sensitive to the initial geometry estimate, the optimized geometry is used iteratively to update the fuzzy point cloud until the geometry has converged. A general flowchart of the full approach for fuzzy geometry fitting is presented in Figure 2.

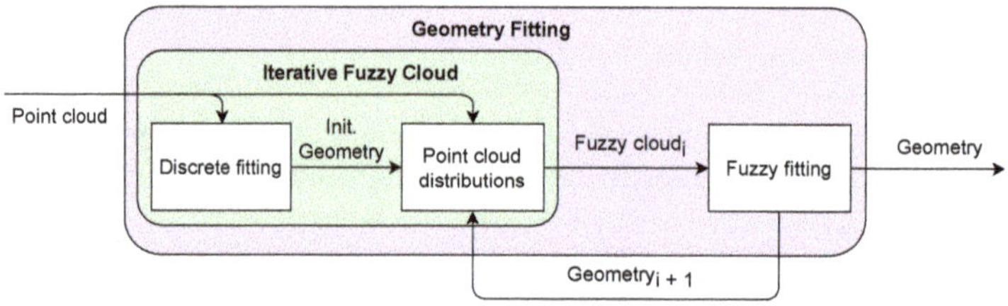

Figure 2. General fuzzy geometry fitting flowchart. Index i denotes fuzzy point cloud iteration.

4. Cylinder Fitting with Laser Scanning Data as an Example

To illustrate the described approach, it is applied to the case of cylinder fitting to laser scanning point cloud data. The circular cross-sectional shape results in a strongly variable incidence angle and thus varying local uncertainty between the points. We further note that most geometry in the built and natural environment can be approximated to be locally of constant cross-section making this an acceptable simplification.

The cylinder is parameterized by its radius r, its azimuth and elevation angles, and its center $\hat{c}_0$. We note that for a given azimuth and elevation angle (i.e., cylinder axis), the cylinder center may be projected onto the cross-sectional reference plane $Q \perp \vec{v}$ s.t. its dimensionality is now two, i.e., $\hat{c}_Q = \text{proj}_Q(\hat{c}_0) \in \mathbb{R}^2$. For the intent of this paper, the cylinder's length l is ignored. The approach to reduce the problem's dimensionality is applicable to any shape of constant cross-section and is given in the Appendix A.

4.1. Data

Using the Gaussian approximation for the power within a laser beam, the standard deviation in the radial direction σ_{radial} given by Equation (6) is a quarter of the beam diameter d_B, as it is commonly defined as covering four standard deviations [8]. It has an initial exit diameter d_0 and increases according to the beam divergence half-angle λ. As the range R between the measured point $\hat{\mu}$ and scanner $\hat{s}$ can safely be assumed to be much greater than its Rayleigh length, the increase in diameter is approximately linear.

$$\begin{aligned} \sigma_{radial}(\hat{\mu} \mid \hat{s}) = \frac{d_B}{4} = \frac{d_0 + 2R\tan(\lambda)}{4} = \\ \frac{d_0}{4} + \frac{1}{2}||\hat{\mu} - \hat{s}||\tan(\lambda). \end{aligned} \tag{6}$$

The standard deviation in the propagation direction σ_{prop} can be seen as a 'smearing out' of the radial uncertainty due to the incidence angle α between the beam and the surface. It follows from the dot product between the vectors from $\hat{\mu}$ to its projection onto the cylinder axis $\hat{p}$ and the scanner location $\hat{s}$, respectively.

$$\alpha(\hat{\mu} \mid \hat{s}, \hat{c}, \vec{v}) = \arccos(\frac{|\langle \hat{\mu} - \hat{p}, \hat{\mu} - \hat{s} \rangle|}{R||\hat{\mu} - \hat{p}||}) \tag{7}$$

Additionally, the propagation uncertainty σ_{prop} includes σ_0 as the base-level range uncertainty of the device:

$$\sigma_{prop}(\hat{\mu} \mid \hat{s}, \hat{c}, \vec{v}) = \sigma_0 + \sigma_{radial}(\hat{\mu} \mid \hat{s})\tan(\alpha). \tag{8}$$

This ensures that $\sigma_{prop} > 0$, and thus that the Mahalanobis distance is finite.

Simulated data are used to evaluate the method. Two hundred random point clouds are sampled from an initial fuzzy point cloud created using the true geometry, which enables us to determine average and standard deviations for the geometry parameters. As an example, a section of a cylinder's fuzzy point cloud from which these point clouds may be sampled is shown in Figure 1.

4.2. Objective

As already mentioned in Section 3.2, the (squared) Mahalanobis distance equals the unitless (squared) Euclidean distance when the distribution's covariance matrix equals the identity matrix. For this reason, as well as the simplification of future calculations, the Gaussian distribution N is transformed by T to a standard normal S, i.e., $T : N \mapsto S$. The Mahalanobis distance between point $\hat{x}$ and the cylinder's two-dimensional representation C is then equal to the Euclidean distance D in the transformed problem, i.e., $M(\hat{x}, \hat{p} \in C) = D(T(\hat{x}), T(\hat{p}) \in T(C)) = D(\hat{a}, \hat{e} \in E)$. The Euclidean distance from any point $\hat{a}$ to the resulting ellipse follows from the method described in [12], where the nearest point on the ellipse may be found using Ferrari's method [13]. This method, however, prohibits an analytical expression for the expected distance, and additionally, it is inherently biased as will be discussed in Section 5.

The circle C is therefore instead approximated by its envelope of tangent lines L with their point $\hat{c} \in C$. After translating and scaling the axes such that the distribution is a standard normal, the distance τ between $\hat{\mu}$ and the line is calculated and the expected squared Mahalanobis distance of this line follows from Equation (9), where we note that the subscripts denote the relevant dimension. The approach is illustrated in Figure 3, where without loss of generality axes x, y are aligned with the projected Gaussian's axes with standard deviation σ_x and σ_y, respectively. Said axes become a, b after transformation to the standard normal.

$$\begin{gathered} \tau = \frac{|c_x(c_x - \mu_x) + c_y(c_y - \mu_y)|}{\sqrt{\sigma_x^2 c_x^2 + \sigma_y^2 c_y^2}}, \\ E[M^2(L \mid \hat{\mu}, \Sigma)] = \tau^2 + 1. \end{gathered} \tag{9}$$

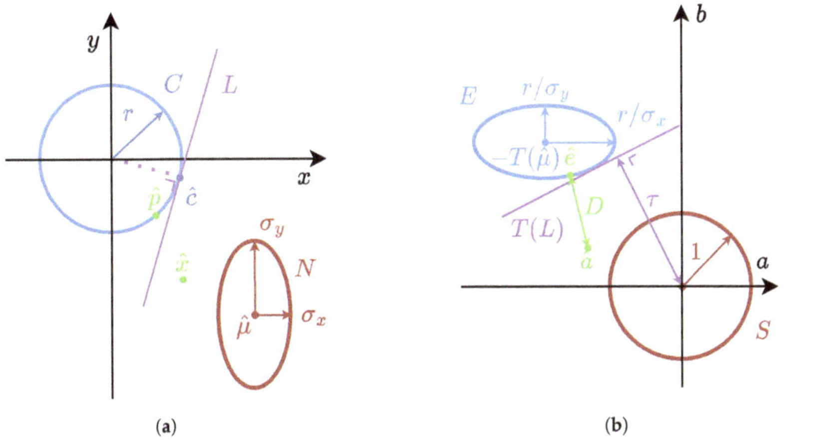

Figure 3. Illustration of the procedure to determine the expected Mahalanobis distance between a Gaussian distribution (**red**) and a tangent line (**purple**) placed on the circle (**blue**) before (**a**) and after transformation to the standard normal (**b**). The Gaussian distribution N is defined by the expected value $\hat{\mu}$ and the standard deviations σ_x and σ_y in the x- and y-directions, respectively.

For the full envelope of t tangent lines, the expected squared Mahalanobis distance of each line is weighted by the distance of the point $\hat{c}$ to the distribution. The contribution of each tangent line to the objective function is illustrated in Figure 4. For clarity, the tangent lines are drawn as finite lines, and only 25 are shown versus the 1000 that were used in the actual computations. Care is taken that the Mahalanobis distance in the denominator is taken to a greater power than the expected Mahalanobis distance $E[M]$, such that far-away tangent lines are weighed less even when $E[M(L_i \mid \hat{\mu}, \Sigma)] \approx M(\hat{c}_i \mid \hat{\mu}, \Sigma)$.

$$E[M^2(C \mid \hat{\mu}, \Sigma)] = \frac{1}{t} \sum_{i=1}^{t} \frac{E[M^2(L_i \mid \hat{\mu}, \Sigma)]}{M^3(\hat{c}_i \mid \hat{\mu}, \Sigma)} \tag{10}$$

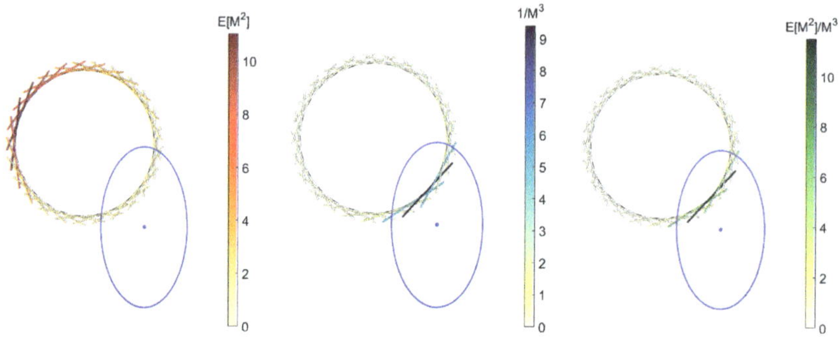

Figure 4. Illustration of the contributions of tangent lines in the envelope to (**a**) the expected Mahalanobis distance; (**b**) the inverse cubed Mahalanobis distance; and (**c**) the weighted expected Mahalanobis distance to the Gaussian distribution shown in blue (1σ). Note that the tangent lines are in reality infinite and all variables are dimensionless.

4.3. Comparison with Least Squares

Geometry is often reconstructed from point cloud data using Euclidean distance least squares, i.e., for cylinder reconstruction [10,14]. Euclidean least squares (ELS) are assessed both with the full point cloud and using RANSAC, in which case it is abbreviated as RELS [15]. As the least squares fitting of a cylinder is a non-linear problem, an iterative Gauss–Newton approach is used to find the optimum. Similarly, an iterative interior-point algorithm is used to minimize the objective function using the envelope approach with the expected Mahalanobis (EM) distance described in this article. It is dependent on the initial geometry estimate, which came from ELS in which case the method is abbreviated as EM. If RANSAC was also used for the initial geometry estimate, the method is abbreviated to REM.

The approaches are compared for a cylinder with a radius and length of 5 and 25 cm, respectively, with the cylinder axis parallel to the z-axis. Scanner specifications correspond to the Faro Focus Plus series [16], however, at half the resolution (i.e., a quarter of the number of points). The scanner is located 50 and/or 100 m away from the cylinder. When two scanners are simulated, they are 90 degrees apart as seen from the cylinder.

The relative geometry fitting error (ν) and uncertainty (σ) are shown in Table 1. To increase the interpretability of the results, the values are divided by the true radius for the center and radius. As the vector is of unit length, no normalization is used there. For the purposes of this article, the location of the center along the cylinder axis (z-axis) and length are not relevant. Additionally, the error in the vector is fully described by the error of its x and y components. The dependent pairwise Student's t-test is used to determine whether or not the difference between the methods is statistically significant and the resulting probability values p are given in Table 2.

Table 1. Relative cylinder fitting error ν and uncertainty σ. The cylinder (solid blue dot) is located at the origin.

		Center$_x$		Center$_y$		Vector$_x$		Vector$_y$		Radius	
		$\nu(\%)$	$\sigma(\%)$	$\nu(\%)$	$\sigma(\%)$	$\nu(\%)$	$\sigma(\%)$	$\nu(\%)$	$\sigma(\%)$	$\nu(\%)$	$\sigma(\%)$
	Scanner at (0, 50, 0)										
	ELS	1.41	5.82	24.2	17.2	−0.82	4.19	0.64	5.16	6.58	9.80
	RELS	1.82	7.10	18.1	17.6	−1.06	4.24	0.90	2.89	7.07	12.7
	EM	−0.92	12.8	9.90	27.4	−0.49	6.79	0.53	6.55	−1.07	16.2
	REM	−1.08	11.4	8.32	26.1	−0.19	6.59	0.55	5.27	−0.35	16.0
	Scanner at (0, 100, 0)										
	ELS	12.6	57.1	52.0	101	−8.12	20.5	6.85	15.6	32.8	90.6
	RELS	11.4	55.7	59.8	107	−6.47	21.2	6.53	16.1	27.6	93.6
	EM	3.87	47.3	41.6	120	−1.76	20.0	1.99	15.6	14.5	103
	REM	1.15	51.3	49.7	133	−4.17	21.6	2.18	16.2	18.9	114
	Scanners at (0, 100, 0) and (50, 0, 0)										
	ELS	9.70	12.0	9.36	6.82	−0.89	2.54	0.89	3.69	11.4	6.55
	RELS	7.25	12.3	7.88	7.50	−0.64	2.22	1.47	4.10	9.80	8.53
	EM	1.40	14.2	4.05	11.0	0.06	3.58	−0.19	6.99	4.28	10.2
	REM	2.05	16.4	4.36	11.2	0.05	4.35	0.20	6.16	3.80	11.3

Table 2. Probability values p according to the dependent pairwise Student's t-test. The number of digits is such that $p < 0.005$, i.e., ten times lower than our limit for statistical significance, is shown as 0.00.

	Center$_x$	Center$_y$	Vector$_x$	Vector$_y$	Radius
Scanner at (0, 50, 0)					
p (ELS vs. EM)	0.01	0.00	0.41	0.76	0.00
p (RELS vs. REM)	0.00	0.00	0.04	0.34	0.00
Scanner at (0, 100, 0)					
p (ELS vs. EM)	0.01	0.31	0.00	0.00	0.00
p (RELS vs. REM)	0.01	0.22	0.26	0.00	0.03
Scanners at (0, 100, 0) and (50, 0, 0)					
p (ELS vs. EM)	0.00	0.00	0.00	0.01	0.00
p (RELS vs. REM)	0.00	0.00	0.02	0.00	0.00

The computational times of the 200 iterations are given in Table 3 and are meant only to give an idea of the relative computational time between the different methods. The time for the approach outlined in this article EM excludes the time needed for the initial geometry fit. Computations were performed with MATLAB R2023b in parallel on a workstation with a 2.70 GHz 6-core processor. There is no meaningful difference in memory usage between the different methods. RANSAC was limited to a maximum of a 100 point cloud subsets to evaluate. Meanwhile, the expected Mahalanobis distance approach used

1000 tangent lines and a maximum of 10 point cloud distribution updates. The number of tangent lines has relatively little effect on the computational time; however, the time scales linearly with the number of distribution updates.

Table 3. Computational time in seconds for the three scanning situations and tested methods.

	ELS	RELS	EM
Scanner at (0, 50, 0)	10.65 s	909.8 s	1993 s
Scanner at (0, 100, 0)	11.45 s	953.9 s	576.0 s
Scanners at (0, 100, 0) and (50, 0, 0)	10.50 s	527.2 s	2783 s

5. Discussion

In this paper, we proposed a method to reconstruct geometry by transforming a point cloud into a fuzzy point cloud consisting of distributions of known form. We, however, note that in reality the distributions cannot be known exactly. For the laser scanning example, several factors such as vibration and wind will affect the uncertainty of each point but are unknown. Furthermore, as the scanners are made by commercial companies the hit-registration algorithm is not public, and thus, its uncertainty is similarly unknown. Additionally, while the effect of the incidence angle is taken into account, the cylinder example assumes a fully smooth surface. The roughness that exists in reality may be modeled statistically, for instance, as a Gaussian where the standard deviation and auto-correlation govern the height and planar distribution, respectively [17].

We note that the fuzzy geometry fitting approach such as described in this paper has an implicit dependence of the uncertainty on the geometry and is therefore an advancement over other methods found in literature [1–3,5]. It however has the inherent disadvantage that it requires a robust initial geometry estimate. If the initial estimate is highly erroneous the initial fuzzy point cloud estimate is likely to be erroneous as well and convergence of the iterative uncertainty updating procedure described in Section 3.3 is not guaranteed.

An aspect that is not considered in this paper is that a lack of points does not necessarily mean a lack of geometry but may instead be due to occlusion. The absence of evidence is thus not evidence of absence if the laser beam's trajectory is obstructed. This can happen either due to the geometry itself (i.e., the back half of a cylinder) or due to other objects. We are unable to distinguish between areas devoid of points due to a lack of geometry and areas where points might have been were the area not obscured. The effect of occlusion on complex geometrical objects such as trees is significant and incorporating this into any analysis may greatly improve the results [18].

In this paper, the expected Mahalanobis distance $E[M]$ is chosen over the expected Euclidean distance $E[D]$. Section 3.2 mentioned the general benefits that it is unitless, scale-invariant, and anisotropic. The scale dependence of $E[D]$ induces bias that is easiest to explain when the distributions lie on the geometry. Generally speaking, $E[D] \propto |\Sigma|$; thus, the objective function would be biased to points with greater uncertainty, i.e., greater covariance. Similarly, bias due to isotropy is easiest to explain with the case of a line L where $\hat{\mu} \in L$. $E[D]$ is then minimal when the fitted line L_{fit} is parallel to the principal axis PC_σ of the covariance matrix. This bias decreases with the number of points when the coincidence of the points to the fitted line and the parallelity of the principal axes $PC_\Sigma \parallel L_{\text{fit}}$ are mutually exclusive.

Bias is introduced for curved geometry, as the distance to a convex shape is always smaller from the inside than outside. In other words, this means that for a circle C with radius r as an example with $\hat{\mu} \in C$, $P_{\text{out}} > P_{\text{in}}$ leading to $r_{\text{fit}} > r$ if the expected distance is taken to the curved shape directly. This further motivates our choice for approximating the shape by an envelope of tangent lines. An alternative approach is given in [5] which includes a term in the objective function that compensates for this bias. Such an approach was tested by the authors; however, it requires an accurate estimate of the geometry and was found to reduce robustness.

The interior-point algorithm used to minimize the expected Mahalanobis distance cylinder fitting objective requires a locally convex space to determine the update step, i.e., a positive-definite Hessian, which is not generally the case for our optimization problem. Instead, a positive-definite approximation of the Hessian is then used, meaning that the optimization problem the optimizer solves can deviate from the one specified. As such then, the optimizer sometimes finds a solution that may not be locally convex per the real Hessian. An alternative approach that does not approximate the Hessian may thus yield different results.

The approach described in this article provides an alternative to the least squares approach commonly used, and the average absolute error is lower for the geometry parameters of each of the three scanning situations. It is, however, important to note that the relative uncertainty in these results is often higher, and therefore, the performance difference is not uniformly statistically significant ($p < 0.05$). Additionally, the computational time of least squares without RANSAC was significantly smaller. It is important to note that for our approach the computational time scales roughly linearly with the number of points and the number of point cloud distribution updates. The requirement for such updates may, however, be relaxed when the number of points is higher or the noise level is lower, for instance, for the scanning case at 50 m. For simplicity, this was not taken into account in this study.

RANSAC in combination with least squares was implemented as a simpler approach to deal with noisy point cloud data; however, it does not consistently outperform least squares over the full point cloud and does not provide a benefit as an initial geometry fitting approach.

We further note that the cylinder vector has to be determined accurately for the cross-sectional circle fitting approaches of both least squares and our approach to be evaluated. As expected, the highest vector fitting error was found at 100 m due to the reduced quality and quantity of data, with the highest absolute error for least squares of 8.1% and for our approach of 4.2%, which is deemed acceptable by the authors.

The radius of cylinders scanned by laser data is commonly overestimated [14,19,20], likely because the aforementioned curvature-induced bias means points are on average located outside the surface. Least squares consistently overestimate the radius, and to a greater extent when the level of noise is greater (6.6% to 32.8%). The outlined approach is able to overcome this bias to a large extent with a statistically significant difference in error from −1.1% to 18.9%.

The case of two scanners at different ranges and different orientations was chosen in particular to test the proposed method. Ideally, combining a high- and low-quality data set would lead to a better result even if the average data quality deteriorates. The performance of the proposed method is, however, not uniformly better compared with using exclusively the higher quality data. The difference with respect to least squares (with and without RANSAC) is, however, statistically significant for all variables. It is further interesting to note that for least squares the error is roughly equal in the direction towards either scanner, while with our method it is clearly lower in the direction towards the closer (more certain) scanner.

Finally, it is of interest to see how the approach performs for other geometric shapes, and more work is needed to evaluate the approach more generally, such as for shapes of non-constant cross-sections or with asymmetrical curvature.

6. Conclusions

We presented a fuzzy point cloud data concept, which is a cloud of distributions instead of points, and a conceptual approach to using fuzzy point clouds in geometric shape fitting. The objective function consists of the average expected distance from the data points to the shape using the Mahalanobis distance instead of the Euclidean distance. Fuzzy point clouds model the uncertainty in the measurements and incorporate the possibility that the uncertainty can vary significantly in size and shape from point to point. This is the situation in laser scanning measurements in a complex environment such as forests. We demonstrated the approach with cylinder fitting to simulated laser scanner data, and the results show that the approach has a consistently lower average error than least squares fitting, with and without RANSAC.

Author Contributions: Conceptualization, V.B.V., P.R. and M.Å.; methodology, V.B.V., P.R. and M.Å.; software, V.B.V.; validation, V.B.V.; formal analysis, V.B.V.; investigation, V.B.V.; resources, P.R.; data curation, V.B.V.; writing—original draft preparation, V.B.V.; writing—review and editing, V.B.V. and P.R.; visualization, V.B.V.; supervision, P.R. and M.Å.; project administration, P.R.; funding acquisition, P.R. All authors have read and agreed to the published version of the manuscript.

Funding: This work was supported by the Research Council of Finland (Flagship of Advanced Mathematics for Sensing Imaging and Modelling grant 359185, Centre of Excellence of Inverse Modelling and Imaging grant 353090).

Institutional Review Board Statement: Not applicable.

Informed Consent Statement: Not applicable.

Data Availability Statement: The MATLAB R2023b code is available online at https://github.com/InverseTampere/Cylinder_Fitting, accessed on 29 December 2024.

Conflicts of Interest: The authors declare no conflicts of interest.

Appendix A

Due to the constant cross-sectional shape, the problem can be simplified by projecting the three-dimensional (3D) Gaussians onto the cross-sectional plane, where the z-axis is parallel to the cylinder axis, $Q \perp \vec{v} \equiv \vec{z}$, resulting in a two-dimensional (2D) Gaussian.

The exponential term of the probability density function for an arbitrary Gaussian $N(\hat{\mu}, \Sigma)$ can be rewritten to an ellipsoidal parametrization such that it can be directly expressed as a function of the Cartesian coordinate system, in this case, x, y, z. The ellipsoidal representation is shown in Equations (A1)–(A3) using the precision matrix $P = \Sigma^{-1}$. The indices denote those dimensions' entry; thus, P_{xx}, P_{yy}, P_{zz}, P_{xy}, P_{xz}, and P_{yz} are elements of the precision matrix P.

$$\begin{aligned} f(\hat{x}) &= \frac{e^{-\frac{1}{2}(\hat{x}-\hat{\mu})^T \Sigma^{-1} (\hat{x}-\hat{\mu})}}{\sqrt{8\pi^3 |\Sigma|}} \\ &= K e^{-\frac{1}{2}(\hat{x}-\hat{\mu})^T \Sigma^{-1} (\hat{x}-\hat{\mu})} \\ &= K e^{-\frac{1}{2}(P_{xx}x^2 + P_{yy}y^2 + P_{zz}z^2) + Ax + By + Cz - P_{xy}xy - P_{xz}xz - P_{yz}yz + D} \\ K &= \frac{1}{\sqrt{8\pi^3 |\Sigma|}}, \end{aligned} \tag{A1}$$

The constants A, B, C, and D are given by Equations (A2) and (A3):

$$\begin{bmatrix} A \\ B \\ C \end{bmatrix} = P\hat{\mu} \tag{A2}$$

$$D = -\frac{1}{2}(P_{xx}\hat{\mu}_x^2 + P_{yy}\hat{\mu}_y^2 + P_{zz}\hat{\mu}_z^2) - (P_{xy}\hat{\mu}_x\hat{\mu}_y + P_{xz}\hat{\mu}_x\hat{\mu}_z + P_{yz}\hat{\mu}_y\hat{\mu}_z) \tag{A3}$$

The marginal distribution $f(x,y)$ on the plane is the integral of the probability density with respect to z, which due to the ellipsoidal parametrization can be derived easily as shown in Equation (A4). The expected value and precision matrix of the projected 2D Gaussian are denoted by μ_Q and Q, respectively.

$$f(x,y) = \int_{\mathbb{R}} f(\hat{x})dz = K\sqrt{\frac{2\pi}{P_{zz}}}e^{-\frac{1}{2}(\hat{x}-\hat{\mu}_Q)^T Q(\hat{x}-\hat{\mu}_Q)} \tag{A4}$$

The dimensionality reduction is also evident by the parametrization of a 2D ellipse in the exponential with constants α through ω.

$$\begin{aligned}
&-\frac{1}{2}(\hat{x}-\hat{\mu}_Q)^T Q(\hat{x}-\hat{\mu}_Q) \\
&= \frac{1}{2}(\frac{P_{xz}^2}{P_{zz}} - P_{xx})x^2 \\
&+ \frac{1}{2}(\frac{P_{yz}^2}{P_{zz}} - P_{yy})y^2 + (\frac{P_{xz}P_{yz}}{P_{zz}} - P_{xy})xy \\
&+ (A - C\frac{P_{xz}}{P_{zz}})x + (B - C\frac{P_{yz}}{P_{zz}})y + \frac{C^2}{2P_{zz}} + D \\
&= \alpha x^2 + \beta y^2 + \gamma xy + \delta x + \epsilon y + \omega
\end{aligned} \tag{A5}$$

From this, it follows that the precision matrix equals $Q = -\begin{bmatrix} 2\alpha & \gamma \\ \gamma & 2\beta \end{bmatrix}$.

Its inverse is the 2D covariance matrix

$$\Sigma_Q = Q^{-1} = \frac{1}{\gamma^2 - 4\alpha\beta}\begin{bmatrix} 2\beta & -\gamma \\ -\gamma & 2\alpha \end{bmatrix} = \psi\begin{bmatrix} 2\beta & -\gamma \\ -\gamma & 2\alpha \end{bmatrix}. \tag{A6}$$

Its two eigenvectors $v_{1,2}$ form the projected Gaussian axes, and its eigenvalues $\lambda_{1,2}$ are the variances along said axes given by (A7) and (A8), respectively.

$$\vec{v}_{1,2} = \frac{1}{\sqrt{\psi^2\gamma^2 + (2\psi\beta - \lambda_{1,2})^2}}\begin{bmatrix} \psi\gamma \\ 2\psi\beta - \lambda_{1,2} \end{bmatrix} \tag{A7}$$

$$\sigma_{1,2}^2 = \lambda_{1,2} = \psi(\alpha + \beta \pm \sqrt{\alpha^2 + \beta^2 + \gamma^2 - 2\alpha\beta}) \tag{A8}$$

References

1. Keksel, A.; Ströer, F.; Seewig, J. Bayesian approach for circle fitting including prior knowledge. *Surf. Topogr. Metrol. Prop.* **2018**, *6*, 035002. [CrossRef]
2. Keksel, A.; Eli, B.; Eifler, M.; Seewig, J. Bayesian analysis of uncertainties in circle, straight-line and ellipse fitting considering a-priori knowledge- comparative analysis with total-least-squares approaches. *Surf. Topogr. Metrol. Prop.* **2024**, *12*, 015015. [CrossRef]
3. Zhao, M.; Jia, X.; Ma, L.; Shi, Y.; Jiang, J.; Li, Q.; Yan, D.M.; Huang, T. A Bayesian Approach Toward Robust Multidimensional Ellipsoid-Specific Fitting. *IEEE Trans. Pattern Anal. Mach. Intell.* **2024**, *46*, 10106–10123. [CrossRef]
4. McLachlan, G.J. Mahalanobis distance. *Resonance* **1999**, *4*, 20–26. [CrossRef]
5. Faion, F.; Zea, A.; Hanebeck, U.D. Reducing bias in Bayesian shape estimation. In Proceedings of the 17th International Conference on Information Fusion (FUSION), Salamanca, Spain, 7–10 July 2014; IEEE: Piscataway, NJ, USA, 2014; pp. 1–8.
6. Calders, K.; Adams, J.; Armston, J.; Bartholomeus, H.; Bauwens, S.; Bentley, L.P.; Chave, J.; Danson, F.M.; Demol, M.; Disney, M.; et al. Terrestrial laser scanning in forest ecology: Expanding the horizon. *Remote Sens. Environ.* **2020**, *251*, 112102. [CrossRef]
7. Pesci, A.; Teza, G.; Bonali, E.; Casula, G.; Boschi, E. A laser scanning-based method for fast estimation of seismic-induced building deformations. *ISPRS—J. Photogramm. Remote Sens.* **2013**, *79*, 185–198. [CrossRef]

8. Hartzell, P.J.; Gadomski, P.J.; Glennie, C.L.; Finnegan, D.C.; Deems, J.S. Rigorous error propagation for terrestrial laser scanning with application to snow volume uncertainty. *J. Glaciol.* **2015**, *61*, 1147–1158. [CrossRef]
9. Sun, W.; Wang, J.; Yang, Y.; Jin, F.; Sun, F. Accurate deformation analysis based on point position uncertainty estimation and adaptive projection point cloud comparison. *Geocarto Int.* **2023**, *38*, 2175916. [CrossRef]
10. Raumonen, P.; Kaasalainen, M.; Åkerblom, M.; Kaasalainen, S.; Kaartinen, H.; Vastaranta, M.; Holopainen, M.; Disney, M.; Lewis, P. Fast automatic precision tree models from terrestrial laser scanner data. *Remote Sens.* **2013**, *5*, 491–520. [CrossRef]
11. Tang, P.; Huber, D.; Akinci, B.; Lipman, R.; Lytle, A. Automatic reconstruction of as-built building information models from laser-scanned point clouds: A review of related techniques. *Autom. Constr.* **2010**, *19*, 829–843. [CrossRef]
12. Eberly, D. *Distance from a Point to an Ellipse, an Ellipsoid, or a Hyperellipsoid*; Geometric Tools: Redmond, WA, USA, 2020.
13. Jury, E.; Mansour, M. Positivity and nonnegativity conditions of a quartic equation and related problems. *IEEE Trans. Autom. Control* **1981**, *26*, 444–451. [CrossRef]
14. Michałowska, M.; Rapiński, J.; Janicka, J. Tree position estimation from TLS data using hough transform and robust least-squares circle fitting. *Remote Sens. Appl. Soc. Environ.* **2023**, *29*, 100863. [CrossRef]
15. Fischler, M.A.; Bolles, R.C. Random sample consensus: A paradigm for model fitting with applications to image analysis and automated cartography. *Commun. ACM* **1981**, *24*, 381–395. [CrossRef]
16. FARO. *Faro Focus Laser Scanner Tech Sheet*; FARO: Lake Mary, FL, USA, 2020.
17. van Ginneken, B.; Stavridi, M.; Koenderink, J.J. Diffuse and specular reflectance from rough surfaces. *Appl. Opt.* **1998**, *37*, 130–139. [CrossRef] [PubMed]
18. Wan, P.; Wang, T.; Zhang, W.; Liang, X.; Skidmore, A.K.; Yan, G. Quantification of occlusions influencing the tree stem curve retrieving from single-scan terrestrial laser scanning data. *For. Ecosyst.* **2019**, *6*, 1–13. [CrossRef]
19. Forsman, M.; Börlin, N.; Olofsson, K.; Reese, H.; Holmgren, J. Bias of cylinder diameter estimation from ground-based laser scanners with different beam widths: A simulation study. *ISPRS—J. Photogramm. Remote Sens.* **2018**, *135*, 84–92. [CrossRef]
20. Abegg, M.; Bösch, R.; Kükenbrink, D.; Morsdorf, F. Tree volume estimation with terrestrial laser scanning—Testing for bias in a 3D virtual environment. *Agric. For. Meteorol.* **2023**, *331*, 109348. [CrossRef]

Journal of
Imaging

MDPI

Article

Arbitrary Optics for Gaussian Splatting Using Space Warping

Jakob Nazarenus [1,*], Simin Kou [2], Fang-Lue Zhang [2] and Reinhard Koch [1]

[1] Department of Computer Science, Kiel University, 24118 Kiel, Germany; rk@informatik.uni-kiel.de
[2] School of Engineering and Computer Science, Victoria University of Wellington, Wellington 6012, New Zealand; simin.kou@vuw.ac.nz (S.K.); fanglue.zhang@vuw.ac.nz (F.-L.Z.)
* Correspondence: jna@informatik.uni-kiel.de

Abstract: Due to recent advances in 3D reconstruction from RGB images, it is now possible to create photorealistic representations of real-world scenes that only require minutes to be reconstructed and can be rendered in real time. In particular, 3D Gaussian splatting shows promising results, outperforming preceding reconstruction methods while simultaneously reducing the overall computational requirements. The main success of 3D Gaussian splatting relies on the efficient use of a differentiable rasterizer to render the Gaussian scene representation. One major drawback of this method is its underlying pinhole camera model. In this paper, we propose an extension of the existing method that removes this constraint and enables scene reconstructions using arbitrary camera optics such as highly distorting fisheye lenses. Our method achieves this by applying a differentiable warping function to the Gaussian scene representation. Additionally, we reduce overfitting in outdoor scenes by utilizing a learnable skybox, reducing the presence of floating artifacts within the reconstructed scene. Based on synthetic and real-world image datasets, we show that our method is capable of creating an accurate scene reconstruction from highly distorted images and rendering photorealistic images from such reconstructions.

Keywords: 3D reconstruction; novel view synthesis; 3D Gaussian Splatting; camera models

1. Introduction

Based on the foundational work in the field of 3D reconstruction, in recent years, the topic of Novel View Synthesis (NVS) has gained momentum due to the rapid progress of learning-based methods [1]. Using a differentiable scene representation in combination with a differentiable rendering pipeline, these methods have achieved state-of-the-art results in recreating photorealistic scenes from images [2]. As the underlying models were trained to encode view-dependent effects, the models outperform renderings from classical 3D reconstruction models with respect to visual fidelity, enabling the recreation of complex lighting scenarios. Since then, more recent contributions have mitigated one of the most significant downsides of learning-based NVS methods: their training and rendering times [3,4]. One method in particular, 3D Gaussian Splatting (3DGS), established a significant improvement over the current state of the art by introducing an efficient scene representation and rendering pipeline [5]. 3DGS enables real-time rendering and training scenes in several minutes while achieving a very high degree of detail. Its efficiency and visual quality have led to its adoption in VR, game engines, and Web frameworks [6–8]. One main reason for the success of 3DGS, its differentiable Gaussian rasterizer, comes with a significant limitation: its underlying pinhole camera model. This simplification works well as an approximation for many real-world cameras, requiring an additional preprocessing step to undistort the images. However, this undistortion is not well suited for all types of lenses, such as fisheye lenses. Due to their wide Field of View (FoV), they have a high information density and are able to capture scenes with only a few frames. Due to these reasons, they have gained popularity in smartphones and action cameras. The same reasons make fisheye images an interesting candidate for NVS tasks, with a high level of overlap in the images providing depth information for scene reconstruction. Although this does not pose an issue for ray tracing-based methods [2], the underlying 3D Gaussian raster prevents the use of fisheye images with 3DGS. Undistorting these images as pinhole images is possible in principle but usually results in cropping into the images and discarding significant information from the peripheral view, potentially significantly reducing the accuracy of the scene reconstruction. This necessitates the introduction of novel methods that combine the efficiency of 3DGS with the flexibility of using different camera models. Concurrently with our work, an extension of the Gaussian rasterizer for equidistant and panoramic fisheye images, termed Fisheye-GS, has been published [9].

With this contribution, we propose to solve the problem of training a 3DGS model with an arbitrary camera model. We achieve this by using a space-warping module to enable the unmodified pinhole rasterizer to render views for arbitrary optics, including fisheye optics with polynomial distortion, as commonly used to represent real-world cameras [10]. Our proposed space-warping module shifts the scene's Gaussians according to a predefined distortion function, emulating an arbitrary camera lens. The scale and rotation of the Gaussians are determined as tangential approximations of the actual distortion by leveraging the Jacobian of the distortion function with a subsequent orthogonalization. This approach allows for seamless integration into the existing 3DGS pipeline, as no modifications are made to the rasterization module, enabling the method to work with future versions of the rasterizer. In addition, to reduce floating artifacts in outdoor scenes, we enforce a learnable skybox that is directly integrated into the underlying model of the scene.

This paper is structured as follows. Section 2 reviews the recent literature on NVS methods and their applicability to models without pinhole cameras. Section 3 derives the space-warping module for the Gaussian rasterizer, as well as two explicit distortion functions for the common OpenCV and Blender polynomial fisheye camera models. Section 4 presents the results of extensive experiments on synthetic and real-world datasets, compares our method with other methods, and demonstrates the reasonableness of our design decisions through several ablations, which are subsequently discussed in Section 5, followed by a brief conclusion in Section 6.

2. Related Work

In this section, we review the related literature that impacts our research, from traditional Structure from Motion (SfM) to modern methods such as 3DGS. We further position our proposed method among existing methods.

2.1. *SfM*

These methods apply feature-matching algorithms to find correspondences between images. This leads to an initial pose estimation, enabling a dense matching of overlapping views to find a dense depth model of the scene [11,12]. Although classical methods cannot compete with modern learning-based methods with respect to the visual quality of rendered images, their underlying robust pose estimation algorithms are still widely used for the initialization of more recent methods. 3DGS, in particular, relies on the sparse COLMAP model to initialize its Gaussian model for fast convergence [5]. Although there are extensions of learning-based 3D reconstruction methods avoiding the use of classical preprocessing methods for initialization [13–15], we consider this beyond the scope of our article and utilize the robust COLMAP pose estimation for initialization of our model.

2.2. *Neural Radiace Fields (NeRF)*

The main contribution accelerating the development of learning-based NVS methods was the introduction of Neural Radiace Fields (NeRF). Combining an Multilayer Perceptron (MLP) as an implicit differentiable scene representation with a differentiable ray tracer led to a significant step forward in the visual quality of rendered images. In particular, the ability of the underlying MLP to capture view-dependent effects such as reflections enabled a previously unseen level of realism. One major drawback of this method is the inefficiency of the ray tracer and the scene representation, requiring 1–2 days per scene for optimization [2].

Initiated by this contribution, several subsequent papers have attempted to improve upon the initial NeRF. Proposed approaches include anti-aliasing [16], appearance changes [17], deformations [18–20], and learned backgrounds [21], the latter being conceptually adopted by our proposed method to reduce floating artifacts introduced by sky textures. Another direction of developments is related to the underlying scene representation. Several papers have successfully demonstrated that the optimization and rendering times can be significantly reduced by choosing a more explicit scene representation, such as octrees [4], feature grids [4], factorized feature planes [19], or hash functions [3]. These explicit methods have achieved real-time rendering speeds and reduced optimization times ranging from hours to minutes [3].

2.3. 3DGS

This contribution adopts a similar approach as the preceding explicit methods by representing the scene as a set of 3D Gaussians, each with a 3D position, rotation, 3D scale, opacity, and directional radiance [5]. Three-dimensional Gaussians have been proven to be an efficient graphical primitive with the capability of representing very fine structures. The authors who proposed 3DGS went one step further by simultaneously replacing the previous ray-tracing module with a more efficient differentiable Gaussian rasterizer. Its main benefit is not rendering an image pixel by pixel using individual rays but, instead, rasterizing the whole image at one time. The proposed rasterizer comes with the significant limitation of being restricted to pinhole images, necessitating an undistortion preprocessing step, which prevents the use of wide-angle fisheye images. Eliminating this limitation is the main contribution of our paper by introducing a flexible lightweight space-warping module that enables the emulation of arbitrary camera lenses using the unmodified pinhole rasterizer.

Concurrently with our work, an extension of the Gaussian rasterizer for equidistant and panoramic fisheye images, termed Fisheye-GS, has been published [9]. Our approach conceptually differs in the following aspects: While Fisheye-GS utilizes a modified Gaussian rasterizer, we keep the original rasterizer intact and represent the camera lens using a spatial distortion of the scene without changing. This allows users to implement new camera optics without needing to write additional CUDA code and presents a more modular approach so that the existing rasterizer can be more easily replaced by future versions. In contrast to Fisheye-GS, which focuses on equidistant fisheye images, we derive distortion functions for the common OpenCV polynomial fisheye camera model, eliminating the need for an additional preprocessing step for real-world datasets such as ScanNet++ [22]. In Section 4.2, we provide a qualitative and quantitative comparison between Fisheye-GS and our proposed method.

Another very recently proposed method for scene reconstructions for non-pinhole images is 3D Gaussian ray tracing [23]. Similarly to the initial NeRF, this method relies on a ray-tracing algorithm for rendering, circumventing the issues associated with pinhole rasterization. As is typical for ray-tracing algorithms, this method enables various camera models and effects, such as reflections, refraction, and depth of field. We acknowledge the contribution of the authors; however, we consider 3D Gaussian ray tracing not to be directly comparable to our proposed method, as our main focus lies in enabling non-pinhole optics for the 3DGS rasterization process.

Simultaneously to our submission, a preprint proposing UniGaussian has been published, utilizing a the compression of 3D Gaussians as a similar concept to the warping function introduced in our contribution [24]. Although we utilize an ordered Gram–Schmidt orthogonalization for the computation of the distorted scale and rotation of the 3D Gaussians, UniGaussian relies on an eigendecomposition for this purpose. While acknowledging the work of the authors who proposed UniGaussian, we cannot provide quantitative comparisons with their proposed method, as their source code is not yet publicly available.

3. Methods and Data

In this section, we first introduce our approach of emulating arbitrary camera models using a space-warping function. Subsequently, we discuss the specifics of fisheye lenses and, finally, describe our approach of enforcing a learned skybox to reduce floating artifacts. In addition, we describe the synthetic and real-world datasets used in the evaluation of our method.

3.1. Space-Warping Function

In 3DGS, a scene is represented by a set of 3D Gaussians. Each Gaussian is fully described by the following properties: position, rotation, scale, opacity, and spherical harmonics coefficients. We denote the original pinhole rendering function as $\mathcal{P}_{\text{pinhole}} : \mathbb{R}^3 \to \mathbb{R}^2$, mapping 3D points to pixel coordinates within the rendered image. As illustrated in Figure 1, we emulate an arbitrary rendering function ($\mathcal{P}_{\text{arb}} : \mathbb{R}^3 \to \mathbb{R}^2$) using a space-warping function ($\mathcal{W} : \mathbb{R}^3 \to \mathbb{R}^3$) that satisfies

$$\mathcal{P}_{\text{arb}}(x) = \mathcal{P}_{\text{pinhole}}(\mathcal{W}(x)) \quad \forall x \in \mathbb{R}^3. \tag{1}$$

This enables us to keep the underlying rendering function ($\mathcal{P}_{\text{pinhole}}$) unchanged while obtaining the desired behavior of $\mathcal{P}_{\text{arb}}$. Assuming that $\mathcal{W}$ exists, this emulation is correct for the projection of points $x \in \mathbb{R}^3$. However, distortion of a Gaussian also distorts its scale and rotation, as shown in Figure 2b. As Figure 2c illustrates, we compute the Jacobian, which provides a local linear approximation of the distortion function:

$$J_{\mathcal{W}}(x) = \begin{bmatrix} \frac{\partial \mathcal{W}_1}{\partial x_1} & \frac{\partial \mathcal{W}_1}{\partial x_2} & \frac{\partial \mathcal{W}_1}{\partial x_3} \\ \frac{\partial \mathcal{W}_2}{\partial x_1} & \frac{\partial \mathcal{W}_2}{\partial x_2} & \frac{\partial \mathcal{W}_2}{\partial x_3} \\ \frac{\partial \mathcal{W}_3}{\partial x_1} & \frac{\partial \mathcal{W}_3}{\partial x_2} & \frac{\partial \mathcal{W}_3}{\partial x_3} \end{bmatrix}. \tag{2}$$

The resulting axes are, in general, not orthogonal, necessitating an additional orthogonalization step. As shown in Figure 2d, for a distorted set of axes ($A \in \mathbb{R}^{3\times3}$), we find an ordering permutation (π) of the axis and compute a Gram–Schmidt orthogonalization of the ordered axes, resulting in orthogonal axes $\tilde{A} \in \mathbb{R}^{3\times3}$:

$$A = [a_1, a_2, a_3], \tag{3}$$

$$\pi \in \left\{ \sigma \in S_3 \;\middle|\; |a_{\sigma(1)}| \geq |a_{\sigma(2)}| \geq |a_{\sigma(3)}| \right\}, \tag{4}$$

$$b_1 = a_{\pi(1)}, \quad b_2 = a_{\pi(2)} - b_1 \frac{a_{\pi(2)}^\top b_1}{b_1^\top b_1}, \quad b_3 = a_{\pi(3)} - b_1 \frac{a_{\pi(3)}^\top b_1}{b_1^\top b_1} - b_2 \frac{a_{\pi(3)}^\top b_2}{b_2^\top b_2}, \tag{5}$$

$$\tilde{A} = \left[b_{\pi^{-1}(1)}, b_{\pi^{-1}(2)}, b_{\pi^{-1}(3)} \right], \tag{6}$$

where S_3 represents the symmetric group, with π being chosen from a constrained set whose elements (σ) all fulfill the ordering constraint.

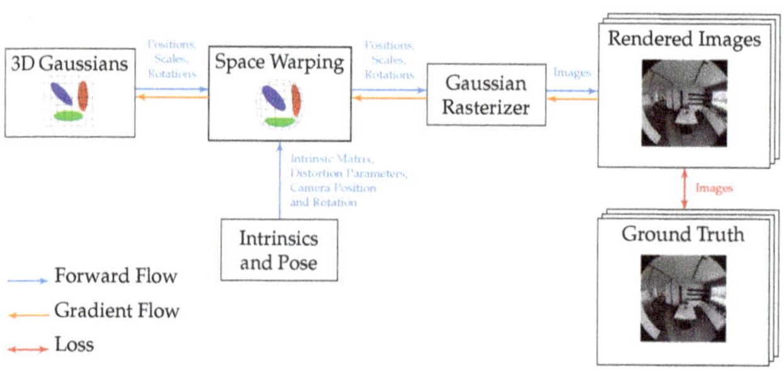

Figure 1. Pipeline of our proposed method. Before being forwarded to the pinhole Gaussian rasterizer, we apply a space-warping module to the position, rotation, and scale to emulate the distortion of a lens specified by the camera's intrinsics.

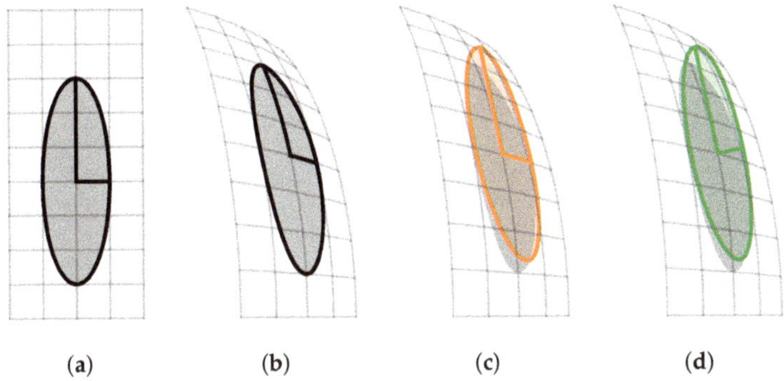

Figure 2. Distortion of scale and rotation. The four images show the steps in the distortion pipeline from left to right. An undistorted Gaussian (**a**) is non-linearly distorted (**b**). This distortion is linearly approximated using Jacobian $J_{\mathcal{W}}$ (**c**), with a subsequent orthogonalization of the axes (**d**). For (**c**,**d**), the gray area shows the true distorted Gaussian to visualize the approximation error.

3.2. Fisheye Cameras

Fisheye cameras usually distort the image in relation to the polar angle. For this reason, it is practical to express the warping function ($\mathcal{W}$) in terms of a warping function ($\mathcal{W}_{\text{sph}}$) that operates in spherical coordinates. For this purpose, we define the following two transformations:

$$\mathcal{T}_{\text{cart}\to\text{sph}}\left(\begin{bmatrix} x \\ y \\ z \end{bmatrix}\right) = \begin{bmatrix} \sqrt{x^2+y^2+z^2} \\ \arccos\left(z/\sqrt{x^2+y^2+z^2}\right) \\ \arctan\left(y/x\right) \end{bmatrix}, \tag{7}$$

$$\mathcal{T}_{\text{sph}\to\text{cart}}\left(\begin{bmatrix} r \\ \theta \\ \varphi \end{bmatrix}\right) = \begin{bmatrix} r\sin\theta\cos\varphi \\ r\sin\theta\sin\varphi \\ r\cos\theta \end{bmatrix}. \tag{8}$$

This enables us to write the warping function as

$$\mathcal{W}(x) = \left(\mathcal{T}_{\text{sph}\to\text{cart}} \circ \mathcal{D} \circ \mathcal{T}_{\text{cart}\to\text{sph}}\right)(x). \tag{9}$$

To approximate D for a given scene's camera, we sample a set of corresponding 3D–2D points (X_i, x_i) and compute their corresponding polar angles as

$$(\mathcal{T}_{\text{cart}\to\text{sph}}(X_i)_\theta, \mathcal{T}_{\text{cart}\to\text{sph}}(\mathcal{P}^{-1}_{\text{pinhole}}(x_i))_\theta). \tag{10}$$

As the distortion only depends on the polar angle, we sample these points at equidistant polar angles ($\theta \in [0, \pi]$) for $\varphi = 0$ and $r = 1$. Using pairs of polar angles, we fit the coefficients of an 8th-degree polynomial. These coefficients need to be recomputed for any change in the camera model or its specific intrinsic parameters. In our pipeline, this is performed automatically at the start of each optimization run.

With $\mathcal{W}$ fully defined by the coordinate transformations and the polynomial polar angle distortion function, we compute their combined Jacobian using the SymPy computer algebra system [25]. The resulting Jacobian is provided in detail in Appendix A. Furthermore, in Appendix B, we demonstrate the application of our proposed method for non-fisheye lenses with an exemplary implementation of an orthographic camera model.

3.3. Skybox

In classical SfM, depth can only be computed for points that contain enough features to be matched between different views [11]. For this reason, reconstruction methods tend to struggle with areas with minimal texture, such as uniformly colored walls or skies. This issue can be mitigated by regularizing the depth through the use of pre-learned monocular depth [26] or other prior knowledge about the scene. In our case, reconstructing the sky in open scenes seemed to be most problematic, resulting in inconsistent depth with floating sky artifacts, which reduced the overall rendering quality. We chose to mitigate this issue, similarly to Nerf++ [21], by enforcing a learned skybox at a long distance. We implemented this by placing 1000 isotropic Gaussians on a Fibonacci sphere at a distance twice as far as the furthest point within the sparse initialization point cloud. During optimization, the movement of these Gaussians is restricted to the surface of the initial sphere. In addition, they are prevented from pruning and from having their opacity reset, ensuring a consistent background being learned during optimization. Compared to other methods [21], instead of creating a separate model for the background, we fully integrate the background Gaussians into a single model, enabling a seamless rendering process.

3.4. Datasets

We decided to evaluate our proposed method on two different datasets—one synthetic and one real-world dataset. The synthetic dataset (Blender) eliminates most sources of errors and allows us to focus solely on the performance of the model under perfect conditions, while the real-world dataset (ScanNet++) enables comparisons with existing methods and provides convincing indications about the applicability of the method.

3.4.1. Synthetic Blender Dataset

We chose 6 publicly available blender scenes containing 3 indoor and 3 outdoor scenes [27]:

- Archiviz;
- Barbershop;
- Classroom;
- Monk;
- Pabellon;
- Sky.

For each scene, we rendered 100 photorealistic, ray-traced frames along a predefined trajectory with a resolution of 1024×1024 px. Each 8th image was taken as a test image; all other images were part of the training set. We chose a simulated fisheye camera with an FoV of 180° and a sensor size of 32 mm. Blender's camera model is configured by specifying five coefficients, which were chosen with the following values [28]:

$$(1.0 \cdot 10^{-5}, -8.7 \cdot 10^{-2}, -3.5 \cdot 10^{-6}, 3.5 \cdot 10^{-6}, -2.6 \cdot 10^{-8}).$$

In Blender's camera model, these coefficients model a polynomial mapping radial distances on the camera sensor to camera rays. The values for the coefficients were chosen as a large negative linear component to achieve fisheye distortion with a large FoV. Furthermore, we chose small coefficients for higher orders to avoid artifacts caused by a non-injective distortion function while avoiding zero-valued coefficients to preserve the projection's complexity. For each scene, we computed our own polynomial representation of the distortion function according to the samples presented in Equation (10).

3.4.2. ScanNet++ Dataset

This dataset is provided by the Technical University of Munich and consists of several indoor scenes with images, camera poses, and point clouds [22]. For comparability with Fisheye-GS, we chose the following 6 scenes:

- Bedroom (e8ea9b4da8);
- Kitchen (bb87c292ad);
- Office Day (4ba22fa7e4; the corresponding scene from the Fisheye-GS paper is currently not available, so we chose a similar scene);

- Office Night (8d563fc2cc);
- Tool Room (d415cc449b);
- Utility Room (0a5c013435).

Each scene consists of a varying number of fisheye images (147–406) with a resolution of 1752×1168 px. Again, each 8th image was taken as a test image; all other images were part of the training set. For each scene, the authors provide the COLMAP poses, intrinsics, and a sparse point cloud. For the Blender dataset, we converted each scene's intrinsic parameters to our custom distortion function using the samples described in Equation (10).

4. Results

This section contains the results of our proposed method for synthetic and real-world datasets, as well as comparisons with Fisheye-GS. We further show several ablations to validate the design choices of our method. The experiments were carried out on a system with an *AMD EPYC 72F3*, 256 GB of memory, and an NIVIA A100 GPU. However, the overall system requirements are lower, identical to 3DGS.

4.1. Synthetic Blender

As a first experiment, we ran our method on the six synthetic blender scenes for 30,000 iterations. The per-scene results, reporting the number of Gaussians, Peak Signal-to-Noise Ratio (PSNR), Structural Similarity Index Measure (SSIM) [29], and Learned Perceptual Image Patch Similarity (LPIPS) [30], are shown in Table 1. As indicated by the metrics and the qualitative samples in Figures 3–5, the rendered images achieve photorealistic quality with only minor artifacts. As seen in the rendered *Barbershop* image, the method struggles with reflections in the mirror. This is expected because the spherical harmonics do not provide sufficient capacity for encoding the full reflected room within the Gaussians of the mirror. A detailed illustration of this problem is shown in Figure 3.

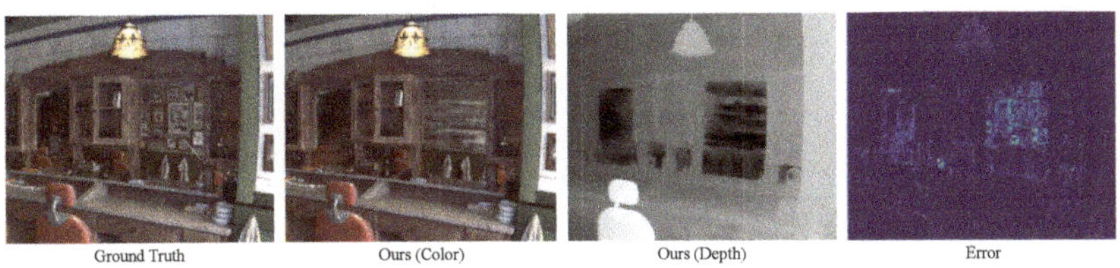

Figure 3. Reconstruction results for the synthetic *Classroom* scene.

Table 1. Experimental results of the proposed method on the synthetic Blender dataset. For each metric, the arrows indicate if lower or higher results are preferred.

Scene	#Gaussians↓	PSNR↑	SSIM↑	LPIPS↓
Archiviz	604,504	38.36	0.979	0.064
Barbershop	573,016	36.97	0.979	0.048
Classroom	676,173	35.56	0.972	0.094
Monk	470,918	33.74	0.964	0.061
Pabellon	502,938	33.91	0.910	0.209
Sky	118,730	42.29	0.989	0.036

The rendered images clearly show a high level of local reconstruction error in the region of the mirror. As the rendered depth shows, within the mirror, there are Gaussians on the mirror's surface, as well as Gaussians behind the mirror. Due to their limited expressiveness, the Gaussians on the mirror's surface lead to blurry horizontal artifacts. The other Gaussians represent a mirrored room behind the mirror, which is heavily undersampled, preventing a photorealistic reconstruction. According to our understanding, this issue is inherent to 3DGS, as it does not model any secondary rays. Furthermore, there are minor artifacts visible in the shady regions of the *Monk* scene. An outlier is the *Pabellon* scene, which shows significantly worse metrics than all other methods. In this scene, the method struggles to render the texture of the pool, causing blurry artifacts.

Figure 4. Results of our proposed method on synthetic Blender scenes (*Archiviz*, *Barbershop*, and *Classroom*). Red rectangles indicate areas in which our method produced reconstruction artifacts. Zoom in for details.

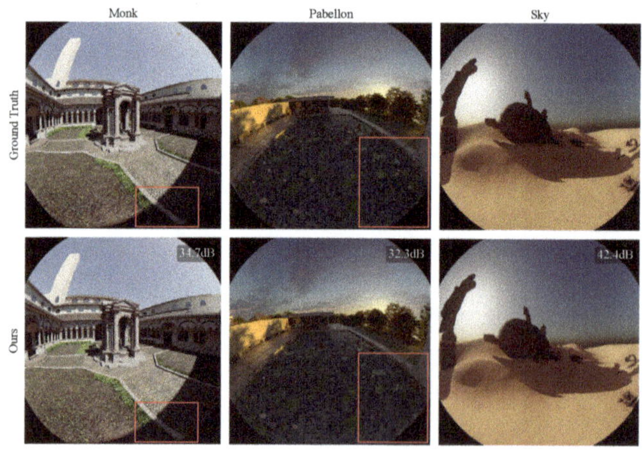

Figure 5. Results of our proposed method on synthetic Blender scenes (*Monk*, *Pabellon*, and *Sky*). Red rectangles indicate areas in which our method produced reconstruction artifacts. Zoom in for details.

4.2. ScanNet++

To further confirm the synthetic results, we performed a second experiment on the real-world Scannet++ dataset for 30,000 iterations, using our proposed method and Fisheye-GS. As the authors who proposed Fisheye-GS thoroughly demonstrated, the use of classical 3DGS for the reconstruction of fisheye scenes performs significantly worse than Fisheye-GS, so in this paper, we did not compare our method against 3DGS and relied on Fisheye-GS as a benchmark. We did not optimize a separate skybox in this experiment because of the absence of outdoor scenes. As our method is directly applicable to OpenCV fisheye images, the pre-processing step to convert the images to equidistant projections was only performed for Fisheye-GS. In Table 2, we report the number of Gaussians, PSNR, SSIM, and LPIPS for each experimental run. For better comparability, we also report the mean of the normalized differences for each metric. Qualitative samples are shown in Figures 6 and 7. When comparing the mean results, our method outperforms Fisheye-GS on every metric except SSIM. However, the increases in PSNR and LPIPS are relatively small, and the main improvement of our method over Fisheye-GS is the reduction in the number of Gaussians and, thus, the model size, which shows a relative difference of 42.18%.

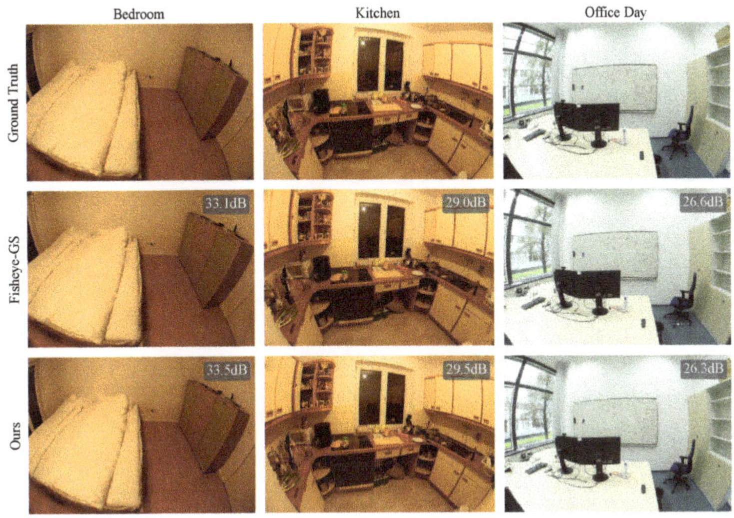

Figure 6. Results of our proposed method and Fisheye-GS on ScanNet++ scenes (*Bedroom*, *Kitchen*, and *Office Day*). Zoom in for details.

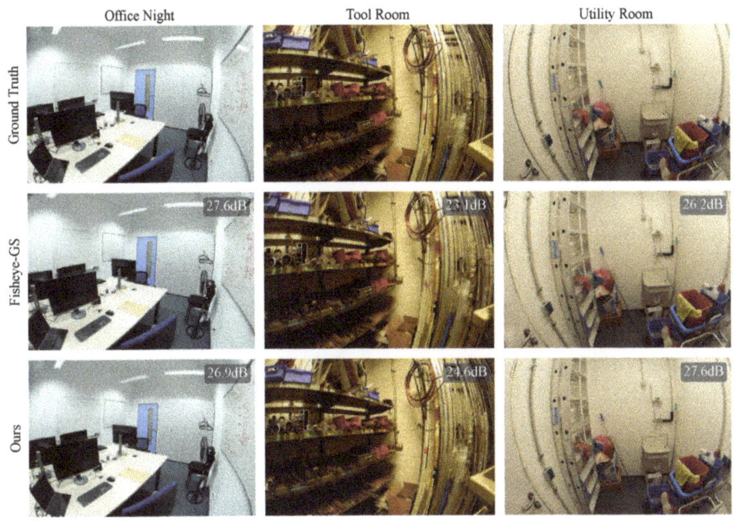

Figure 7. Results of our proposed method and Fisheye-GS on ScanNet++ scenes (*Office Night*, *Tool Room*, and *Utility Room*). Zoom in for details.

4.3. Ablations

To test the design decisions for our proposed method, we chose to conduct experiments to validate the Jacobian distortion function, the learned skybox, and the number of coefficients used for the polar distortion polynomial.

Table 2. Experimental results of the proposed method on the ScanNet++ dataset. For each dataset, the best result is highlighted in bold.

Scene	Method	#Gaussians↓	PSNR↑	SSIM↑	LPIPS↓
Bedroom	Fisheye-GS	345,084	32.09	**0.947**	0.192
	Ours	**239,923**	32.09	0.946	**0.191**
Kitchen	Fisheye-GS	593,270	30.75	**0.935**	**0.179**
	Ours	**392,722**	**30.80**	0.928	0.188
Office Day	Fisheye-GS	669,351	25.92	**0.888**	0.182
	Ours	**480,055**	**26.22**	0.885	**0.176**
Office Night	Fisheye-GS	684,649	**26.30**	**0.907**	**0.176**
	Ours	**448,746**	26.27	0.899	0.179
Tool Room	Fisheye-GS	2,647,082	**27.01**	**0.856**	0.222
	Ours	**1,479,128**	26.95	0.851	**0.221**
Utility Room	Fisheye-GS	749,024	28.06	0.915	0.165
	Ours	**453,970**	**29.20**	**0.923**	**0.155**
Relative Mean	Fisheye-GS	42.81%	−0.83%	**0.31%**	0.76%
	Ours	**−42.81%**	**0.83%**	−0.31%	**−0.76%**

4.3.1. Jacobian Distortion

To qualitatively visualize the importance of this design component, we created a scene consisting of Gaussians along the edges of a cube. Each Gaussian was stretched along the axis of its corresponding edge. As Figure 8 shows, disabling Jacobian distortion leads to blurred edges of the cube, since the Gaussians are not rotated according to the spatial distortion. To confirm this observation quantitatively, we performed an optimization on the *Utility Room* scene, showing overall improved visual metrics with the Jacobian distortion enabled, as shown in Table 3.

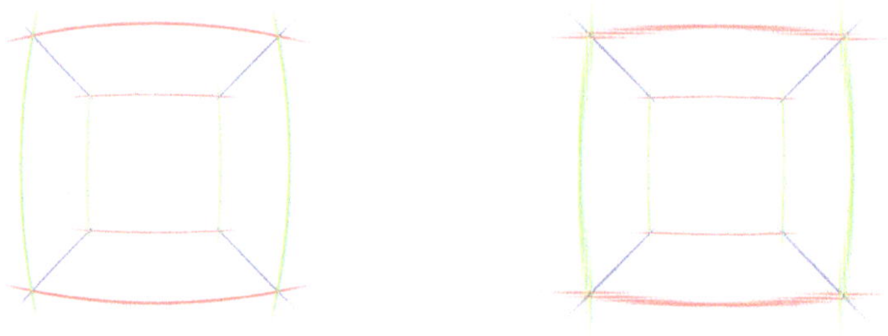

Figure 8. Renderings of a cube with Gaussians along the edges. The left rendering has the scale and rotation adjusted according to the Jacobian; for the right rendering, scale and rotation were left unmodified.

Table 3. Experimental results for evaluation of the Jacobian distortion for rotation and scale of the Gaussians.

Scene	Jacobian	#Gaussians↓	PSNR↑	SSIM↑	LPIPS↓
Utility Room	Enabled	455,653	**29.07**	**0.923**	**0.155**
	Disabled	**387,457**	28.80	0.918	0.171

4.3.2. Degree of the Distortion Polynomial

To test the effect of varying the number of polynomial coefficients within the polar distortion function, we performed experiments on the *Utility Room* scene with an increasing number of coefficients. As shown in Table 4 and Figure 9, all metrics improve with an increasing number of coefficients, with only minor improvements from sixth degree upward. This is to be expected, as increasing the number of coefficients decreases the approximation error, with diminishing returns when the approximation error approaches the pixel size of the rendered image. For the chosen polynomial degree of eight, we computed the sensitivity as the mean absolute deviation from the results for higher and lower polynomial degrees and found the values of $\Delta\text{PSNR} = 1.29 \cdot 10^{-1}$, $\Delta\text{SSIM} = 1.75 \cdot 10^{-4}$, and $\Delta\text{LPIPS} = 1.97 \cdot 10^{-4}$. This demonstrates the overall robustness of the proposed method for the chosen polynomial degree.

Table 4. Experimental results for evaluation of the number of polynomial coefficients in the polar distortion polynomial.

Scene	Degree	#Gaussians↓	PSNR↑	SSIM↑	LPIPS↓
Utility Room	2	537,376	27.65	0.903	0.170
	4	665,298	28.90	0.916	0.156
	6	472,092	29.03	**0.923**	**0.154**
	8	**456,368**	29.27	**0.923**	**0.154**
	10	456,941	**29.29**	**0.923**	**0.154**

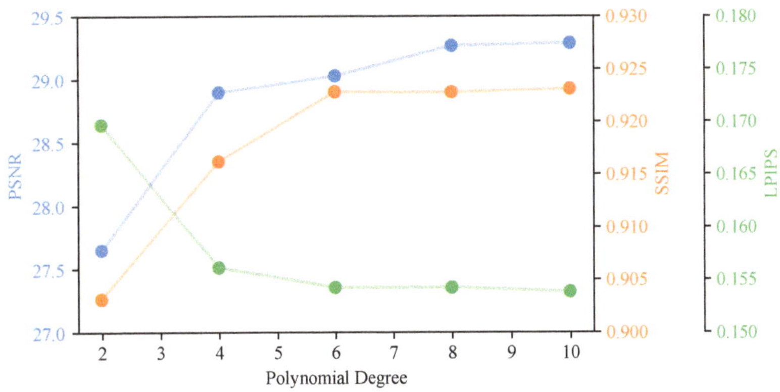

Figure 9. Evaluation metrics for the *Utility Room* scene for varying degrees of the polynomial polar distortion function.

4.3.3. Learnable Skybox

For validation of the learned skybox, we performed two optimizations on the synthetic outdoor *Monk* scene—one with the skybox enabled and one with it disabled. As Table 5 shows, all visual metrics improve significantly with the enabled feature. In addition, Figure 10 illustrates the reason for this discrepancy. With the learned skybox disabled, the sky is represented using a set of Gaussians in close proximity to the camera, which creates floating sky artifacts and hides parts of the scene. The enforcement of the learned skybox successfully mitigates this issue.

Table 5. Experimental results for evaluation of the learned skybox.

Scene	Skybox	#Gaussians↓	PSNR↑	SSIM↑	LPIPS↓
Monk	Disabled	**389,608**	31.92	0.958	0.074
	Enabled	471,256	**33.63**	**0.963**	**0.062**

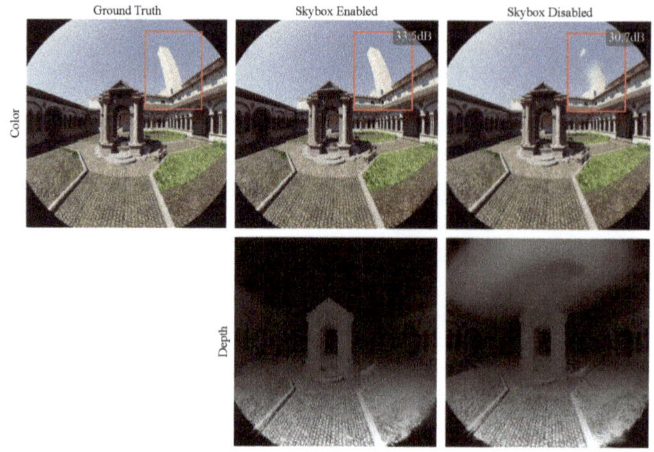

Figure 10. Results for our proposed model trained on synthetic data with the learned skybox enabled (middle) and disabled (right).

4.4. Performance

To measure the performance of the model, we rendered 100 random views for each model when trained on the ScanNet++ dataset and measured the inference latency. As Table 6 shows, the current version of our method achieves interactive frame rates for most scenes, whereas it shows high latencies for the *Tool Room* scene. This scene contains significantly more Gaussians than the others, indicating a possible bottleneck for rendering speeds. This behavior is expected, as our space-warping module leads to an additional computational effort linear to the number of Gaussians being processed. We acknowledge the existence of several projects focusing on the acceleration of 3DGS [8,31]; however, we consider this process of optimization to be outside of the scope of this paper.

Table 6. Average rendering latencies for our proposed model when trained on the ScanNet++ dataset.

Scene	Latency/ms	#Gaussians
Bedroom	27.4 ± 3.5	239,923
Kitchen	33.8 ± 4.8	392,722
Office Day	35.9 ± 3.5	480,055
Office Night	38.2 ± 3.6	448,746
Tool Room	120.8 ± 18.9	1,479,128
Utility Room	37.9 ± 5.3	453,970

5. Discussion

With the photorealistic synthetic rendering results, we can confirm that our method is a valid approach for extending 3DGS to arbitrary camera models. Although there are still some reconstruction artifacts in shaded parts of the scene and highly reflective surfaces, the metrics indicate a high visual quality. Furthermore, despite a significant discrepancy between synthetic and real-world results, our method matched or outperformed the Fisheye-GS method on the real-world dataset. From a quantitative viewpoint, the most significant improvement of our space-warping approach is the reduction in the number of Gaussians, enhancing the overall space efficiency of the reconstructed model. Qualitatively, our method processes ScanNet++ OpenCV fisheye images directly, removing the need for additional preprocessing steps. Additionally, extensive ablations supported the design decisions in our method, such as the learned skybox, the Jacobian distortion, and the number of polynomial coefficients. The learned skybox, in particular, effectively mitigates floating artifacts and improves the overall visual quality of outdoor scenes.

However, several open questions remain. One is the generalization capability of our model. To compare our method with Fisheye-GS, we adopted the same eighth image selection strategy for the test set. To further explore generalization, future research could include rendering views from more distinct viewpoints or depth comparisons with ground-truth data. Another limitation is the real-time capability of our model. While it achieves interactive frame rates for most scenes, performance in larger scenes suggests a need for further optimizations to reduce rendering latency.

6. Conclusions and Outlook

Our method demonstrates significant advancements in extending 3DGS to arbitrary camera models, particularly in terms of space efficiency and preprocessing requirements. The ability to process fisheye images without additional steps and the improved visual quality achieved through the learned skybox showcase the potential of our approach. Future research directions include optimizing camera parameters to reduce the reliance on robust and precise calibration processes. The trainability of polynomial lens distortion coefficients could be explored to enhance adaptability. Building on the modularity of our method, it could integrate seamlessly with existing extensions for 3DGS, such as for dynamic scene reconstruction [32], reconstruction without known poses [15], or more compact scene representations [33]. Finally, we are currently developing the warping module as a *gsplat* component. This step may enable the integration of our method into real-time applications, broadening its practical utility and impact.

Author Contributions: Conceptualization, J.N. and S.K.; methodology, J.N.; software, J.N. and S.K.; validation, J.N.; data curation, J.N. and S.K.; writing—original draft preparation, J.N.; writing—review and editing, J.N., S.K., F.-L.Z. and R.K.; visualization, J.N.; supervision, F.-L.Z. and R.K.; project administration, F.-L.Z. and R.K.; funding acquisition, F.-L.Z. and R.K. All authors have read and agreed to the published version of the manuscript.

Funding: The authors acknowledge the financial support of Catalyst: Leaders Julius von Haast Fellowship (23-VUW-019-JVH).

Institutional Review Board Statement: Not applicable.

Informed Consent Statement: Not applicable.

Data Availability Statement: We provide all source code and the synthetic blender dataset in the project's GitHub repository: https://github.com/jna-358/warped-gaussians, accessed on 16 Decemeber 2024. The ScanNet++ dataset requires an application submitted through its publisher's website for download (https://kaldir.vc.in.tum.de/scannetpp, accessed on 16 Decemeber 2024).

Conflicts of Interest: The authors declare no conflicts of interest. The funders had no role in the design of the study; in the collection, analyses, or interpretation of data; in the writing of the manuscript; or in the decision to publish the results.

Abbreviations

The following abbreviations are used in this manuscript:

NVS	Novel View Synthesis
3DGS	3D Gaussian Splatting
FoV	Field of View
SfM	Structure from Motion
NeRF	Neural Radiace Fields
MLP	Multilayer Perceptron
PSNR	Peak Signal-to-Noise Ratio
SSIM	Structural Similarity Index Measure
LPIPS	Learned Perceptual Image Patch Similarity

Appendix A. Derivation of the Jacobian

As we introduced in Equation (9), we decided to model the space-warping function ($\mathcal{W}(x)$) as the composition of a coordinate transformation to spherical coordinates, a polynomial distortion of the polar angle, followed by transformation back to Cartesian coordinates:

$$\mathcal{W}(x) = \left(\mathcal{T}_{\text{sph}\rightarrow\text{cart}} \circ \mathcal{D} \circ \mathcal{T}_{\text{cart}\rightarrow\text{sph}}\right)(x). \tag{A1}$$

Jacobian $J_{\mathcal{W}}$ is, thus, formed by the product the three Jacobians corresponding to these functions:

$$J_{\mathcal{W}}(x) = \left[J_{\text{sph}\rightarrow\text{cart}}(\mathcal{D}(\mathcal{T}_{\text{cart}\rightarrow\text{sph}}(x)))\right]\left[J_{\mathcal{D}}(\mathcal{T}_{\text{cart}\rightarrow\text{sph}}(x))\right]\left[J_{\text{cart}\rightarrow\text{sph}}(x)\right]. \tag{A2}$$

From the definition of $\mathcal{T}_{\text{cart}\rightarrow\text{sph}}$ in Equation (8), we can derive its Jacobian as

$$J_{\text{cart}\rightarrow\text{sph}}\left(\begin{bmatrix}x\\y\\z\end{bmatrix}\right) = \begin{bmatrix} \frac{x}{\sqrt{x^2+y^2+z^2}} & \frac{y}{\sqrt{x^2+y^2+z^2}} & \frac{z}{\sqrt{x^2+y^2+z^2}} \\ \frac{xz}{(x^2+y^2+z^2)^{3/2}\sqrt{1-\frac{z^2}{x^2+y^2+z^2}}} & \frac{yz}{(x^2+y^2+z^2)^{3/2}\sqrt{1-\frac{z^2}{x^2+y^2+z^2}}} & \frac{-\sqrt{1-\frac{z^2}{x^2+y^2+z^2}}}{\sqrt{x^2+y^2+z^2}} \\ \frac{-y}{x^2+y^2} & \frac{x}{x^2+y^2} & 0 \end{bmatrix}.$$

The Jacobian is undefined at the pole, leading to numerical instabilities at points close to the central camera ray. In our implementation, we discarded these most central points. As the corresponding Gaussians are observed as non-central points from other perspectives, this does not cause degradation of the reconstruction performance. The polar distortion function ($\mathcal{D}$) consists of an identity function for the radius and the azimuthal angle and an eighth-degree polynomial function with coefficients $c_0, \ldots, c_8$:

$$\mathcal{D}\left(\begin{bmatrix}r\\\theta\\\varphi\end{bmatrix}\right) = \begin{bmatrix}r\\\sum_{i=0}^{8} c_i\theta^i\\\varphi\end{bmatrix}, \tag{A3}$$

which results in the following Jacobian:

$$J_{\mathcal{D}}\left(\begin{bmatrix}r\\\theta\\\varphi\end{bmatrix}\right) = \begin{bmatrix}1 & 0 & 0\\0 & \sum_{i=1}^{8} c_i i\theta^{i-1} & 0\\0 & 0 & 1\end{bmatrix}. \tag{A4}$$

The final Jacobian corresponding to the transformation back to Cartesian coordinates can be derived from Equation (8) as

$$J_{\text{sph}\rightarrow\text{cart}}\left(\begin{bmatrix}r\\\theta\\\varphi\end{bmatrix}\right) = \begin{bmatrix}\sin\theta\cos\varphi & r\cos\varphi\cos\theta & -r\sin\varphi\sin\theta\\\sin\varphi\sin\theta & r\sin\varphi\cos\theta & r\sin\theta\cos\varphi\\\cos\theta & -r\sin\theta & 0\end{bmatrix}.$$

Appendix B. Other Camera Models

To demonstrate the application of our proposed method to other camera models, we considered an orthographic projection model. Within this camera model, a 3D point $(x, y, z)^\top$ is projected to the pixel position $(u, v)^\top$ as

$$u(x) = \frac{wx}{s} + \frac{w}{2}, \tag{A5}$$

$$v(y) = \frac{hy}{s} + \frac{h}{2}, \tag{A6}$$

where w and h are the image width and height in pixels, while s represents the orthographic scale. This projection is equivalent to an initial warping using

$$\mathcal{W}\left(\begin{bmatrix} x \\ y \\ z \end{bmatrix}\right) = \begin{bmatrix} xz \\ yz \\ z, \end{bmatrix} \tag{A7}$$

followed by a perspective projection with the following intrinsic matrix:

$$K = \begin{bmatrix} w/s & 0 & w/2 \\ 0 & r/s & r/2 \\ 0 & 0 & 1 \end{bmatrix}, \tag{A8}$$

Differentiating the warping function yields

$$J_{\mathcal{W}}\left(\begin{bmatrix} x \\ y \\ z \end{bmatrix}\right) = \begin{bmatrix} z & 0 & x \\ 0 & z & y \\ 0 & 0 & 1 \end{bmatrix}. \tag{A9}$$

To test this camera model within our proposed pipeline, we generated a synthetic dataset based on the public *Lego* scene [34], which has been used within several NVS publications [2]. We utilized Blender's orthographic camera model with its built-in ray-tracing renderer. We rendered 100 frames and took every eighth frame as a validation frame. After optimizing for 30,000 iterations, for the validation set, we obtained a PSNR of 36.45, an SSIM score of 0.978, and an LPIPS value of 0.03. Qualitative samples of the validation views are shown in Figure A1. The demonstrated samples show a photorealistic reconstruction quality with a consistent rendered depth. This example illustrates the capability of our proposed method to be applied to a variety of camera models, including non-fisheye lenses.

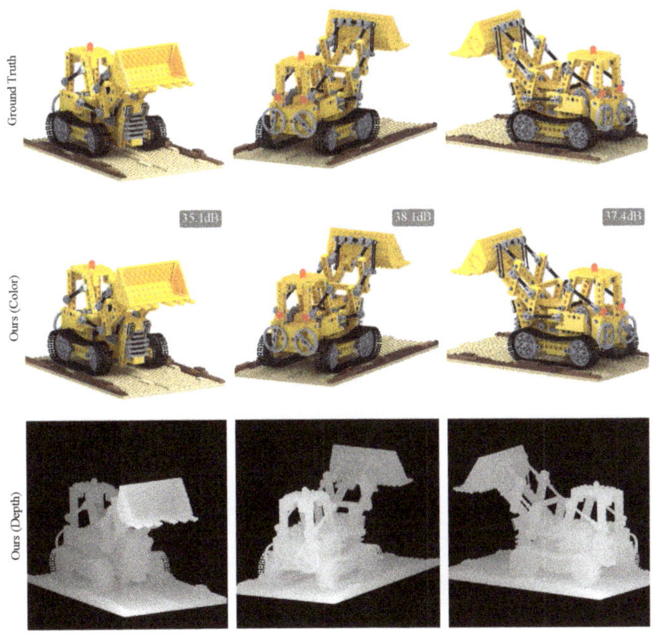

Figure A1. Results for three validation views optimized on our synthetic orthographic dataset.

Appendix C. Additional Scenes

For the demonstration of our method's generalization capabilities, we picked an additional five scenes from the ScanNet++ dataset. These scenes were chosen to contain objects not contained in the six main evaluation scenes to ensure a broad coverage of reconstructed objects:

- Bathtub (1c876c250f);
- Conference Room (1b75758486);
- Electricity Room (1c4b893630);
- Plant (06a3d79b68);
- Printer (1b9692f0c7).

As shown in Table A1 and Figure A2, our proposed method achieves photorealistic reconstruction quality on the additional scenes, demonstrating its overall generalization capability.

Table A1. Experimental results of the proposed method on the five additional ScanNet++ scenes.

Scene	#Gaussians↓	PSNR↑	SSIM↑	LPIPS↓
Bathtub	353,599	32.62	0.956	0.155
Conference Room	901,825	27.79	0.917	0.174
Electrical Room	1,405,169	31.88	0.946	0.129
Plant	1,046,335	28.03	0.871	0.182
Printer	498,382	30.49	0.917	0.199

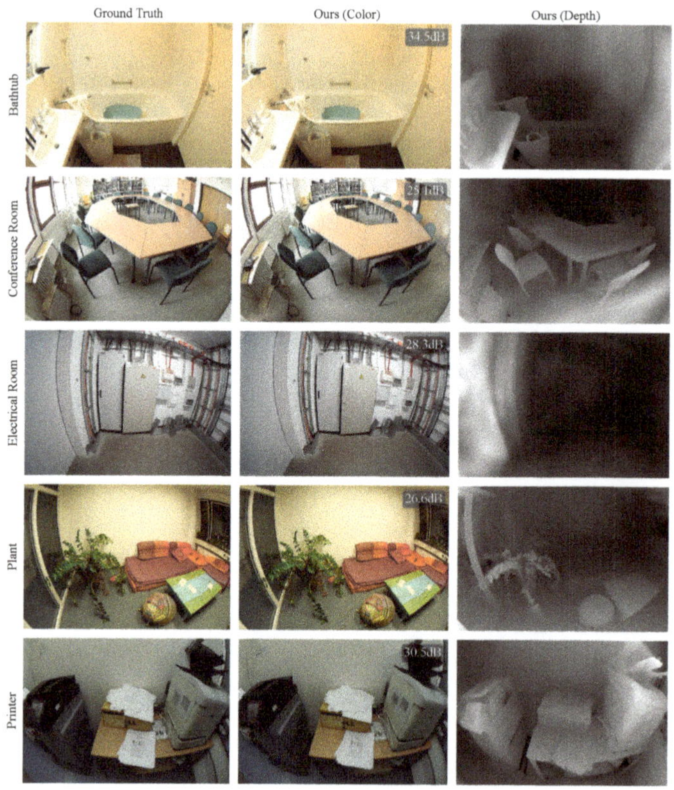

Figure A2. Results for the five additional real-world scenes from the ScanNet++ dataset.

References

1. Gao, K.; Gao, Y.; He, H.; Lu, D.; Xu, L.; Li, J. Nerf: Neural radiance field in 3d vision, a comprehensive review. *arXiv* **2022**, arXiv:2210.00379.
2. Mildenhall, B.; Srinivasan, P.P.; Tancik, M.; Barron, J.T.; Ramamoorthi, R.; Ng, R. Nerf: Representing scenes as neural radiance fields for view synthesis. *Commun. ACM* **2021**, *65*, 99–106. [CrossRef]
3. Müller, T.; Evans, A.; Schied, C.; Keller, A. Instant neural graphics primitives with a multiresolution hash encoding. *ACM Trans. Graph. (TOG)* **2022**, *41*, 1–15. [CrossRef]
4. Fridovich-Keil, S.; Yu, A.; Tancik, M.; Chen, Q.; Recht, B.; Kanazawa, A. Plenoxels: Radiance Fields without Neural Networks. In Proceedings of the 2022 IEEE/CVF Conference on Computer Vision and Pattern Recognition (CVPR), New Orleans, LA, USA, 18–24 June 2022; pp. 5491–5500. [CrossRef]
5. Kerbl, B.; Kopanas, G.; Leimkühler, T.; Drettakis, G. 3D Gaussian Splatting for Real-Time Radiance Field Rendering. *ACM Trans. Graph.* **2023**, *42*, 139:1–139:14. [CrossRef]
6. GitHub-xverse-Engine/XV3DGS-UEPlugin: A Unreal Engine 5 (UE5) Based Plugin Aiming to Provide Real-Time Visulization, Management, Editing, and Scalable Hybrid Rendering of Guassian Splatting Model. Available online: https://github.com/xverse-engine/XV3DGS-UEPlugin (accessed on 16 Decemeber 2024).
7. GitHub-clarte53/GaussianSplattingVRViewerUnity: A VR Viewer for Gaussian Splatting Models Developped as Native Plugin for Unity with the Original CUDA Rasterizer. Available online: https://github.com/clarte53/GaussianSplattingVRViewerUnity (accessed on 16 Decemeber 2024).
8. Ye, V.; Li, R.; Kerr, J.; Turkulainen, M.; Yi, B.; Pan, Z.; Seiskari, O.; Ye, J.; Hu, J.; Tancik, M.; et al. gsplat: An open-source library for Gaussian splatting. *arXiv* **2024**, arXiv:2409.06765.
9. Liao, Z.; Chen, S.; Fu, R.; Wang, Y.; Su, Z.; Luo, H.; Ma, L.; Xu, L.; Dai, B.; Li, H.; et al. Fisheye-GS: Lightweight and Extensible Gaussian Splatting Module for Fisheye Cameras. *arXiv* **2024**, arXiv:2409.04751.
10. OpenCV-Fisheye Camera Model. Available online: https://docs.opencv.org/4.x/db/d58/group__calib3d__fisheye.html (accessed on 16 Decemeber 2024).
11. Schönberger, J.L.; Frahm, J.M. Structure-from-Motion Revisited. In Proceedings of the Conference on Computer Vision and Pattern Recognition (CVPR), Las Vegas, NV, USA, 27–30 June 2016.
12. Moulon, P.; Monasse, P.; Perrot, R.; Marlet, R. OpenMVG: Open multiple view geometry. In Proceedings of the International Workshop on Reproducible Research in Pattern Recognition, Cancún, Mexico, 4 December 2016; Springer: Berlin/Heidelberg, Germany, 2016; pp. 60–74.
13. Meng, Q.; Chen, A.; Luo, H.; Wu, M.; Su, H.; Xu, L.; He, X.; Yu, J. GNeRF: GAN-based Neural Radiance Field without Posed Camera. In Proceedings of the 2021 IEEE/CVF International Conference on Computer Vision (ICCV), Montreal, BC, Canada, 11–17 October 2021; pp. 6331–6341. [CrossRef]
14. Bian, W.; Wang, Z.; Li, K.; Bian, J.W. NoPe-NeRF: Optimising Neural Radiance Field with No Pose Prior. In Proceedings of the 2023 IEEE/CVF Conference on Computer Vision and Pattern Recognition (CVPR), Vancouver, BC, Canada, 17–24 June 2023; pp. 4160–4169. [CrossRef]
15. Fu, Y.; Liu, S.; Kulkarni, A.; Kautz, J.; Efros, A.A.; Wang, X. COLMAP-Free 3D Gaussian Splatting. In Proceedings of the IEEE/CVF Conference on Computer Vision and Pattern Recognition (CVPR), Seattle, WA, USA, 16–22 June 2024; pp. 20796–20805.
16. Barron, J.T.; Mildenhall, B.; Tancik, M.; Hedman, P.; Martin-Brualla, R.; Srinivasan, P.P. Mip-NeRF: A Multiscale Representation for Anti-Aliasing Neural Radiance Fields. In Proceedings of the IEEE/CVF International Conference on Computer Vision, Montreal, BC, Canada, 11–17 October 2021; pp. 5855–5864.
17. Martin-Brualla, R.; Radwan, N.; Sajjadi, M.S.M.; Barron, J.T.; Dosovitskiy, A.; Duckworth, D. NeRF in the Wild: Neural Radiance Fields for Unconstrained Photo Collections. In Proceedings of the IEEE/CVF Conference on Computer Vision and Pattern Recognition, Nashville, TN, USA, 20–25 June 2021; pp. 7210–7219.
18. Pumarola, A.; Corona, E.; Pons-Moll, G.; Moreno-Noguer, F. D-NeRF: Neural Radiance Fields for Dynamic Scenes. In Proceedings of the IEEE/CVF Conference on Computer Vision and Pattern Recognition, Nashville, TN, USA, 20–25 June 2021; pp. 10318–10327.
19. Fridovich-Keil, S.; Meanti, G.; Warburg, F.; Recht, B.; Kanazawa, A. K-Planes: Explicit Radiance Fields in Space, Time, and Appearance. *arXiv* **2023**, arXiv:2301.10241.
20. Cao, A.; Johnson, J. HexPlane: A Fast Representation for Dynamic Scenes. In Proceedings of the 2023 IEEE/CVF Conference on Computer Vision and Pattern Recognition (CVPR), Vancouver, BC, Canada, 17–24 June 2023; pp. 130–141. [CrossRef]
21. Zhang, K.; Riegler, G.; Snavely, N.; Koltun, V. NeRF++: Analyzing and Improving Neural Radiance Fields. *arXiv* **2020**, arXiv:2010.07492. [CrossRef]
22. Yeshwanth, C.; Liu, Y.C.; Nießner, M.; Dai, A. ScanNet++: A High-Fidelity Dataset of 3D Indoor Scenes. In Proceedings of the International Conference on Computer Vision (ICCV), Paris, France, 2–3 October 2023.
23. Moenne-Loccoz, N.; Mirzaei, A.; Perel, O.; de Lutio, R.; Esturo, J.M.; State, G.; Fidler, S.; Sharp, N.; Gojcic, Z. 3D Gaussian Ray Tracing: Fast Tracing of Particle Scenes. *ACM Trans. Graph. SIGGRAPH Asia* **2024**, *43*, 1–19. [CrossRef]
24. Ren, Y.; Wu, G.; Li, R.; Yang, Z.; Liu, Y.; Chen, X.; Cao, T.; Liu, B. UniGaussian: Driving Scene Reconstruction from Multiple Camera Models via Unified Gaussian Representations. *arXiv* **2024**, arXiv:2411.15355.
25. Meurer, A.; Smith, C.P.; Paprocki, M.; Čertík, O.; Kirpichev, S.B.; Rocklin, M.; Kumar, A.; Ivanov, S.; Moore, J.K.; Singh, S.; et al. SymPy: Symbolic computing in Python. *PeerJ Comput. Sci.* **2017**, *3*, e103. [CrossRef]

26. Yang, L.; Kang, B.; Huang, Z.; Xu, X.; Feng, J.; Zhao, H. Depth Anything: Unleashing the Power of Large-Scale Unlabeled Data. In Proceedings of the IEEE/CVF Conference on Computer Vision and Pattern Recognition (CVPR), Seattle, WA, USA, 16–22 June 2024.
27. Blender Demo Files. Available online: https://www.blender.org/download/demo-files/ (accessed on 16 Decemeber 2024).
28. Blender Cameras-Fisheye Lens Polynomial. Available online: https://docs.blender.org/manual/en/latest/render/cycles/object_settings/cameras.html (accessed on 16 Decemeber 2024).
29. Wang, Z.; Bovik, A.C. Mean squared error: Love it or leave it? A new look at signal fidelity measures. *IEEE Signal Process. Mag.* **2009**, *26*, 98–117. [CrossRef]
30. Zhang, R.; Isola, P.; Efros, A.A.; Shechtman, E.; Wang, O. The Unreasonable Effectiveness of Deep Features as a Perceptual Metric. In Proceedings of the IEEE/CVF Conference on Computer Vision and Pattern Recognition (CVPR), Salt Lake City, UT, USA, 18–23 June 2018.
31. Feng, G.; Chen, S.; Fu, R.; Liao, Z.; Wang, Y.; Liu, T.; Pei, Z.; Li, H.; Zhang, X.; Dai, B. Flashgs: Efficient 3d gaussian splatting for large-scale and high-resolution rendering. *arXiv* **2024**, arXiv:2408.07967.
32. Wu, G.; Yi, T.; Fang, J.; Xie, L.; Zhang, X.; Wei, W.; Liu, W.; Tian, Q.; Wang, X. 4d gaussian splatting for real-time dynamic scene rendering. In Proceedings of the IEEE/CVF Conference on Computer Vision and Pattern Recognition, Seattle, WA, USA, 16–22 June 2024; pp. 20310–20320.
33. Morgenstern, W.; Barthel, F.; Hilsmann, A.; Eisert, P. Compact 3d scene representation via self-organizing gaussian grids. In Proceedings of the European Conference on Computer Vision, London, UK, 15–16 January 2025; Springer: Berlin/Heidelberg, Germany, 2025; pp. 18–34.
34. Blend Swap-Lego 856 Bulldozer. Available online: https://blendswap.com/blend/11490 (accessed on 16 Decemeber 2024).

Journal of *Imaging*

MDPI

Article

A Mathematical Model for Wind Velocity Field Reconstruction and Visualization Taking into Account the Topography Influence

Guzel Khayretdinova [1,2,*] and Christian Gout [1,2,3]

1 National Institute for Applied Sciences (INSA Rouen Normandie), Laboratoire de Mathématiques de l'INSA, LMI-UR 3226, 76000 Rouen, France
2 CNRS, Normandie Mathématiques, FR CNRS 3335, 76000 Rouen, France
3 Department of Mathematics, University of Hawai'i at Manoa, 2565 McCarthy Mall, Keller Hall, Honolulu, HI 96822, USA
* Correspondence: guzel.khayretdinova@insa-rouen.fr

Abstract: In this paper, we propose a global modelling for vector field approximation from a given finite set of vectors (corresponding to the wind velocity field or marine currents). In the modelling, we propose using the minimization on a Hilbert space of an energy functional that includes a fidelity criterion to the data and a smoothing term. We discretize the continuous problem using a finite elements method. We then propose taking into account the topographic effects on the wind velocity field, and visualization using a free library is also proposed, which constitutes an added value compared to other vector field approximation models.

Keywords: vector flow visualization; current/wind velocity field approximation; wind velocity field modelling

1. Introduction

Vector field approximation has many applications, such as to predict wind turbine production; in oceanography, to study marine currents; and more generally in a computer sciences framework. As introduced in [1], in order to approximate a vector field, several approaches have been developed: a finite element method interpolation (see [2]), PDE-based methods, kriging methods, a Lagrange interpolation method and spline and Rational Basis Function (RBF) approximations (see [3–8]). These approaches have drawbacks, particularly when a small number of data are available and the approximation's result is qualitatively insufficient. In this work, we are precisely in a case where the number of data from anemometers is considered low in comparison to the large study area. It is, therefore, necessary to propose a mathematical model allowing for this type of data to be processed via a robust energy functional minimization. The major methodological contribution of this work consists in the modelling using D^m splines, as well as the contribution of adding the topography effect into the numerical results; most of the methods proposed in the literature do not integrate this important aspect, but because there are little available data, it is important to include as much information in the model as possible to generate a realistic wind field, and taking the topography into account is quite simple and brings significant added value. Other methods (like [3–5]) only focus on the modelling and the mathematical aspects of the approximation without adding more information than the input dataset. To our knowledge, our approach is the first to offer such a global framework.

In this paper, we first give the considered vector field approximation model, using a D^m spline operator, rigorously introduced in [1,7,9–11]. The dataset consists of a finite set of vectors $(xi, yi, wi)_i$, where (xi, yi) locates the point in 2D and wi = (Ui, Vi) $\in R^2$ gives the direction and speed of the wind at location (xi, yi). A minimization problem is introduced, leading to a variational problem whose solution is the searched for wind vector field. We give the discretization using a classic finite element method. We then give details on how to integrate into the model the effect of the topography on the obtained wind velocity field; this last part greatly improves upon previous models (given in [1]), where topographic effects are not taken into account. We show the effects of this topographic influence on the synthetic dataset given by an explicit function. We then give numerical examples for a real dataset, including a specific tool for visualization. A global review of this complete approximation framework is given in Figure 1.

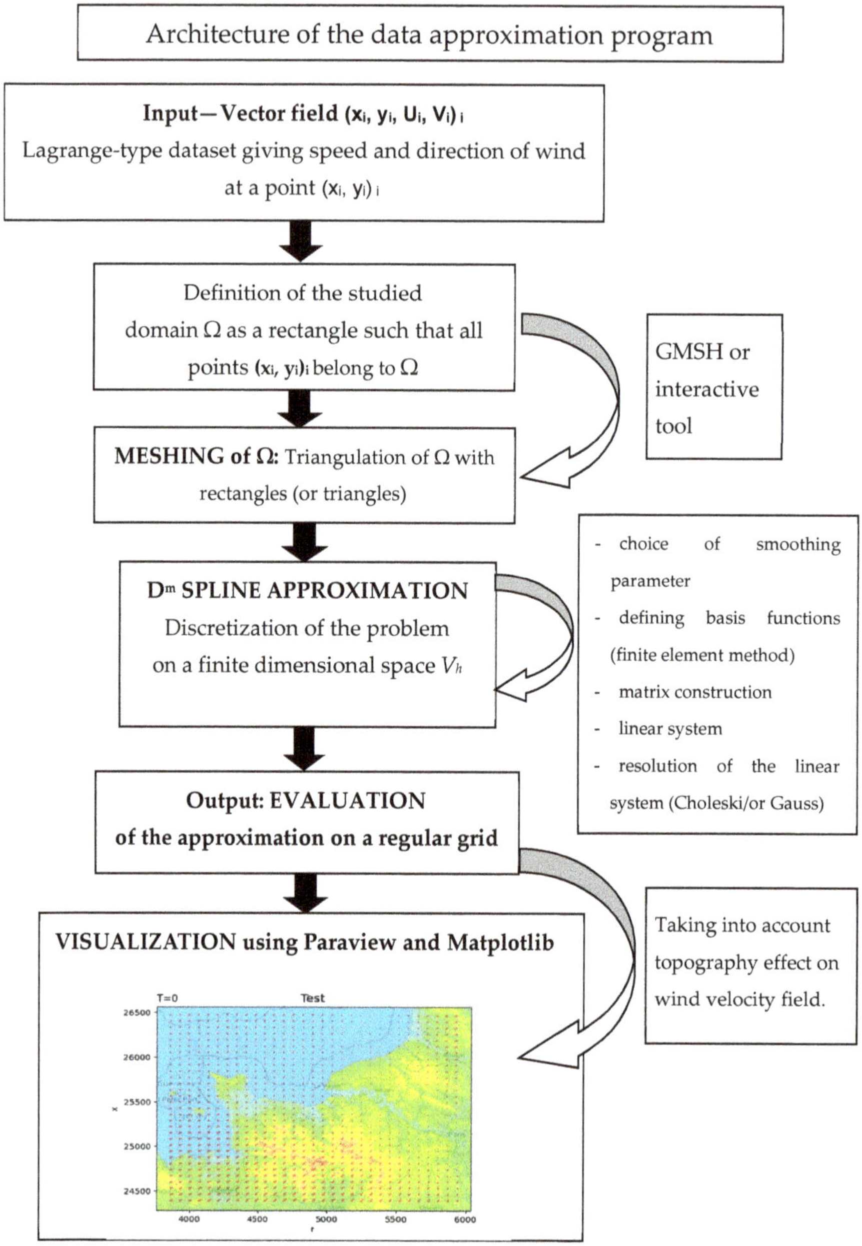

Figure 1. Global view of the approximation framework.

2. Mathematical Modelling

The mathematical modelling of our approximation problem is constructed using a D^m spline operator as follows: For all $v \in H^{m+1}(\Omega)$, we introduce the energy functional consisting of two terms. The first one is the data fidelity criterion, while the second one is the smoothing parameter:

$$J_\varepsilon(v) = \langle \rho(v-w) \rangle^2 + \varepsilon |v|^2_{m+1,\Omega,R} \tag{1}$$

where Ω is an open subset corresponding to the studied zone. $w = (w_1, w_2, \ldots, w_N) \in \left(R^2\right)^N$ is the vector field dataset, ρ is a linear operator linked to the dataset, $|\bullet|^2_{m+1,\Omega,IR}$ is the usual semi-norm of the usual Sobolev space $H^{m+1}(\Omega)$ as defined in [1], $\langle \bullet \rangle$ is the Euclidean norm in R^2 and ε is a smoothing parameter generally equal to 10^{-6} in many applications (see [1,4,7,8] for more details). We recall that $H^{m+1}(\Omega)$ is a space of functions belonging to $L^2(\Omega)$ and their $(m+1)$ derivatives. We also introduce an ordered set of N points $(x_i)_i$ of Ω, where we suppose as known the value of the wind velocity field. The linear operator ρ is given by $\rho(v) = (v(x_1), v(x_2), \ldots v(x_N)) \in \left(R^2\right)^N$. We use the D^m spline (see Gout et al. [1] and Arcangéli et al. [7] for a complete study of this approximation operator) approximation framework to solve this problem. We call σ_ε^d the smoothing spline on Ω relative to ρ, which is the unique solution of the minimization problem:

$$\begin{cases} \text{find } \sigma_\varepsilon \in H^{m+1}(\Omega), \text{ such that for any } v \in H^{m+1}(\Omega): \\ \qquad J_\varepsilon(\sigma_\varepsilon) \le J_\varepsilon(v). \end{cases} \tag{2}$$

We can use the Lax–Milgram theorem to establish the uniqueness of this minimization problem, since the solution σ_ε^d of this minimization problem is the solution of the following variational problem

$$\begin{cases} \text{find } \sigma_\varepsilon \in H^{m+1}(\Omega), \text{ such that for any } v \in H^{m+1}(\Omega): \\ \qquad \left\langle \rho\sigma_\varepsilon^d, \rho v \right\rangle + \varepsilon\left(\sigma_\varepsilon^d, v\right)_{m+1,\Omega,R} = \langle w, \rho v \rangle. \end{cases} \tag{3}$$

where $(\bullet,\bullet)_{m+1,\Omega,R}$ denotes the semi-norm of $H^{m+1}(\Omega)$. To apply the Lax–Milgram theorem, we recall that all the hypotheses of this theorem are satisfied, since $H^{m+1}(\Omega)$ is a Hilbert space, and we also have as follows:

- $a(u,v) = \langle \rho u, \rho v \rangle + \varepsilon(u,v)_{m+1,\Omega,R}$ is a bilinear form, being the sum of scalar products.
- $a(u,v)$ is continuous on $\left(H^{m+1}(\Omega)\right)^2$ because $|a(u,v|) \le \max(1,\varepsilon)\|u\|_{m+1,\Omega}\|v\|_{m+1,\Omega}$, using the Cauchy–Schwarz inequality and the norm equivalence between $\left(\langle \rho v, \rho v \rangle + (v,v)_{m+1,\Omega,R}\right)^{1/2}$ and $\|v\|_{m+1,\Omega}$.
- $a(v,v)$ is elliptic on $H^{m+1}(\Omega)$ since $a(v,v) \ge \min(1,\varepsilon)\|v\|^2 m+1,\Omega$.
- $\langle w, \rho v \rangle$ is a continuous linear form.

We now propose a discretization of the variational problem using a finite element discretization (see [1,9,12] for more details on such a discretization). We recall that the main idea of the finite element method is to replace the (Hilbert) space $H^{m+1}(\Omega)$ used to define the variational Equation (3) by a finite dimensional subspace V_h. Of course, we have $V_h \subset H^{m+1}(\Omega)$. The functions belonging to V_h. are piecewise polynomials, and the bases of the functions for the space V_h are constructed such that they have small support. For any real $h > 0$, let T_h be a triangulation of Ω by n-simplices or n-rectangles K with diameter $h_K \le h$. We classically approximate the space $H^{m+1}(\Omega)$ by the space $V_{h,}$, a finite dimensional space included in $H^{m+1}(\Omega)$ and admitting a polynomial basis of polynomial functions $(\Phi_j)_j$. We write the solution of $\sigma^d_{\varepsilon,h}$ on the basis of V_h as

$$\sigma^d_{\varepsilon,h} = \sum_{j=1}^{\dim V_h} \beta_j \Phi_j \; : \; \beta_j \in R. \tag{4}$$

Note that the polynomials $(\Phi_j)_j$ are given, since they are computed following the chosen generic finite element. The generic finite element we choose here is the Bogner–Fox–Schmit (BFS) rectangle of class C^1, where a function of V^h. is completely determined by its four values (value, values of the two first derivatives and value of the twist derivative) at a nodal point (see Appendix A). The choice of the BFS finite element is due to their capability to easily tessellate rectangular domains and to guarantee a final approximation of class C^1.

From (4), we have to find $(\beta_j)_j$ in order to find the solution $\sigma^d_{\varepsilon,h}$ of our approximation problem. We can now give the discretization of problem (3) using (4):

$$\begin{aligned} &Find\ (\beta_j)_j \in R^{\dim V_h}\ such\ \ that\ \forall k = 1, \ldots, \dim V_h, \\ &\sum_{i=1}^{N} \left\langle \sum_{j=1}^{\dim V_h} \beta_j.\Phi_j, v_h(x_i) \right\rangle + \varepsilon \left(\sum_{j=1}^{\dim V_h} \beta_j.\Phi_j, v_h \right)_{m+1,\Omega,R^n} = \sum_{i=1}^{N} \langle w_i, v_h(x_i) \rangle. \end{aligned} \tag{5}$$

Equation (5) leads to the following linear system, taking as the test function v_h all the basis functions Φ_k: $k = 1, \ldots, dim\ V_h$

$$\sum_{i=1}^{N} \sum_{j=1}^{\dim V_h} \beta_j.\Phi_j, \Phi_k(x_i) + \varepsilon \sum_{j=1}^{\dim V_h} \beta_j.(\Phi_j, \Phi_k)_{m+1,\Omega,R^n} = \sum_{i=1}^{N} \langle w_i, \Phi_k(x_i) \rangle. \tag{6}$$

We finally have to solve the following linear system to find the unknown real values $(\beta_j)_j$:

$$\left((A)^T A + \varepsilon R\right)\beta = (A)^T w, \text{with } A = \left[\Phi_j(x_i)\right]_{1 \le i \le N, 1 \le j \le \dim V_h} \text{ and } R = \left[(\Phi_j, \Phi_j)_{m+1,\Omega,R}\right]_{1 \le i,j \le \dim V_h}. \tag{7}$$

For the numerical simulation, we take m = 1, and we use the Bogner–Fox–Schmit finite element with the basis function as a polynomial of degree 3 (see [7] for more details). The modelling we have introduced in this section permits the approximation of a wind velocity field on all Ω from a finite set of data given, for instance, by several anemometers (as illustrated in the numerical section of this work—Figure 10 and Figure 11).

3. Taking into Account the Topography

In Section 2, we proposed a framework to approximate a wind velocity field from a finite set of measures. It is, of course, well known that the topography plays an important role in wind field velocity variations. Obstacles modify air flows due to pressure forces (see Figure 2). The wind slows down upstream of an obstacle, and accelerates downstream of it.

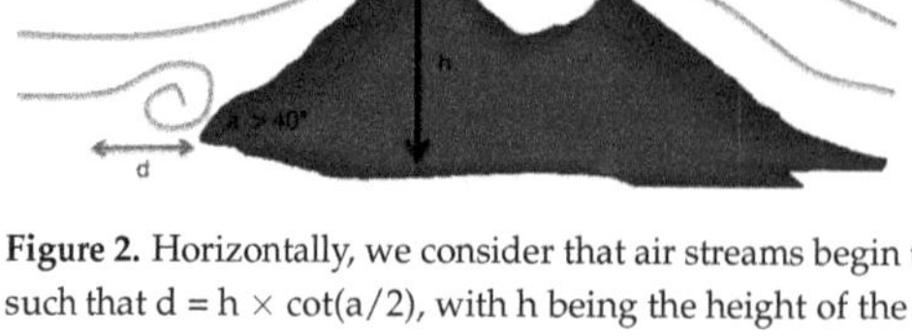

Figure 2. Horizontally, we consider that air streams begin to rise upstream of an obstacle at a distance such that d = h × cot(a/2), with h being the height of the obstacle and *a* the angle of the slope.

Since the approach introduced in Section 2 does not take into account physical considerations, we propose here a way to approximate the topography's influence by post processing the approximated wind field obtained from the model given in Section 2. More precisely, let us consider a wind vector field on N points as

$$W' = (w'_1, \ldots, w'_N),$$

where each wind vector w'_i belongs to R x R, with given coordinates (x_i, y_i) for each w'_i, according to the topographic configuration around it as follows: $W'_i = c_i(\theta)\, w_i$, where W'_i is the adjusted wind vector and $c_i(\theta)$ is the coefficient computed from the topographic configuration at point (x_i, y_i), depending on the wind direction θ. This approximation holds for local topographic effects. It cannot take into account large-scale effects, such as Venturi effects in valleys or straits. To compute the topographic coefficient c_i for a given wind direction θ at point (x_i, y_i), we used the formulas given in parts 1–4 of [13], depending on the slope Φ

$$\begin{cases} 1, & \Phi < 0.05 \\ 1 + 2s\Phi, & 0.05 < \Phi < 0.3 \\ 1 + 0.6s, & \Phi > 0.3 \end{cases} \tag{8}$$

where s is the characteristic coefficient of the obstacle, depending on its features (see Table 1 and Figure 3).

Table 1. Features used to compute the characteristic coefficient s of an obstacle (source: [13]).

Variables	Definition
s	Orographic location factor
Φ	Upwind slope H/L_u in the wind direction (see Figure 3)
L_e	Effective length of the upwind side
L_u	Length of the upwind side
L_d	Length of the downwind side
H	Effective height of the obstacle
x	Horizontal distance between point (x,y) and the top of the obstacle
z	Height of the considered point (x,y)

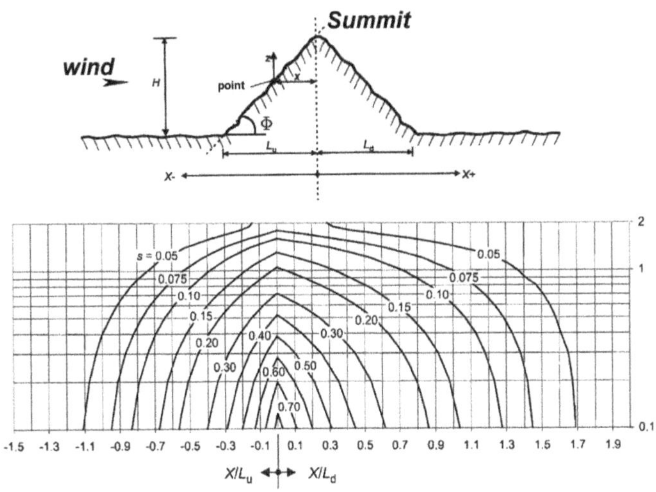

Figure 3. Top: considered parameters for hills and ridges (source: [13]). **Bottom**: corresponding values of parameter s.

The effective length L_e is computed as follows (type of slope $\Phi = H/L_u$):

$$L_e = \begin{cases} L_u, & 0.05 < \Phi < 0.3 \\ \frac{H}{0.3}, & \Phi > 0.3. \end{cases} \tag{9}$$

We also have to compute the value of the orographic location factor s used in (8). As shown in [13], the value of s is related to the ratio H/L_e. More precisely, for an upwind section, for ranges $-1.5 \leq \frac{x}{L_u} \leq 0$ and $0 \leq \frac{z}{L_e} \leq 2$ we take $s = A\exp\left(B\frac{x}{L_u}\right)$, where

$$A = 0.1552\left(\frac{z}{L_e}\right)^4 - 0.8575\left(\frac{z}{L_e}\right)^3 + 1.8133\left(\frac{z}{L_e}\right)^2 - 1.9115\left(\frac{z}{L_e}\right) + 1.0124,$$

and $B = 0.3542\left(\frac{z}{L_e}\right)^2 - 1.0577\left(\frac{z}{L_e}\right) + 2.6456$. Note that when $\frac{x}{L_u} \leq -1.5$ or $2 \leq \frac{z}{L_e}$ we take $s = 0$.

For a downwind section, as shown in [13], we take $s = A\left(\log\frac{x}{L_e}\right)^2 + B\log\frac{x}{L_e} + C$, with

$$A = -1.342\left(\log\frac{z}{L_e}\right)^3 - 0.822\left(\log\frac{z}{L_e}\right)^2 + 0.4609\log\frac{z}{L_e} - 0.0791,$$

$$B = -1.0196\left(\log\frac{z}{L_e}\right)^3 - 0.891\left(\log\frac{z}{L_e}\right)^2 + 0.5343\log\frac{z}{L_e} - 0.1156,$$

$$\text{and } C = 0.803\left(\log\frac{z}{L_e}\right)^3 + 0.4236\left(\log\frac{z}{L_e}\right)^2 - 0.5738\log\frac{z}{L_e} + 0.1606.$$

To compute the topographic coefficients of a domain $D = [0, 1]^2 \times R^2$, we consider a regular grid D_h of D of step h. Then, for each point $(x_i, y_i) = (ih, jh)$ in D_h, we compute a coefficient for each wind direction θ. We split up the compass wind into eight directions θ_j from 0 to 360°, by steps of 45 degrees (see Figure 4).

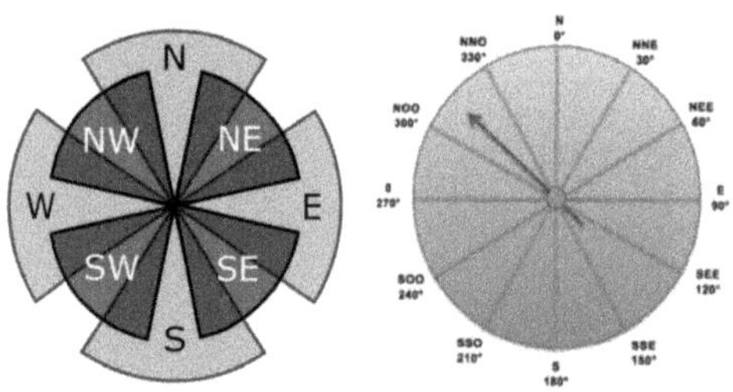

Figure 4. Usual examples of compass wind and wind rose.

Once the collection of topographic coefficients is computed using (8), we use it to adjust the approximated wind field, selecting coefficients according to the direction of each wind vector and using them on the obtained approximating wind velocity field.

4. Numerical Examples

In this section, we give several numerical examples, including the computation of the topographic coefficients (using a given function f) and the approximation of a vector field from a finite set of vectors giving the direction and speed of the wind.

4.1. Computation of the Topographic Coefficients

In order to illustrate the proposed methodology on synthetic data given by an explicit given function f, we simulate an obstacle (hills) in domain D using the basic 2D function f defined as follows:

$$\begin{aligned} f(x,y) = &\tfrac{3}{4}\exp\left(-\tfrac{1}{4}(-9x-2)^2 - \tfrac{1}{4}(9y-2)^2\right) \\ &+\tfrac{3}{4}\exp\left(-\tfrac{1}{49}(-9x+1)^2 - \tfrac{1}{4}(9y+1)^2\right) \\ &+\tfrac{1}{2}\exp\left(-\tfrac{1}{4}(-9x-7)^2 - \tfrac{1}{4}(9y-3)^2\right) \\ &-\tfrac{1}{5}\exp\left(-\tfrac{1}{4}(-9x-4)^2 - \tfrac{1}{4}(9y-7)^2\right) \end{aligned} \tag{10}$$

We define the discretized domain as D_h, with $h = 1/n$. The obstacle is obtained by computing f for every couple point (x_i, y_i), where $i, j = 0, \ldots, n$. For $n = 40$, we obtain the following obstacle (see Figure 5), and we give on this image the computed topographic coefficients obtained using (10) (considering an arbitrary wind direction indicated by the red arrow).

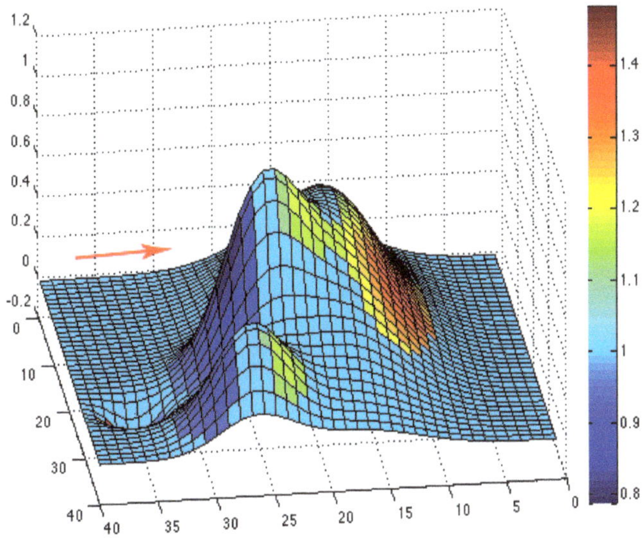

Figure 5. Example of an obstacle given by the function f in (10); the arrow gives the considered wind direction (eastern wind). We also give the colormap of the topographic coefficients associated with the east wind direction.

The simulated topographical data given by the function f in (10) are then used to compute the topographic coefficients as described in the previous subsection. For each of the eight wind directions, we can plot the color map of the computed topographic coefficients on the obstacle. The associated colors go from dark, for zones where the topography slows down the wind flow ($c < 1$), to white, for zones where the topography accelerates the wind flow ($c > 1$). For instance, we have plotted color maps for situations where the wind comes from the northeast (see Figure 6), west (see Figure 7) and southeast (see Figure 8). For each figure, the arrow indicates the direction of the wind.

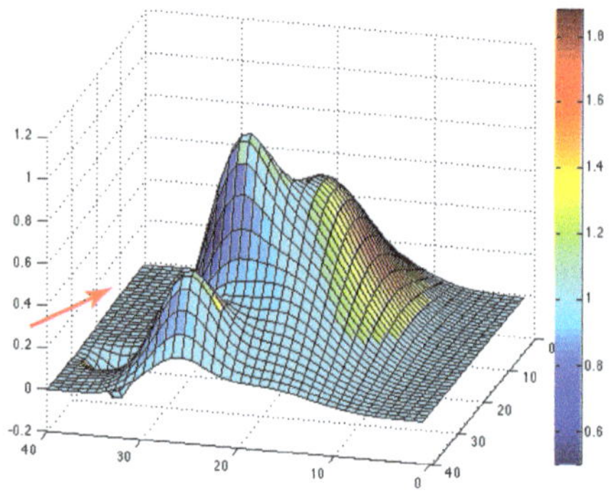

Figure 6. Color map of topographic coefficients associated with a northeast wind direction. The arrow gives the considered wind direction.

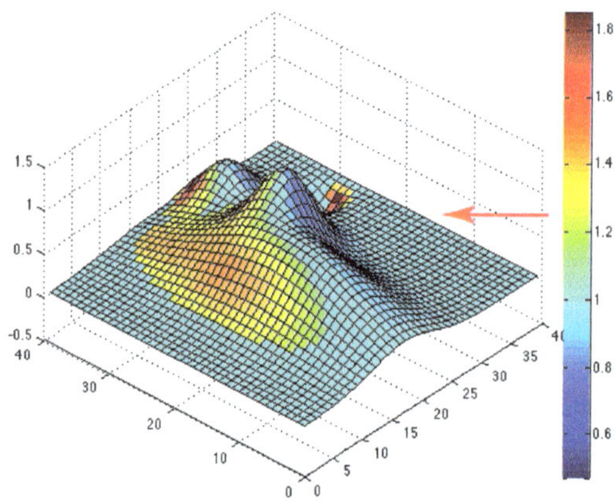

Figure 7. Color map of topographic coefficients associated with a west wind direction. The arrow gives the considered wind direction.

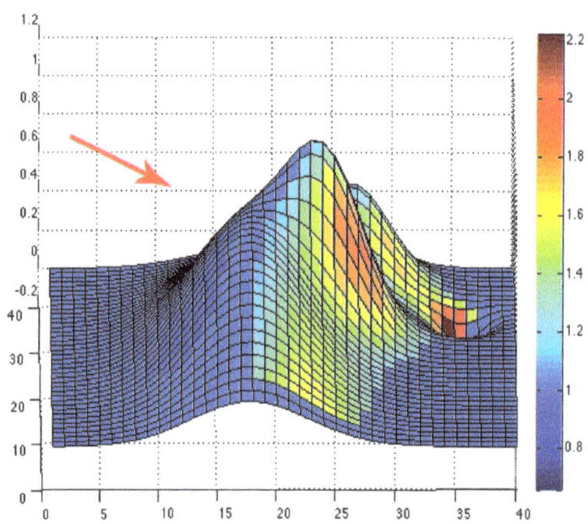

Figure 8. Color map of topographic coefficients associated with a southeast wind direction. The arrow gives the considered wind direction.

Note that the function given in (10) permits illustrating the topographic effect in cases where we consider hills and ridges (as in Figure 3). We have tested the topographic effect on more vertical hills and the results were satisfying. For the case of cliffs or escarpments (Figure 9), the computation is slightly different but the reasoning is analogous, and the values of the corresponding orographic parameters are given in Figure 10. An improvement could be to propose calculations to take into account the influence of buildings (vertical walls of buildings with a rectangular plan, the influence of the angle of roofs, etc.) or of vegetation (trees, etc.), especially if we want to reconstruct the wind on a micro-scale.

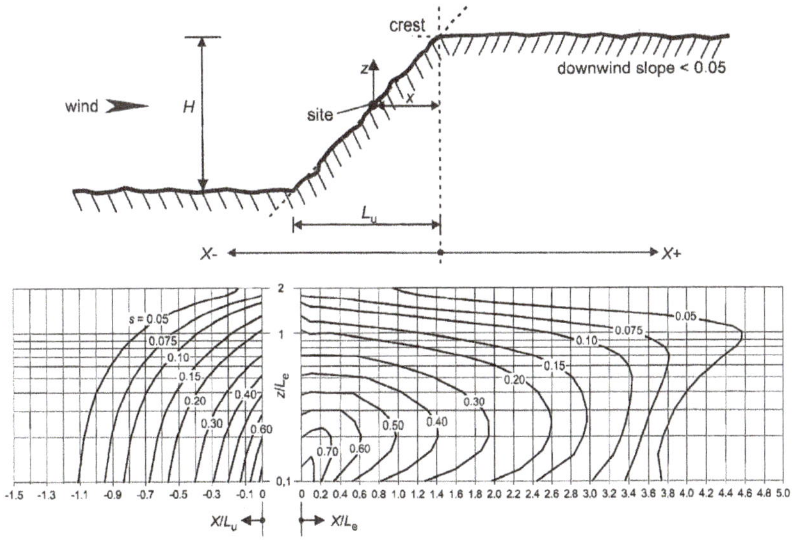

Figure 9. Top: case of cliffs or escarpments [13]. **Bottom**: equivalent method to compute orographic coefficients (as we did for hills and ridges).

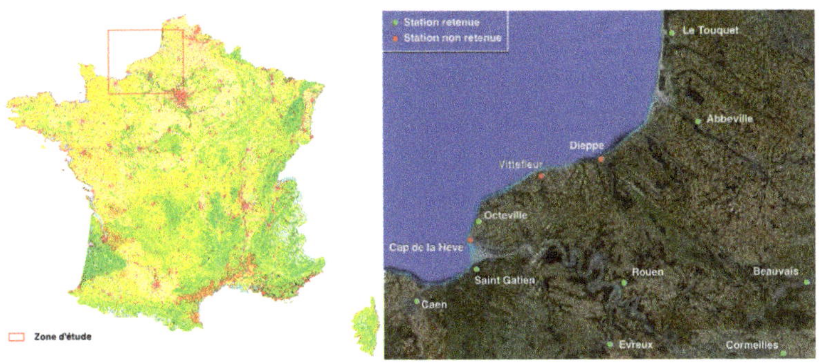

Figure 10. Studied zone (northwest France). Anemometers located in Caen, Octeville, Rouen, Beauvais, Abbeville and Le Touquet were selected.

We now give several numerical experiments on real datasets. The experiments are performed on a 2.21 GHz Athlon with 1.00 GB of RAM.

We focus on the data for wind vector fields acquired in northwest France; the dataset takes into account eight weather stations (Meteo France, Figures 10–12).

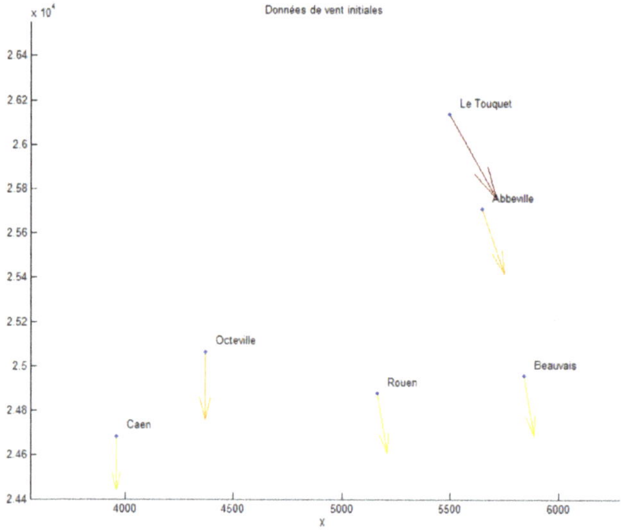

Figure 11. Example of a wind dataset for a given time step. The location is northwest France; the data are from anemometers located at six different airports.

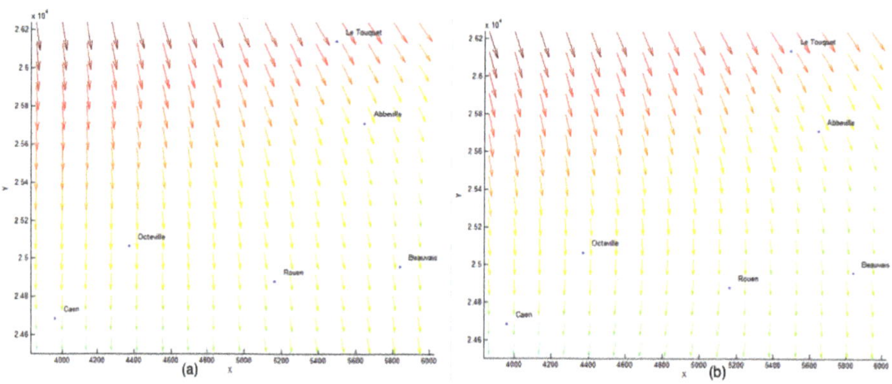

Figure 12. We give two different approximations using the model given in Section 2 of the wind velocity field using a 4×4 finite element grid (**a**) and a 3×3 finite element grid (**b**). Colors indicate wind speed (same colormap as on Figure 11).

4.2. Numerical Simulations of the Global Algorithm

In this subsection, from a set of six velocity wind data, we give the approximation obtained by the method in Section 2, and we then compute the topographic coefficients of the studied zone using the method given in Section 3.

Here is some information about the numerical examples:

- Dataset: six anemometers located at six airports giving the direction and speed of the wind, see Figures 10 and 11;
- Parameter $\varepsilon = 0.000001$;
- Generic finite element: Bogner–Fox–Schmit of class C^1 (See Appendix A);
- Studied domain: $[3500, 6000] \times [2.44, 2.62]$;
- Meshing: 4×4 rectangles and 3×3 rectangles. The results are given in Figure 12.

The choice of the finite element meshing is crucial, and it must be linked to the number of data we have in the input. For a grid of 3×3 rectangles, we have 9 rectangles, 16 nodes and, as we have four basis functions per node with the BFS finite element of class C^1, the dimension of the space V_h is equal to 64 (while it is 100 with a 4×4 meshing, leading to 16 rectangles and 25 nodes). As we do not have a large amount of data, we choose a low number of rectangles in our mesh. If we choose a finer grid, the approximation error increases.

We compute the quadratic error given by the following quotient

$$Quad\_Error = \left(\frac{\sum_{i=1}^{N} \left\langle \sigma^d_{\varepsilon,h}(a_i) - w_i \right\rangle_2^2}{\sum_{i=1}^{N} \langle w_i \rangle_2^2} \right)^{1/2},$$

where $\langle \bullet \rangle_2$ denotes the Euclidean scalar norm. In all our different tests, the quadratic errors is of 10^{-3} and 10^{-4} orders, which is considered as very good in the context of vector field approximation. We then compute this obtained approximation using the topography of the considered zone (see Figures 7 and 13 for the result).

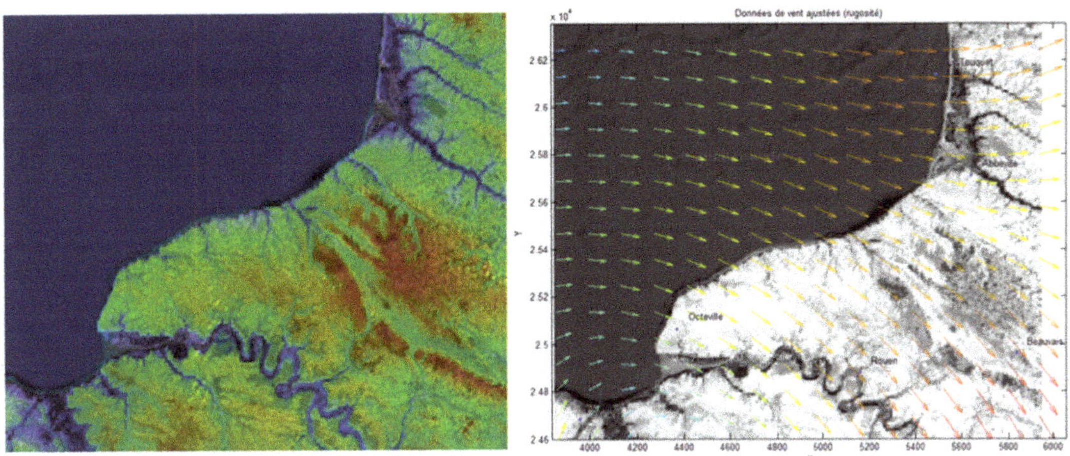

(a) (b)

Figure 13. Topographic map of the studied zone (Normandy Region, France) (**a**). Wind vector field (approximated from the six different Meteo France locations at airports) on the topographic map (**b**).

In order to show this method on more complicated datasets, we consider the wind conditions over 90 h; we have the value of the wind vector field at each Meteo France station every 3 h (total of 30 datasets). We apply the previous method for each time step. We then obtain the approximated wind velocity field over the 90 h. We have to propose a way to visualize such datasets.

5. Visualization

To obtain a simulation on time using a free library, we first propose using Matplotlib using Python. The following code was developed at INSA Rouen Normandie by the authors (and thanks to H. Merelle from the Applied Math. Department for his help). The main advantage of this code is that it gives a complete framework from the input (dataset) to the numerical simulation, including the approximation using the spline functions, finite element methods and the topography influence.

Algorithm for visualization using Matplolib [14].

Here is the list of files necessary for correct processing and the different steps of the proposed method linked to the flowcharts given in Figure 1:

- Initial Input: dataset (xi, yi, (Ui, Vi))i.
- Definition of Ω—meshing of Ω with rectangles (as we use rectangular finite elements): the number of subdivisions is linked to the number of data; in the examples here, it is 3×3 or 4×4 subdivisions in x and y.
- D^m spline approximation: the output is the evaluation of the vector field on each point of a fine grid of Ω.
- Computation of topography effect on the vector field: output.txt file.
- Script_visualization.py
- The "output.txt" file (in the same folder).

The purpose of the program is to visualize a vector flow from text files, with the possibility of adding a background (topography, etc.).

Data conditions:

- For an animation:

- The "output.txt" file is of the form

  X1 Y1 U1_1 V1_1 U1_2 V1_2...

  X2 Y2 U2_1 V2_1 U2_2 V2_2...

  with X1 and Y1 being the first coordinates, followed by U1_1 V1_1 U1_2 V1_2...; the different sizes of the vectors are a function of time.

- To execute in a terminal under Ubuntu, we use the Python script_visualization.py, with the following instructions:
  - The title: it represents the file name (when exporting) and the title of the figure.
  - The size of the vector arrows: the bigger they are, the smaller the vectors appear.
  - The number of images: if your output file is of the form "Animation", in this case you will have the following question, "*Enter the number of frames per second*"; it determines the frame rate per second.
  - For the background: "O" for accept or "N" otherwise.
  - For the name of the image you must give the file extension: here is a non-exhaustive list of usable formats: [name].png, [name].jpg, [name].jpeg and [name].gif.
  - To display the result: This command is only used to show you the result. The result is still saved even if you do not display it.
  - Data output: For an animation, you can find the animation in the folder in the form [title].gif, and for a fixed image, you can find the rendering in the folder in the form [title].png.

In order to show a numerical simulation, we give a simulation for all of the Normandy region (wind velocity field using the Meteo France dataset) using this Python script and using Matplotlib.

Examples of the obtained visualizations on a sequence (time) of an approximated velocity vector field (test in the Normandy region, France) and marine current (Seine River at Rouen, France) are given in Figures 14 and 15.

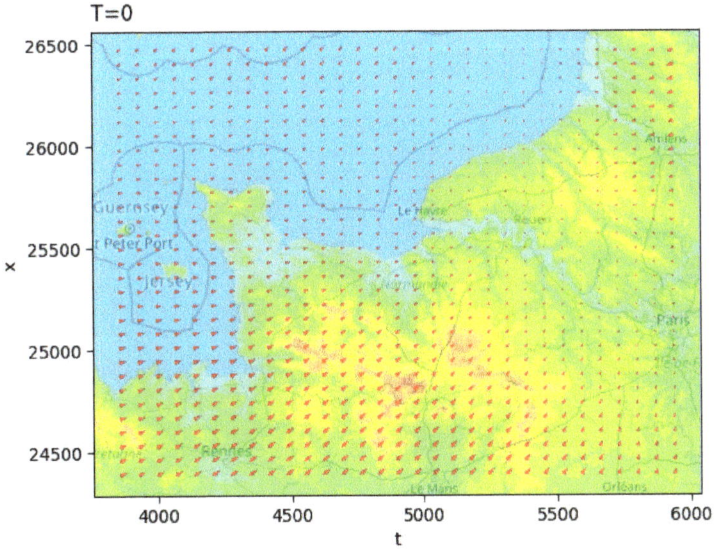

Figure 14. Visualization of a vector flow in Normandy, including the topography effect using Matplotlib (http://lmi.insa-rouen.fr/images/contenu/Movies/Test.gif accessed on 1 November 2024).

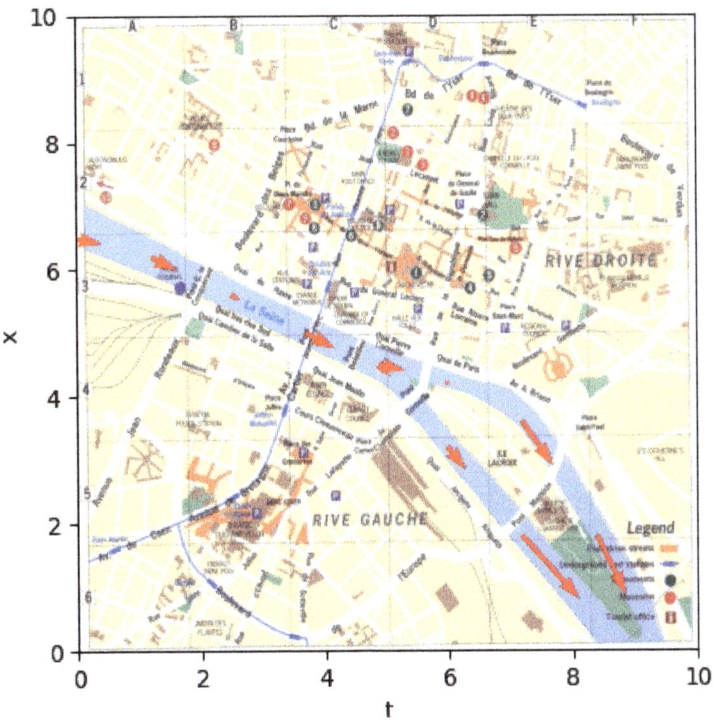

Figure 15. Example of visualization of marine currents in Rouen, France. Arrows indicate directions and speed (following length of the arrow) of the current. Visualization is performed using Matplotlib. (http://lmi.insa-rouen.fr/images/contenu/Movies/Rouen.gif accessed on 1 November 2024).

6. Discussion and Future Directions

About the approximation method using D^m splines, note that a theoretical study of the error in the approximation method following the used finite element mesh is a work in progress. To do that, we use previous results obtained from smoothing spline approximations from a finite set of points, as performed in [10,11]. Another development, linked to the approximation part, will be to propose an automatic meshing of the domain and different choices for the used generic finite element (based on triangles, etc.).

Another goal consists in improving both the modelling and visualization. About the modelling, the goal will be to include new kind of datasets; nowadays, it is possible to obtain wind datasets from Lidar located on wind turbines (see Figure 16). This dataset gives the wind velocity field with a specific geometry: along a spiral. It makes the computation much more difficult because it requires a specific finite elements meshing, which makes the process much less automatic.

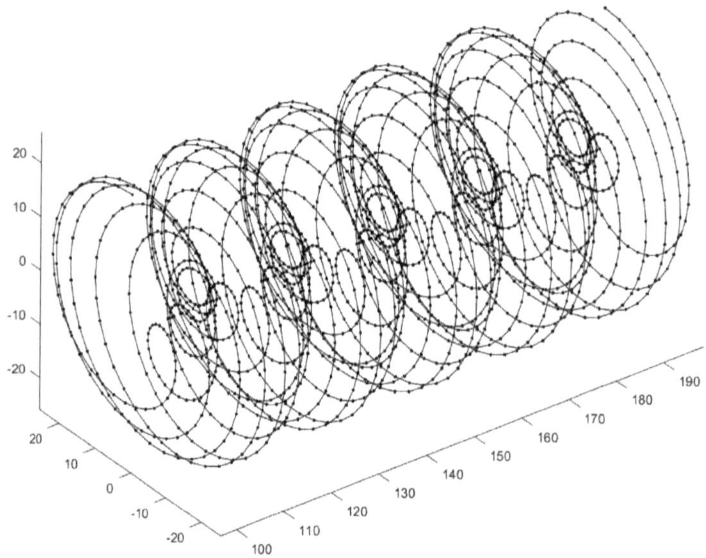

Figure 16. Example of the location of data in a Lidar dataset.

We also plan to add a smooth visualization based on texture using the Matplotlib library. Another crucial point consists in the effect of using a smooth visualization of the flow using streamlines. Vector field data are produced by scientific experiments and numerical simulations, which are now widely used to study complex dynamic phenomena, using a robust method to visualize steady flow field with both line representations and textures. In [15], the authors specify that "a streamline is a line tangential to the vector field at any point. Covering an image with a set of streamlines is a very good way to visualize the flow features" (see Figure 17 and [15] for more details).

In order to show this method on more complicated datasets, we considered the wind conditions over 90 h. We have the value of the wind vector field at each Meteo France station every 3 h (total of 30 datasets equivalent to the one we show in Section 4).

We then computed the obtained vector field using the modelling proposed in this work with the help of B. Jobard [15], and we obtained the movie given in [16]. This result is smooth and promising. But improvements have to be made to propose a tool able to treat the whole process with the same software, and to maybe try other approximation methods (like the one in [17], and to mix this approximation/visualization tool with an image processing framework, or the one in [18–20], using radial basis functions; this is ongoing work). Moreover, it is also crucial to develop an algorithm with which to approximate more complex datasets, like Lidar ones (instead of anemometers) to compute wind velocity fields.

Figure 17. Different streamlines to visualize a vector flow can be used (Short streamlines on the left and long streamlines in the middle. On the right, this is an image we can obtain using the "streamline algorithm" of [15]).

7. Conclusions

In this work, we successfully proposed a global tool, from vector field approximation to visualization. We proposed a method to obtain a visualization of a vector field from a sparse dataset, after computing its numerical approximation using a mathematical model using energy minimization and finite elements for the discretization. Note that we also integrated the topography effect into the modelling of a wind velocity field approximation method. To our knowledge, this is the first global approach for such numerical simulations from a dataset with few data.

As stressed in Section 6, several developments should occur in the future in order to improve this global approach. Many potential applications exist, from velocity wind approximation for wind turbine energy modelling, current simulation for modelling the morphodynamics of coastal zones and control theory for vehicle navigation (cars, submarines, etc.).

Author Contributions: Conceptualization, G.K. and C.G.; methodology, G.K. and C.G.; software, G.K. and C.G.; validation, G.K. and C.G.; formal analysis, G.K. and C.G.; investigation, G.K. and C.G.; resources, G.K. and C.G.; writing—original draft preparation, G.K. and C.G.; writing—review and editing, G.K. and C.G.; visualization, G.K. and C.G.; project administration, C.G.; funding acquisition, C.G. All authors have read and agreed to the published version of the manuscript.

Funding: This work was funded in parts by Région Normandie and ERDF fund via the I-demo SCALE Op project under convention 00152289 (BPI France and France 2030 AAP). This work was also supported in parts by the project DEFHY3GEO funded by Région Normandie and ERDF fund under convention 21E05300. This work is also supported by the CNRS and Labex AMIES for the PEPS2 M2Si2 project (2023–2025) under Labex AMIES convention ANR-10-LABX-0002-01.

Institutional Review Board Statement: Not applicable.

Informed Consent Statement: Not applicable.

Data Availability Statement: The original contributions presented in this study are included in the article. Further inquiries can be directed to the corresponding author.

Acknowledgments: The authors thank T. Cousin, H. Merelle, and D. Apprato for fruitful exchanges about the visualization of velocity vector fields. C. Gout thanks M. Chyba (University of Hawai'i at Manoa) for her invitation (July/August 2024) where this work was conducted in part, C. Gout also thanks CNRS and FR CNRS 3335 for the CNRS delegation he got in 2024.

Conflicts of Interest: The authors declare no conflicts of interest.

Appendix A. Bogner–Fox–Schmit Finite (BFS) Element of Class C^1

We give some results here about the BFS finite element, keeping the usual notations introduced in this field (see [21,22] for more details). We introduce the BFS of class C^1 we use in this work. We consider a rectangle K, a set of polynomial functions P and a set of degrees of freedom $\sum$ defined as

- K is the rectangle defined by the four points (x_i,y_i), (x_{i+1},y_i), (x_i,y_{i+1}) and (x_{i+1},y_{i+1}) (see Figure A1).
- $P = Q_3(R^2) = \left\{ q(x,y) = \sum_{0\leq i,j\leq 3} \alpha_{ij}x^i y^j, \alpha_{ij} \in R \right\}$.
- $\Sigma = \left\{ \phi_{kl} : p \mapsto p(x_k,y_l); \phi_{kl}^{(1)} : p \mapsto \frac{\partial p}{\partial x}(x_k,y_l); \phi_{kl}^{(2)} : p \mapsto \frac{\partial p}{\partial y}(x_k,y_l), \phi_{kl}^{(3)} : p \mapsto \frac{\partial^2 p}{\partial x \partial y}(x_k,y_l) \right\}$.

It easy to check that *dim P* = *card* $\sum$ = 16 and that $\sum$ is P-unisolvant. Thus, the triplet (K, $\sum$, P) defines a finite element. The 16 elements of $\sum$ are called the "degrees of freedom" of the considered finite element.

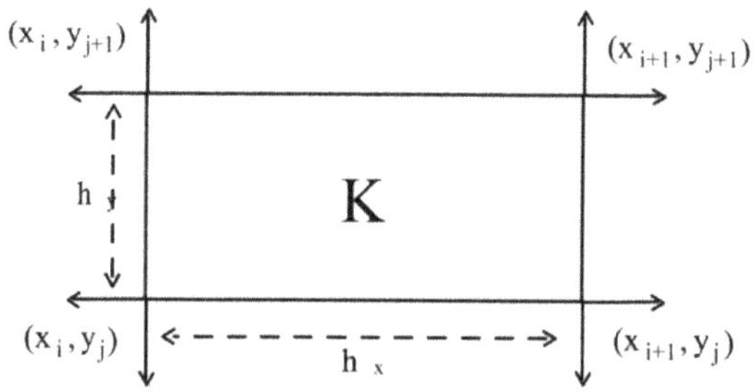

Figure A1. BFS rectangle finite element.

We now have to define the basis function of (K, $\sum$, P). To do that, we first work on a reference finite element corresponding to the rectangle K = [0, 1] × [0, 1]. Then, an affine transformation gives all the basis functions for all the degrees of freedom of the meshing.

Each basis function is a polynomial belonging to P, with a value of 1 for one of the 16 degrees of freedom, and value of 0 for the 15 others.

For example, on the rectangle of reference [0, 1] × [0, 1], the four basis functions at point (0, 0) are

$$\begin{aligned} \phi_{00}(x,y) &= (2x+1)(x-1)^2(2y-1)(y-1)^2. \\ \phi_{00}^{(1)}(x,y) &= (2x+1)(x-1)^2 y(y-1)^2. \\ \phi_{00}^{(2)}(x,y) &= x(x-1)^2(2y+1)(y-1)^2. \\ \phi_{00}^{(3)}(x,y) &= x(x-1)^2 y(y-1)^2. \end{aligned} \tag{A1}$$

We find the 16 basis functions for the 12 other degrees of freedom.

The finite element method then uses the basis function of the "reference" finite element to compute all the basis functions corresponding to the finite element mesh, with four basis functions for each node of the meshing. To do that, we just have to apply a diagonal affine mapping W, such that W([0, 1] × [0, 1]) = K, transforming each vertex of the reference element into one vertex of K (Figure A2). Then, the basis functions of K are trivially obtained using the mapping W and the basis function of the element of reference (see p. 57 of [22] for example).

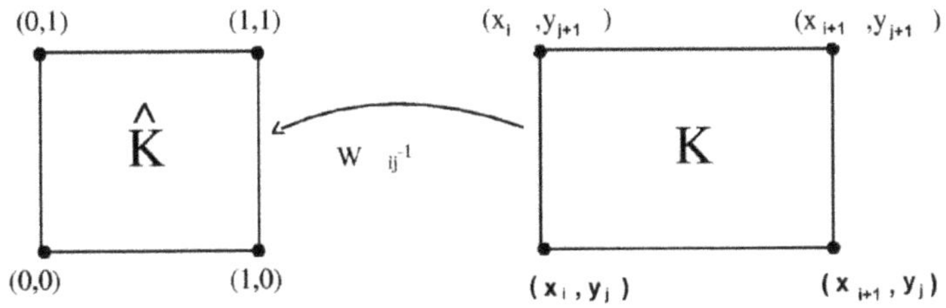

Figure A2. Affine transformation to compute the basis function of any point using the basis function of the reference finite element.

One of the main advantages of a finite element basis is that these basis functions have a very small support; thus, the matrix of the linear system we obtain is sparse (diagonal and positive definite!). A global introduction to the finite element method can be found in [23].

References

1. Gout, C.; Lambert, Z.; Apprato, D. *Data Approximation: Mathematical Modelling and Numerical Simulations*; EDP Sciences: Les Ulis, France, 2019; 168p, ISBN 978-2759823673.
2. Dzhabrailov, A.S.; Klochkov, Y.V.; Marchenko, S.S.; Nikolaev, A.P. The finite element approximation of vector field in curvilinear coordinates. *Russ. Aeronaut.* **2007**, *50*, 115–120. [CrossRef]
3. Benbourhim, M.N.; Bouhamidi, A. Approximation of vector fields by thin plate splines with tension. *J. Approx. Theory* **2005**, *136*, 198–229. [CrossRef]
4. Benbourhim, M.N.; Bouhamidi, A. Pseudo-polyharmonic vectorial approximation for div-curl and elastic semi-norms. *Numer. Math.* **2008**, *109*, 333–364. [CrossRef]
5. Dodu, F.; Rabut, C. Vectorial interpolation using radial-basis-like functions. *Comput. Math. Appl.* **2002**, *43*, 393–411. [CrossRef]
6. Apprato, D.; Gout, C.; Komatitsch, D. A new method for C^k-surface approximation from a set of curves, with application to ship track data in the Marianas trench. *Math. Geol.* **2002**, *34*, 831–843. [CrossRef]
7. Arcangéli, R.; Torrens, J.J.; Cruz de Silanes, M. *Multidimensional Minimizing Splines: Theory and Applications*; Kluwer Academic Publishers: Boston, NY, USA, 2004; 278p.
8. Gout, C.; Komatitsch, D. C^1—Approximation of seafloor surfaces with large variations. In Proceedings of the IEEE 2000 International Geoscience and Remote Sensing Symposium, Honolulu, HI, USA, 24–28 July 2000; Volume 5, pp. 1836–1838.
9. Le Guyader, C.; Gout, C.; Apprato, D. Spline approximation of gradient field: Applications to wind velocity field. *Math. Comput. Simul.* **2014**, *97*, 260–279. [CrossRef]
10. López de Silanes, M.C.; Arcangéli, R. Approximation error estimates for interpolating and smoothing (m,s)-splines. *Numer. Math.* **1989**, *5*, 449–467. [CrossRef]
11. Lopez de Silanes, M.C.; Apprato, D. Estimations de l'erreur d'approximation sur un domaine borné de Rn par Dm splines d'interpolation et d'ajustement discrètes. *Numer. Math.* **1988**, *3*, 367–376. [CrossRef]
12. Khayretdinova, G.; Chaumont-Frelet, T.; Gout, C.; Kuksenko, S. Image segmentation with a priori conditions: Applications to medical and geophysical imaging. *Math. Comput. Appl.* **2022**, *27*, 26. [CrossRef]
13. Eurocode 1: Actions on Structures. Part 1–4: General Actions—Wind Actions—National Annex to NF EN 1991-1-4:2005—General Actions—Wind Actions. Available online: https://www.boutique.afnor.org/en-gb/standard/nf-en-199114/eurocode-1-actions-on-structures-part-14-gereral-actions-wind-actions/fa104153/25897 (accessed on 1 November 2024).
14. Matplotlib. Available online: https://matplotlib.org/ (accessed on 27 October 2024).
15. Jobard, B.; Lefer, W. In Proceedings of the 9th International Conference in Central Europe on Computer Graphics, Visualization and Computer Vision 2001. Plzen, Czech Republic, 5–9 February 2001. Available online: https://dblp.org/db/conf/wscg/wscg2001.html (accessed on 27 October 2024).
16. Jobard, B.; Gout, C. Wind Field Reconstruction in Normandie. Available online: https://lmi2.insa-rouen.fr/~cgout/GG/phd/imagesPart3/vents_seine_maritime_01.avi (accessed on 27 October 2024).
17. Khayretdinova, G.; Apprato, D.; Gout, C. A level set based model for image segmentation under geometric constraints and data approximation. *J. Imaging* **2024**, *10*, 19p. [CrossRef] [PubMed]
18. Cervantes, D.; Gonzalez Casanova, P.; Gout, C.; Juarez, L.H.; Resendiz, R. Vector field approximation using radial basis functions. *J. Comput. Appl. Math.* **2013**, *240*, 163–173. [CrossRef]
19. González-Casanova, P.; Antonio Muñoz-Gómez, J.; Rodríguez-Gómez, G. Node Adaptive Domain Decomposition Method by Radial Basis Functions. *J. Numer. Methods Partial. Differ. Equ.* **2009**, *25*, 1482–1501. [CrossRef]
20. Cervantes, D.; Gonzalez-Casanova, P.; Gout, C.; Moreles, M. A line search algorithm for wind field adjustment with incomplete data and RBF approximation. *Comp. Appl. Math.* **2018**, *37*, 2519–2532. [CrossRef]

21. Bogner, F.K.; Fox, R.L.; Schmit, L.A. The generation of interelement-compatible stiffness and mass matrices by the use of interpolation formulae. In Proceedings of the Conference on Matrix Methods in Structural Analysis, Dayton, OH, USA, 26–28 October 1965.
22. Ciarlet, P.G. *The Finite Element Method for Elliptic Problems*; North-Holland Publishing Company: Amsterdam, The Netherlands, 1978.
23. Le Dret, H. Finite Element Lecture. Available online: https://www.ljll.fr/ledret/M1English/M1ApproxPDE_Chapter5-1.pdf (accessed on 27 October 2024).

Journal of
Imaging

MDPI

Communication

Multi-Head Attention Refiner for Multi-View 3D Reconstruction

Kyunghee Lee, Ihjoon Cho †, Boseung Yang † and Unsang Park *

Department of Computer Science and Engineering, Sogang University, 35 Baekbeom-ro (Sinsu-dong), Mapo-gu, Seoul 04107, Republic of Korea; kyungheelee9913@gmail.com (K.L.); ihjune9708@gmail.com (I.C.); didqhtmd@gmail.com (B.Y.)

* Correspondence: unsangpark@sogang.ac.kr

† These authors contributed equally to this work.

Abstract: Traditional 3D reconstruction models have consistently faced the challenge of balancing high recall of object edges with maintaining a high precision. In this paper, we introduce a post-processing method, the Multi-Head Attention Refiner (MA-R), designed to address this issue by integrating a multi-head attention mechanism into the U-Net style refiner module. Our method demonstrates improved capability in capturing intricate image details, leading to significant enhancements in boundary predictions and recall rates. In our experiments, the proposed approach notably improves the reconstruction performance of Pix2Vox++ when multiple images are used as the input. Specifically, with 20-view images, our method achieves an IoU score of 0.730, a 1.1% improvement over the 0.719 of Pix2Vox++, and a 2.1% improvement in F-Score, achieving 0.483 compared to 0.462 of Pix2Vox++. These results underscore the robustness of our approach in enhancing both precision and recall in 3D reconstruction tasks involving multiple views.

Keywords: multi-view 3D reconstruction; attention mechanism; multi-head attention; refiner; object boundary prediction

1. Introduction

In recent decades, the methodologies for creating 3D representations from 2D images have undergone significant changes. The field of 3D reconstruction has seen development of numerous geometric theory-based vision algorithms, such as feature extraction (SIFT) [1], structure estimation using epipolar geometry [2], and model creation via Delaunay triangulation [3]. However, these algorithms consume a considerable amount of computing power, making 3D reconstruction a time-consuming process. The advent of deep learning has shifted these paradigms, offering improvements in both precision and recall, contributing to advancements in 3D reconstruction methodologies. Deep learning has particularly revolutionized the refinement of reconstructed shapes, addressing the challenges posed by the complexity of details and textures inherent in 3D objects.

The attention mechanism is a methodology in artificial intelligence that enables models to focus on the critical aspects of data. This mechanism has been adopted in various forms throughout deep learning networks, with the transformer architecture [4] being one of the most prominent examples that uses the multi-head attention mechanism. The transformer architecture has demonstrated its high performance in both natural language processing (NLP) [4] and computer vision tasks [5]. Given the demonstrated effectiveness of transformers and the precise focus provided by the attention mechanism, this approach is expected to effectively address challenges in 3D reconstruction. Specifically, the original Pix2Vox++ model often struggles with accurate boundary prediction and maintaining high recall rates, particularly for complex geometries. Therefore, we incorporated a multi-head attention mechanism into the refiner module to enhance the model's capability for focusing on intricate details, thereby reducing boundary prediction errors and improving overall reconstruction accuracy. Multi-head attention enables the model to focus on distinct spatial and contextual features of the input simultaneously, effectively capturing both fine-grained and broader-scale information. This approach is particularly effective for reconstructing complex 3D structures, ensuring accurate representation of intricate boundaries and fine details.

We applied this concept to the existing deep learning 3D reconstruction model Pix2Vox, with the primary contribution of this research being the integration of a multi-head attention mechanism into the refiner module of Pix2Vox++. This enhancement aims to improve the model's capability to effectively balance the precision and recall of object boundaries, particularly when handling multiple views. Our proposed approach not only enhances boundary accuracy, but also ensures better retention of fine details across different input views, resulting in a superior 3D reconstruction quality. This model takes multiple input images of a single object and reconstructs the 3D object in voxel form. The original architecture of this model employs an encoder–decoder structure along with a U-Net-based refiner. We modified the refiner to incorporate a multi-head attention mechanism for better data flow. With this modification, the modified Pix2Vox++ demonstrates a greater improvement in IoU and F-score values when more than four input images are given. Specifically, our modified model showed fewer errors at the object boundaries and improved recall rates. For example, when using 20-view inputs, our model demonstrated a 2.1% improvement in F-Score, highlighting its robustness in enhancing both precision and recall in multi-view 3D reconstruction tasks.

The remainder of this paper is organized as follows: Section 2 reviews the related work, Section 3 describes the proposed methodology and architecture, and Section 4 presents the experimental results, highlighting how the integration of the multi-head attention mechanism enhances Pix2Vox++ in terms of boundary accuracy and overall reconstruction quality. Finally, Section 5 concludes the paper and discusses future work.

2. Related Works

2.1. 3D Reconstruction

Comprehensive reviews of 3D object reconstruction approaches can be found in [6]. The 3D reconstruction task can be classified by input types, such as single-view 3D reconstruction and multiview 3D reconstruction, categorized by output format, including voxel grids, point clouds, and 3D meshes. The single-view 3D reconstruction task is a long-established, ill-posed, and ambiguous problem. Before learning-based methods, many attempts have been made to address this issue, such as Shape from X [7], where X may represent silhouettes [8], shading [9], or texture [10]. These approaches require strong assumptions and abundant experience in natural images [11], so they are rarely applied in real-world scenarios. With the advent of learning-based methods, many approaches have achieved strong performances. The 3D Variational Autoencoder Generative Adversarial Network (3D-VAE-GAN) [12] combines a generative adversarial network (GAN) [13] and a variational autoencoder (VAE) [14] to generate 3D objects from single-view images. Marrnet [15] reconstructs 3D objects by estimating the depth, surface normal, and silhouettes from 2D images.

Multiview 3D reconstruction tasks have been studied with algorithm-based methods. Structure-from-motion (SfM) and simultaneous localization and mapping (SLAM) algorithms require a collection of RGB images. These algorithms estimate 3D structures through dense feature extraction and matching [1]. However, algorithm-based methods struggle when multiple viewpoints are widely separated. Furthermore, as the input is discrete information, it cannot offer a full surface of an object, which leads to reconstructing incomplete 3D shapes with occluded or hollowed-out areas. With the learning-based method, Pixel2Mesh [16] is the first to reconstruct the 3D shape in a triangular mesh from a single image. Octree Generating Networks (OGN) [17] uses octree to represent high-resolution 3D volumes with a limited memory budget. Matryoshka Networks [18] continuously decomposes a 3D shape into nested shape layers, which outperforms octree-based reconstruction methods. More recently, AttSets [19] used an attentional aggregation module to automatically predict a weight matrix as attention scores for input features. Both 3D Recurrent Reconstruction Neural Network (3D-R2N2) [20] and Learnt Stereo Machines (LSM) [21] are based on Recurrent Neural Network (RNN), resulting in the networks being permutation variants and inefficient for aggregating features from images of long sequence.

Beyond the domain of 3D reconstruction, there are several additional methods exist for representing 3D objects from 2D images. Novel view synthesis (NVS) is one such method for representing a 3D object. It generates novel photorealistic views by interpolating given 2D images [22–24]. Novel view synthesis task has supported 3D reconstruction by generating interpolated images, providing additional data to enhance reconstruction process before Neural Radiance field was introduced. Neural radiance field (NeRF) is a method for synthesizing novel views of complex scenes by optimizing an underlying continuous volumetric scene function using a sparse set of input views [25]. The introduction of Neural Radiance Field marked a turning point, as its exceptional results sparked widespread interest and led a number of following studies. Like the models in 3D reconstruction, some of those variant models [26–28] generated a synthesized image from a single image input. Some studies expanded the NVS task into 3D reconstruction. In other words, some studies of NeRFs produced 3D outputs in the form of voxel [29], point cloud, and polygons, while the original NeRF only generated 2D outputs in the form of image and video.

In recent studies, the emphasis has shifted from developing models that perform 3D reconstruction to NeRF models in NVS that are also related to the recently introduced Gaussian Splatting technique [30]. Although we focus on developing a 3D reconstruction model rather than to recent approaches like NeRF, the concept of refining the U-Net structure is generalizable and could be applied across different deep learning architectures, independent of the specific task.

2.2. Attention Mechanisms

The general form of the attention mechanism is presented below [31].

$$\text{Attention} = f(g(x)|x) \tag{1}$$

Here, $f(g(x), x)$ means processing input x based on the attention $g(x)$, which is consistent with processing critical regions and obtaining detailed information. Almost all existing attention mechanisms can be written into the above formulation. As attention mechanisms have been researched, a number of variants such as 'spatial attention [32]: where to pay attention', 'temporal attention [33]: when to pay attention' and 'channel attention [34]: what to pay attention' have been proposed. With self-attention and multi-head attention, transformer architecture has been applied in natural language processing tasks, showing significantly improved results [4].

Several studies have used attention mechanism and transformer architecture itself for enhancing 3D reconstruction performance. EVoIT [35] reformulated 3D reconstruction problem as a sequence-to-sequence prediction problem and proposed a 3D Volume Transformer framework inspired from the success of transformer. Differently from previous CNN-based networks, EVoIT has an advantage by unifying the two stage feature extraction and view fusion into a single stage. The aention mechanism allows them to explore the view-to-view relationships from multi-view input images. Self-attention ONet [36] is an enhanced version of ONet [37] incorporating the self-attention mechanism into original 3D object reconstruction model. By employing a self-attention mechanism, models could extract global information, ignore unimportant details, and obtain more consistent meshes. METRO [38] is a mesh transformer framework that reconstructs 3D human pose and mesh from a single input image. By leveraging the transformer, it could simultaneously reconstruct 3D human body joints and mesh vertices. VoRTX [39] model for 3D volumetric reconstruction could retain finer details from fusing multi-view information by performing data-dependent fusion using a transformer.

Inspired by these previous studies, we introduce a multi-head attention refiner that incorporates multi-head attention mechanism into the refiner module to recover finer details from coarse volume.

3. Proposed Method

3.1. Pix2Vox++ Network Architecture

The Pix2Vox++ [40] network architecture, initially comprising an encoder, decoder, multi-scale context-aware fusion module and a refiner, has been enhanced in this study. The encoder starts by generating feature maps from input images, which are then processed by the decoder to produce coarse 3D volumes. These volumes are further refined by the multi-scale context-aware fusion module, which selects high-quality reconstructions from all the coarse volumes to create a fused 3D volume.

Our study focuses on improving the existing refiner module, which is crucial for correcting inaccuracies in the fused 3D volume but has limitations in capturing intricate object details and maintaining high recall rates. By incorporating a multi-head attention mechanism into the refiner module, we significantly improved the model's capability to predict precise object boundaries, resulting in more accurate and detailed 3D reconstructions. This integration showcases the effectiveness of attention mechanisms in advancing deep learning models for 3D reconstruction tasks.

3.2. Multi-Head Attention Refiner

Traditional 3D reconstruction refiners often struggle with maintaining high recall of object boundaries, particularly in complex geometries. The multi-head attention refiner (MA-R) is introduced in this study to address these shortcomings. By integrating the attention mechanism, MA-R selectively focuses on critical regions of the 3D volume, improving precision at object boundaries and enhancing the overall model accuracy. Specifically, we aim to mitigate the boundary prediction errors observed in previous refiners, as shown in Figure 1. The modified refiner, enhanced with multi-head attention, allows for better preservation of detailed information, ensuring that object boundaries are more accurately reconstructed.

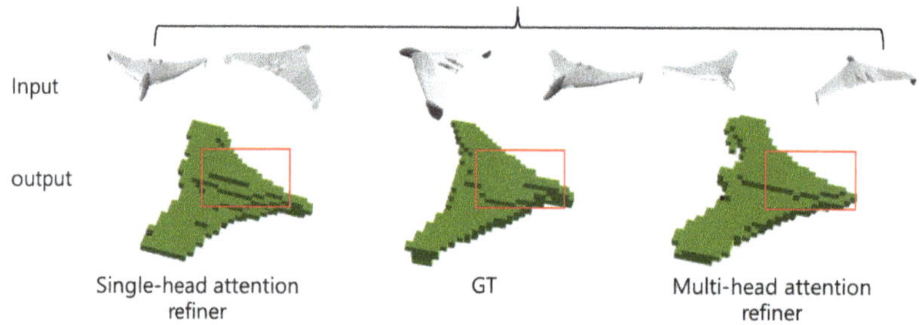

Figure 1. 3D reconstruction of an airplane from multi-view and single-view inputs: This figure presents both the angle-specific reconstruction results of an airplane using multi-view inputs and the results obtained using single-view inputs. The comparison demonstrates the effectiveness of our proposed model in translating multi-view 2D input images into accurate 3D objects.

MA-R helps tackle one of the most persistent challenges in 3D reconstruction—accurately capturing fine details, particularly at object boundaries, without introducing noise or losing important features. By employing multiple attention heads, MA-R processes the 3D volume at various resolutions, ensuring that both fine-grained and large-scale structures are equally well reconstructed. Focusing on the most informative regions of the volume results in more precise and detailed reconstructions, as demonstrated in our experiments.

Equations (2)–(4) represent the mathematical formulations of the mechanism of self-attention in our multi-head attention refiner. Equation (2) indicates a linear transformation to generate the query vector (Q), key vector (K), and value vector (V). Equation (3) represents the computation of attention weights in self-attention. Finally, Equation (4) denotes the final output, which is the product of the attention weights and the value vectors. This integration of the attention mechanism enables the refiner to focus on the most informative parts of the 3D volume, resulting in a more accurate reconstruction, as shown in our output.

$$Q, K, V = \text{Linear}(x) \tag{2}$$

$$g(x) = \text{Softmax}(QK) \tag{3}$$

$$f(g(x), x) = g(x)V \tag{4}$$

The architecture of our proposed model is fundamentally based on Pix2Vox++, incorporating multi-head attention (MHA) specifically within the refiner. In this architecture, multi-view images are first processed through an encoder–decoder structure to generate coarse 3D volumes. The MHA refiner is then applied to these volumes to refine and enhance the details. The MHA mechanism processes each feature map by computing key, query, and value matrices from the input volume. These matrices are used to calculate the attention weights, determine how much focus each region of the volume should receive based on its relevance to the overall 3D structure.

Specifically, the MHA refiner employs several heads to provide attention to different aspects of the 3D volume, allowing the model to capture intricate details that may be missed in single-headed attention. The structure of MHA is depicted in Figure 2. Volumes 16-L, 8-L, and 4-L are processed through the MHA, and their outputs are attached to their corresponding volumes of 32-R, 16-R, and 8-R, respectively. This multi-scale refinement process allows the MHA to focus on fine-grained details at multiple resolutions, ensuring that the final volume output of 32^3 retains high levels of detail and accuracy.

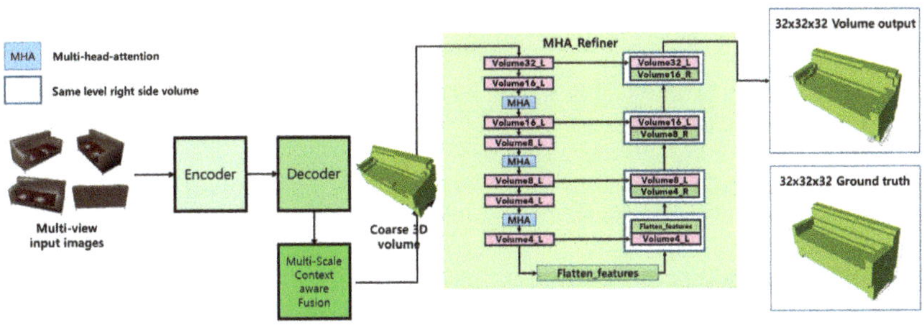

Figure 2. Architectural overview with multi-head self-attention integration in the refiner encoder: This figure demonstrates the integration of multi-head attention within the refiner module of the Pix2Vox++ architecture. The process involves using four-view images and generating a refined 3D volume output of 32 × 32 × 32.

Figure 3 describes how multi-head self-attention is integrated within the encoder section of the refiner. It focuses on the transformation process following layer 3, leading up to the input for layer 4, emphasized by the application of multi-head self-attention. Each layer initially processes the feature map through the convolution 3D layer, batch-normalization 3D layer, LeakyReLu as the activation function, and max pooling 3D layer. In the post-processing layer, the volume undergoes a linear transformation to set the query (Q), key (K), and the value (V). The attention score is generated by the multiplication of the matrix of Q and K. These scores, once normalized with a softmax, stabilize the weights, and their multiplication with V produces an output of the original size, which is fed to the subsequent layers. A unique feature of this model is the repetition of internal attention that extends over several heads.

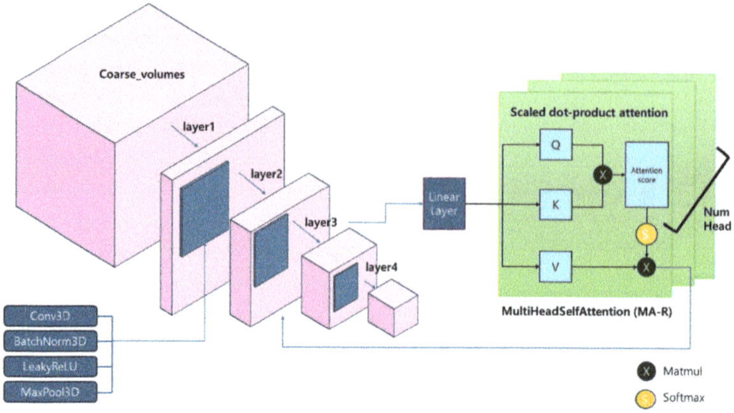

Figure 3. Integration of multi-head self-attention post layer 3: This figure delineates the specific segment within the MA-R refiner where the multi-head self-attention mechanism is employed. It emphasizes the transformation process from the input of the third layer to its output, which then serves as the input for the subsequent fourth layer. This portrays the layered sequential processing and the attention-based enhancement applied within the refiner's encoder, underlining the effectiveness of the multi-head self-attention in refining 3D reconstruction outputs.

In the Section 4, the intersection over union (IoU) and F score will be used as quantitative metrics to demonstrate the effectiveness of the proposed method. Other models will also be compared, with a comparison to Pix2Vox serving as an ablation experiment. Additionally, reconstructed 3D objects will be presented as qualitative results.

4. Experiment

4.1. Datasets

The ShapeNet dataset [41] is a comprehensive and widely used collection of 3D CAD models, organized based on the WordNet taxonomy. It is renowned for its large scale and diversity, encompassing a wide array of object categories, making it a standard benchmark in the field of 3D object reconstruction. ShapeNet provides richly annotated 3D CAD models that are crucial for training and evaluating 3D reconstruction algorithms. In this study, we utilized a subset of the ShapeNet dataset, consisting of approximately 44,000 models across 13 major categories. This selection is aligned with the datasets used in 3D-R2N2 [20], ShapeNetRendering, and ShapeNetVox32. The choice of this subset was driven by our aim to ensure compatibility and facilitate a direct comparison with existing studies, particularly those that have employed 3D-R2N2, a well-established framework in multi-view 3D reconstruction. By using this subset, we aim to benchmark our proposed method against established performances in the field, ensuring that our findings are both relevant and comparable within the current research landscape.

4.2. Evaluation Metrics

To evaluate the quality of the proposed methods, we binarized probabilities at a fixed threshold of 0.3 and calculated IoU as a similarity measure between ground truth and prediction.

$$\mathrm{IoU} = \frac{\sum_{i,j,k} I\left(\hat{P}_{i,j,k} > t\right) I\left(\hat{P}_{i,j,k}\right)}{\sum_{i,j,k} I\left[I\left(\hat{P}_{i,j,k} > t\right) + I\left(P_{i,j,k}\right)\right]} \tag{5}$$

where $\hat{P}_{i,j,k}$ and $P_{i,j,k}$ represent the predicted occupancy probability and the ground truth in the (i, j, k) voxels, respectively. $\mathrm{I}(\cdot)$ is an indicator function and t denotes a threshold. Higher IoU values correspond to a better reconstruction accuracy.

We calculate the F-Score as an additional metric to evaluate the performance of 3D reconstruction results. The F-Score represents the harmonic mean between precision and recall, and is defined as follows:

$$\text{F-Score}(d) = \frac{2 \cdot P(d) \cdot R(d)}{P(d) + R(d)} \quad (6)$$

where $P(d)$ and $R(d)$ are the precision and recall for a given distance threshold d, respectively. These are computed by:

$$P(d) = \frac{1}{n_r} \sum_{r \in \mathcal{R}} \left[\min_{g \in \mathcal{G}} \|g - r\| < d \right] \quad (7)$$

$$R(d) = \frac{1}{n_g} \sum_{g \in \mathcal{G}} \left[\min_{r \in \mathcal{R}} \|g - r\| < d \right] \quad (8)$$

where $\mathcal{R}$ and $\mathcal{G}$ denote the reconstructed and ground truth point clouds, respectively, and n_r and n_g represent the number of points in $\mathcal{R}$ and $\mathcal{G}$, respectively. We used the Marching Cubes algorithm to extract the object surface from the reconstructed voxel. We sampled 8192 points from the surface to compute the F-Score between the prediction and ground truth [42]. Higher F-Scores indicate better reconstruction quality.

4.3. Implementation Details

We trained the proposed methods with batch size of 64 using 224 × 224 RGB images as the inputs. The output data are 32^3 voxels. We implemented our networks in PyTorch 2.4.0+cu121 [43] and trained Pix2Vox++/A using the Adam optimizer [44] with β_1 of 0.9 and β_2 of 0.999. The initial learning rate was set at 0.001 and decayed by 2 after 150 epochs. We trained the networks for 250 epochs, while multiscale context-aware fusion does not apply in single-view reconstruction tasks. All of the experiments were conducted using an NVIDIA A6000 GPU (NVIDIA Corporation, Santa Clara, CA, USA) on a server provided by Sogang University.

4.4. Results

4.4.1. Quantitative Results

This improvement can be attributed to the ability of the MA-R to selectively focus on essential geometric details while minimizing the influence of noise. The multi-head attention mechanism prioritizes high-frequency features, such as sharp edges and object boundaries, which are crucial for preserving the integrity of the 3D reconstruction. Conversely, irrelevant or redundant low-frequency information that could introduce noise is suppressed through the attention mechanism.

As the number of input views increases, the network is better able to refine the object by distinguishing between fine details and noise, leading to more accurate reconstructions, particularly in complex geometries. The improved capability to focus on relevant details directly enhances the model's performance, resulting in increased accuracy, as evidenced by higher IoU and F-Score metrics.

4.4.2. Qualitative Results

We also performed a qualitative evaluation of our proposed method. Figures 4 and 5 present the results of this evaluation.

In Figure 4, we demonstrate the qualitative evaluation of MA-R performance using the sofa and chair datasets. The improvements made by MA-R are emphasized by using color-coded boxes. The red and yellow boxes indicate that the voxels that were present but did not exist in the ground truth have been removed after MA-R application. The blue boxes show areas where parts that were missing in the initial model but present in the ground truth have been improved, resulting in higher-fidelity reproduction. This illustrates the superiority of our MA-R in correcting both deficiencies and excesses in the model, thus enhancing overall accuracy.

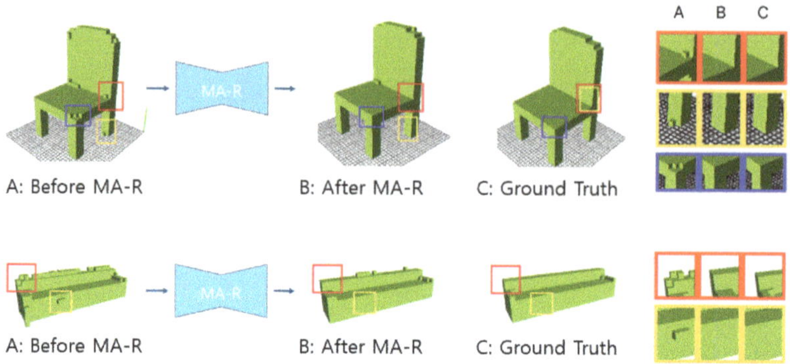

Figure 4. Qualitative demonstration of MA-R performance: Panel A shows the 3D volume before the application of MA-R, Panel B presents the 3D volume after MA-R processing, and Panel C represents the ground truth. The red and yellow boxes highlight areas that were erroneously reproduced in the single head model but have been correctly removed after MA-R, while the blue box indicates a region that was initially missing and has been adequately filled in after MA-R refinement.

In Figure 5, we compare our final MA-R output result with the ground truth meshes and the results from Pix2Vox++. In the case of a chair (as shown in the fourth row of the table), unlike the results from Pix2Vox++, we did not meet the problem of having holes in the middle of the object. In additional comparison examples, it is clear that our outputs demonstrate a higher fidelity to the ground truth mesh compared to the outputs obtained from Pix2Vox++.

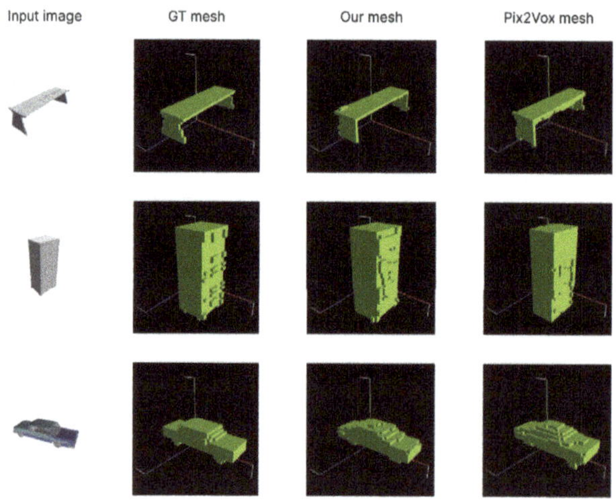

Figure 5. *Cont.*

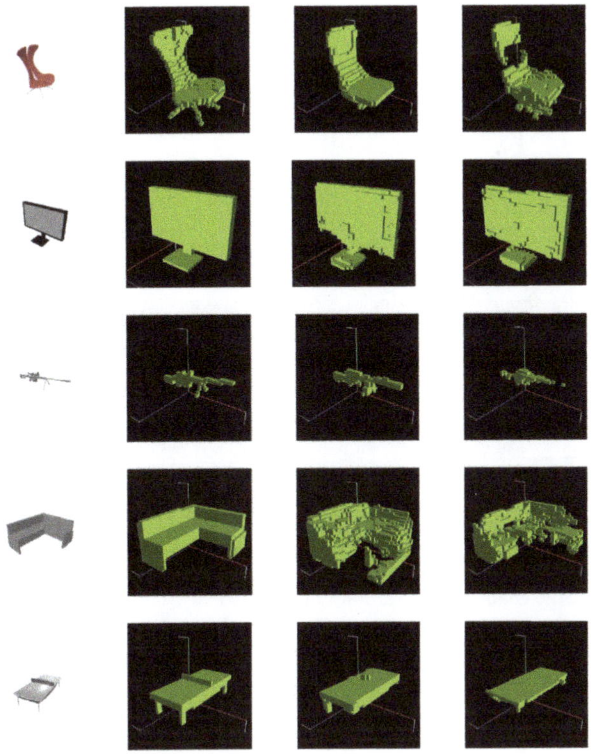

Figure 5. Ground truth meshes and reconstruction results of our model and Pix2Vox++ for various objects: Arranged from top to bottom are the results for bench, cabinet, car, chair, monitor, rifle, sofa, and table.

5. Conclusions

The MA-R method has demonstrated significant improvements in boundary prediction and overall reconstruction accuracy for 3D reconstruction. This is supported by the results in Tables 1 and 2, where our method outperforms the baseline approaches in terms of both IoU and F-Score across multiple view settings. Specifically, at 20 views, our method achieves an IoU score of 0.730 and an F-Score of 0.483, compared to Pix2Vox++'s 0.719 and 0.462, respectively. However, the increased computational complexity of the multi-head attention mechanism, particularly when applied to high-resolution 3D volumes, remains a limitation. This complexity can lead to longer processing times and higher memory consumption, potentially limiting the method's scalability for real-time applications or with very large datasets. Future research will focus on refining the refiner module, experimenting with larger volumetric data beyond the current $32 \times 32 \times 32$ resolution, and exploring more efficient data representations, such as tri-planes, to achieve a better performance with reduced computational costs.

Table 1. Quantitative results of multi-view 3D object reconstruction on ShapeNet at 32^3 resolution, with mean IoU for all categories.

Methods	1 View	2 Views	3 Views	4 Views	5 Views	8 Views	12 Views	16 Views	20 Views
3D-R2N2	0.560	0.603	0.617	0.625	0.634	0.635	0.636	0.636	0.636
AttSets	0.642	0.663	0.670	0.675	0.677	0.685	0.688	0.692	0.693
Pix2Vox++	**0.670**	**0.695**	**0.704**	**0.708**	0.711	0.715	0.717	0.718	0.719
Ours	0.636	0.681	0.699	**0.708**	**0.713**	**0.721**	**0.726**	**0.729**	**0.730**

Table 2. Quantitative results of multi-view 3D object reconstruction on ShapeNet at 32^3 resolution, with mean F-Score for all categories. **Bold** values indicate the highest performance in each category.

Methods	1 View	2 Views	3 Views	4 Views	5 Views	8 Views	12 Views	16 Views	20 Views
3D-R2N2	0.351	0.368	0.372	0.378	0.382	0.383	0.382	0.382	0.383
AttSets	0.395	0.418	0.426	0.430	0.432	0.444	0.445	0.447	0.448
Pix2Vox++	**0.436**	**0.452**	**0.455**	**0.457**	**0.458**	0.459	0.460	0.461	0.462
Ours	0.360	0.414	0.438	0.450	**0.458**	**0.470**	**0.477**	**0.480**	**0.483**

Despite these challenges, the MA-R method holds great promise for advancing the field of 3D reproduction. It effectively addresses persistent issues in multi-view 3D reconstruction, particularly the challenge of maintaining high fidelity at object boundaries. Future research will also focus on improving the performance of the method when applied to larger and more complex 3D volumes, ensuring that the method scales effectively for more detailed reconstructions. By integrating MA-R with emerging techniques such as neural rendering or photogrammetry-based approaches, we can further enhance its capability to handle complex geometries and large-scale environments.

Additionally, refining the balance between noise suppression and detail preservation remains an important area of exploration. While MA-R has shown encouraging results, more sophisticated approaches to identifying and preserving fine details, while minimizing noise, could lead to even more accurate 3D reconstruction. As 3D reconstructions continue to evolve, improving scalability and fidelity for larger, more intricate reconstructions will be key for developing robust, high-performance solutions.

Author Contributions: Conceptualization, K.L.; Methodology, K.L.; Software, K.L.; Validation, K.L.; Writing—original draft preparation, I.C. and B.Y.; Writing—review and editing, I.C., B.Y., K.L. and U.P.; Visualization, K.L.; Supervision and project administration, U.P. All authors have read and agreed to the published version of the manuscript.

Funding: This research was funded by Institute of Information & communications Technology Planning & Evaluation (IITP) grant funded by the Korea government (MSIT) (No.RS-2023-00227871, Development of Digital Object Matching and Management Technology based on Object Identification of Mobile Images and 3D Data Generation).

Institutional Review Board Statement: Not applicable.

Informed Consent Statement: Not applicable.

Data Availability Statement: The data presented in this study are available in ShapeNet at https://www.shapenet.org/, accessed on 11 January 2024. reference number [41]. ShapeNet is a publicly accessible dataset commonly used for 3D object reconstruction research. No new data were generated in this study, and the analysis was conducted using this publicly available dataset.

Conflicts of Interest: The authors declare no conflicts of interest.

References

1. Lowe, D.G. Distinctive image features from scale-invariant keypoints. *Int. J. Comput. Vis.* **2004**, *60*, 91–110. [CrossRef]
2. Hartley, R.; Zisserman, A. *Multiple View Geometry in Computer Vision*, 2nd ed.; Cambridge University Press: Cambridge, UK, 2003.

3. Lee, D.T.; Schachter, B.J. Two algorithms for constructing a Delaunay triangulation. *Int. J. Comput. Inf. Sci.* **1980**, *9*, 219–242. [CrossRef]
4. Vaswani, A.; Shazeer, N.; Parmar, N.; Uszkoreit, J.; Jones, L.; Gomez, A.N.; Kaiser, Ł.; Polosukhin, I. Attention is all you need. In Proceedings of the Advances in Neural Information Processing Systems 30: Annual Conference on Neural Information Processing Systems 2017, Long Beach, CA, USA, 4–9 December 2017; Volume 30, pp. 6000–6010. [CrossRef]
5. Dosovitskiy, A.; Beyer, L.; Kolesnikov, A.; Weissenborn, D.; Zhai, X.; Unterthiner, T.; Dehghani, M.; Minderer, M.; Heigold, G.; Gelly, S.; et al. An image is worth 16x16 words: Transformers for image recognition at scale. *arXiv* **2020**, arXiv:2010.11929.
6. Jin, Y.; Jiang, D.; Cai, M. 3D reconstruction using deep learning: A survey. *Commun. Inf. Syst.* **2020**, *20*, 389–413. [CrossRef]
7. Barron, J.T.; Malik, J. Shape, illumination, and reflectance from shading. *IEEE Trans. Pattern Anal. Mach. Intell.* **2014**, *37*, 1670–1687. [CrossRef] [PubMed]
8. Dibra, E.; Jain, H.; Oztireli, C.; Ziegler, R.; Gross, M. Human shape from silhouettes using generative hks descriptors and cross-modal neural networks. In Proceedings of the IEEE Conference on Computer Vision and Pattern Recognition, Honolulu, HI, USA, 21–26 July 2017; pp. 4826–4836.
9. Richter, S.R.; Roth, S. Discriminative shape from shading in uncalibrated illumination. In Proceedings of the IEEE Conference on Computer Vision and Pattern Recognition, Boston, MA, USA, 7–12 June 2015; pp. 1128–1136. [CrossRef]
10. Witkin, A.P. Recovering surface shape and orientation from texture. *Artif. Intell.* **1981**, *17*, 17–45. [CrossRef]
11. Zhang, Y.; Liu, Z.; Liu, T.; Peng, B.; Li, X. RealPoint3D: An efficient generation network for 3D object reconstruction from a single image. *IEEE Access* **2019**, *7*, 57539–57549. [CrossRef]
12. Wu, J.; Zhang, C.; Xue, T.; Freeman, B.; Tenenbaum, J. Learning a probabilistic latent space of object shapes via 3d generative-adversarial modeling. In Proceedings of the NIPS'16: Proceedings of the 30th International Conference on Neural Information Processing Systems, Barcelona Spain, 5–10 December 2016; Volume 29.
13. Goodfellow, I.; Pouget-Abadie, J.; Mirza, M.; Xu, B.; Warde-Farley, D.; Ozair, S.; Courville, A.; Bengio, Y. Generative adversarial nets. In Proceedings of the NIPS'14: Proceedings of the 27th International Conference on Neural Information Processing Systems, Montreal, QC, Canada, 8–13 December 2014; Volume 27.
14. Kingma, D.P.; Welling, M. Auto-encoding variational bayes. *arXiv* **2013**, arXiv:1312.6114.
15. Wu, J.; Wang, Y.; Xue, T.; Sun, X.; Freeman, B.; Tenenbaum, J. Marrnet: 3d shape reconstruction via 2.5 d sketches. In Proceedings of the 31st Conference on Neural Information Processing Systems (NIPS 2017), Long Beach, CA, USA, 4–9 December 2017; Volume 30.
16. Wang, N.; Zhang, Y.; Li, Z.; Fu, Y.; Liu, W.; Jiang, Y.G. Pixel2mesh: Generating 3d mesh models from single rgb images. In Proceedings of the European Conference on Computer Vision (ECCV), Munich, Germany, 8–14 September 2018; pp. 52–67.
17. Tatarchenko, M.; Dosovitskiy, A.; Brox, T. Octree Generating Networks: Efficient Convolutional Architectures for High-Resolution 3D Outputs. In Proceedings of the 2017 IEEE International Conference on Computer Vision (ICCV), Venice, Italy, 22–29 October 2017; pp. 2107–2115. [CrossRef]
18. Richter, S.R.; Roth, S. Matryoshka networks: Predicting 3d geometry via nested shape layers. In Proceedings of the IEEE Conference on Computer Vision and Pattern Recognition, Salt Lake City, UT, USA, 18–23 June 2018; pp. 1936–1944.
19. Yang, B.; Wang, S.; Markham, A.; Trigoni, N. Robust attentional aggregation of deep feature sets for multi-view 3D reconstruction. *Int. J. Comput. Vis.* **2020**, *128*, 53–73. [CrossRef]
20. Choy, C.B.; Xu, D.; Gwak, J.; Chen, K.; Savarese, S. 3d-r2n2: A unified approach for single and multi-view 3d object reconstruction. In Proceedings of the Computer Vision–ECCV 2016: 14th European Conference, Amsterdam, The Netherlands, 11–14 October 2016; Springer: Berlin/Heidelberg, Germany, 2016; pp. 628–644.
21. Kar, A.; Häne, C.; Malik, J. Learning a multi-view stereo machine. In Proceedings of the NIPS'17: Proceedings of the 31st International Conference on Neural Information Processing Systems, Long Beach, CA, USA, 4–9 December 2017; Volume 30.
22. Gortler, S.; Grzeszczuk, R.; Szeliski, R.; Cohen, M. The lumigraph. In *SIGGRAPH'96: Proceedings of the 23rd Annual Conference on Computer Graphics and Interactive Techniques*; Association for Computing Machinery: New York, NY, USA, 1996; pp. 43–54.
23. Davis, A.; Levoy, M.; Durand, F. Unstructured light fields. In *Proceedings of the Computer Graphics Forum*; Wiley Online Library: Hoboken, NJ, USA, 2012; Volume 31, pp. 305–314.
24. Levoy, M.; Hanrahan, P. Light field rendering. In *Seminal Graphics Papers: Pushing the Boundaries*; ACM, Inc.: New York, NY, USA, 2023; pp. 441–452.
25. Mildenhall, B.; Srinivasan, P.P.; Tancik, M.; Barron, J.T.; Ramamoorthi, R.; Ng, R. Nerf: Representing scenes as neural radiance fields for view synthesis. *Commun. ACM* **2021**, *65*, 99–106. [CrossRef]
26. Dong, P.; Niu, X.; Wei, Z.; Pan, H.; Li, D.; Huang, Z. AutoRF: Auto Learning Receptive Fields with Spatial Pooling. In *Proceedings of the International Conference on Multimedia Modeling*; Springer: Cham, Switzerland, 2023; pp. 683–694.
27. Chen, Y.; Wu, Q.; Zheng, C.; Cham, T.J.; Cai, J. Sem2NeRF: Converting Single-View Semantic Masks to Neural Radiance Fields. In *Computer Vision–ECCV 2022*; Springer Nature: Cham, Switzerland, 2022; pp. 730–748. [CrossRef]
28. Xu, D.; Jiang, Y.; Wang, P.; Fan, Z.; Shi, H.; Wang, Z. SinNeRF: Training Neural Radiance Fields on Complex Scenes from a Single Image. In *European Conference on Computer Vision*; Springer Nature: Cham, Switzerland, 2022.
29. Barron, J.T.; Mildenhall, B.; Verbin, D.; Srinivasan, P.P.; Hedman, P. Mip-NeRF 360: Unbounded Anti-Aliased Neural Radiance Fields. *arXiv* **2022**. [CrossRef]

30. Kerbl, B.; Kopanas, G.; Leimkühler, T.; Drettakis, G. 3D Gaussian Splatting for Real-Time Radiance Field Rendering. *ACM Trans. Graph.* **2023**, *42*, 139. [CrossRef]
31. Guo, M.H.; Xu, T.X.; Liu, J.J.; Liu, Z.N.; Jiang, P.T.; Mu, T.J.; Zhang, S.H.; Martin, R.R.; Cheng, M.M.; Hu, S.M. Attention mechanisms in computer vision: A survey. *Comput. Vis. Media* **2022**, *8*, 331–368. [CrossRef]
32. Mnih, V.; Heess, N.; Graves, A. Recurrent models of visual attention. In Proceedings of the NIPS'14: Proceedings of the 27th International Conference on Neural Information Processing Systems, Montreal, QC, Canada, 8–13 December 2014; Volume 27.
33. Xu, S.; Cheng, Y.; Gu, K.; Yang, Y.; Chang, S.; Zhou, P. Jointly attentive spatial-temporal pooling networks for video-based person re-identification. In Proceedings of the IEEE International Conference on Computer Vision, Venice, Italy, 22–29 October 2017; pp. 4733–4742.
34. Chen, L.; Zhang, H.; Xiao, J.; Nie, L.; Shao, J.; Liu, W.; Chua, T.S. Sca-cnn: Spatial and channel-wise attention in convolutional networks for image captioning. In Proceedings of the IEEE Conference on Computer Vision and Pattern Recognition, Honolulu, HI, USA, 21–26 July 2017; pp. 5659–5667.
35. Wang, D.; Cui, X.; Chen, X.; Zou, Z.; Shi, T.; Salcudean, S.; Wang, Z.J.; Ward, R. Multi-view 3d reconstruction with transformers. In Proceedings of the IEEE/CVF International Conference on Computer Vision, Montreal, QC, Canada, 10–17 October 2021; pp. 5722–5731.
36. Salvi, A.; Gavenski, N.; Pooch, E.; Tasoniero, F.; Barros, R. Attention-based 3D object reconstruction from a single image. In Proceedings of the 2020 International Joint Conference on Neural Networks (IJCNN), Glasgow, UK, 19–24 July 2020; pp. 1–8.
37. Mescheder, L.; Oechsle, M.; Niemeyer, M.; Nowozin, S.; Geiger, A. Occupancy networks: Learning 3d reconstruction in function space. In Proceedings of the IEEE/CVF Conference on Computer Vision and Pattern Recognition, Long Beach, CA, USA, 15–20 June 2019; pp. 4460–4470.
38. Lin, K.; Wang, L.; Liu, Z. End-to-end human pose and mesh reconstruction with transformers. In Proceedings of the IEEE/CVF Conference on Computer Vision and Pattern Recognition, Nashville, TN, USA, 20–25 June 2021; pp. 1954–1963.
39. Stier, N.; Rich, A.; Sen, P.; Höllerer, T. Vortx: Volumetric 3d reconstruction with transformers for voxelwise view selection and fusion. In Proceedings of the 2021 International Conference on 3D Vision (3DV), London, UK, 1–3 December 2021; pp. 320–330.
40. Xie, H.; Yao, H.; Zhang, S.; Zhou, S.; Sun, W. Pix2Vox++: Multi-scale context-aware 3D object reconstruction from single and multiple images. *Int. J. Comput. Vis.* **2020**, *128*, 2919–2935. [CrossRef]
41. Wu, Z.; Song, S.; Khosla, A.; Yu, F.; Zhang, L.; Tang, X.; Xiao, J. 3d shapenets: A deep representation for volumetric shapes. In Proceedings of the IEEE Conference on Computer Vision and Pattern Recognition, Boston, MA, USA, 7–12 June 2015; pp. 1912–1920.
42. Tatarchenko, M.; Richter, S.R.; Ranftl, R.; Li, Z.; Koltun, V.; Brox, T. What do single-view 3D reconstruction networks learn? In Proceedings of the IEEE Conference on Computer Vision and Pattern Recognition (CVPR), Long Beach, CA, USA, 15–20 June 2019.
43. Paszke, A.; Gross, S.; Massa, F.; Lerer, A.; Bradbury, J.; Chanan, G.; Killeen, T.; Lin, Z.; Gimelshein, N.; Antiga, L.; et al. Pytorch: An imperative style, high-performance deep learning library. In Proceedings of the 33rd International Conference on Neural Information Processing Systems, Vancouver, BC, Canada, 8–14 December 2019; Curran Associates Inc.: Red Hook, NY, USA, 2019; p. 721.
44. Kingma, D.P.; Ba, J. Adam: A method for stochastic optimization. *arXiv* **2014**, arXiv:1412.6980.

Journal of
Imaging

MDPI

Article

Three-Dimensional Reconstruction of Indoor Scenes Based on Implicit Neural Representation

Zhaoji Lin [1,*], Yutao Huang [2] and Li Yao [2]

1 School of Computer Science and Engineering, Sanjiang University, Nanjing 210012, China
2 School of Computer Science and Engineering, Southeast University, Nanjing 211189, China; 172210501215@stu.just.edu.cn (Y.H.); yao.li@seu.edu.cn (L.Y.)
* Correspondence: lin_zhaoji@sju.edu.cn; Tel.: +86-137-7669-3202

Abstract: Reconstructing 3D indoor scenes from 2D images has always been an important task in computer vision and graphics applications. For indoor scenes, traditional 3D reconstruction methods have problems such as missing surface details, poor reconstruction of large plane textures and uneven illumination areas, and many wrongly reconstructed floating debris noises in the reconstructed models. This paper proposes a 3D reconstruction method for indoor scenes that combines neural radiation field (NeRFs) and signed distance function (SDF) implicit expressions. The volume density of the NeRF is used to provide geometric information for the SDF field, and the learning of geometric shapes and surfaces is strengthened by adding an adaptive normal prior optimization learning process. It not only preserves the high-quality geometric information of the NeRF, but also uses the SDF to generate an explicit mesh with a smooth surface, significantly improving the reconstruction quality of large plane textures and uneven illumination areas in indoor scenes. At the same time, a new regularization term is designed to constrain the weight distribution, making it an ideal unimodal compact distribution, thereby alleviating the problem of uneven density distribution and achieving the effect of floating debris removal in the final model. Experiments show that the 3D reconstruction effect of this paper on ScanNet, Hypersim, and Replica datasets outperforms the state-of-the-art methods.

Keywords: 3D reconstruction; indoor scene; neural radiance fields; signed distance function; normal prior; mesh

1. Introduction

The goal of 3D reconstruction of indoor scenes is to reconstruct and restore an accurate scene model from 2D images of indoor scenes from multiple angles to reflect the geometry, structure, and appearance characteristics of the actual scene [1]. This process can obtain an explicit 3D model observed from any perspective. This task has been a hot topic and an important task in computer vision and graphics research in recent years, and has broad application prospects in house interior restoration, interior design, virtual reality, augmented reality, indoor navigation, etc. [2].

Unlike object-level reconstruction, indoor environments usually include large and small objects, different materials, and complex spatial layouts, which puts higher demands on feature extraction and scene understanding. Indoor lighting conditions are also complex and changeable, and may include natural light, artificial light, and shadow areas. These factors will affect the quality of the reconstructed model and the difficulty of reconstruction. In addition, occlusion between objects is more common in indoor scenes, which makes it more difficult to obtain complete scene information from a limited perspective. Existing methods often have poor effects on uneven lighting (as shown in Figure 1a) and partial planar texture processing (as shown in Figure 1b). There are many floating debris noises in the air of the reconstructed indoor 3D model (as shown in Figure 1c). The recently proposed neural radiance fields reconstruction method can achieve good results, but the computation is large, and it cannot directly obtain a 3D mesh model. We strengthen surface learning by adding normal priors and use an SDF, a compact and continuous multi-layer perceptron (MLP), to parameterize the representation of the implicit model, and finally obtain a high-quality 3D mesh model. In summary, this paper has the following main contributions:

(1) It proposes a new indoor 3D reconstruction method that combines NeRF and SDF scene expression, which not only preserves the high-quality geometric information of the NeRF, but also uses the SDF to generate an explicit mesh with a smooth surface.
(2) By adding adaptive normal priors to provide globally consistent geometric constraints, the reconstruction quality of planar texture areas and details is significantly improved.
(3) By introducing a new regularization term, the problem of uneven distribution of NeRF density is alleviated, and the effect of removing floating debris is achieved in the final generated model, which improves the look and feel of the visualization results.

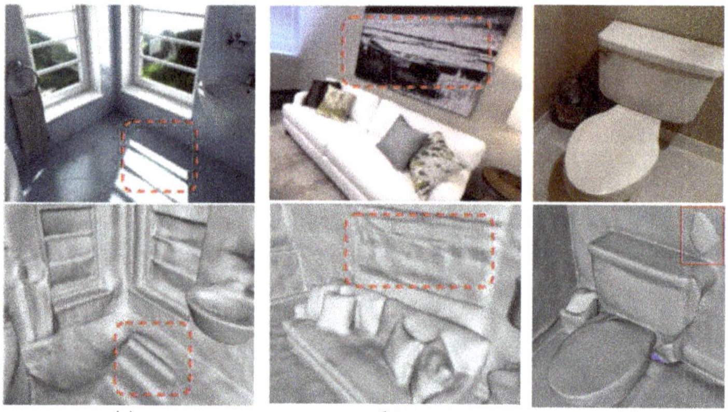

Figure 1. (**a**) Distortion of reconstructed 3D models under uneven lighting conditions enclosed by the red dashed box; (**b**) distortion of 3D reconstruction of smooth planar texture areas enclosed by the red dashed box; (**c**) floating debris noise in red box in 3D reconstruction.

2. Related Works

2.1. Three-Dimensional Reconstruction of Based on Visual SLAM

Simultaneous Localization and Mapping (SLAM) refers to the process of a moving object carrying a sensor to locate itself during movement and to synchronously map the surrounding environment in an appropriate manner. In 2016, Google open-sourced an indoor SLAM library called Cartographer [3], which is still being updated. Its main application area is indoor reconstruction, providing functions such as positioning, mapping, and loop detection. LOAM [4] is a SLAM algorithm based on a 3D laser sensor. Compared with Cartographer, it can solve indoor and outdoor problems, but it does not have loop detection. Later, based on LOAM, researchers proposed LeGOLOAM [5], a new algorithm derived from the LOAM framework. By introducing a global map and loop detection module, the positioning accuracy and robustness are improved, and at the same time, the entire algorithm is made more lightweight. The advantage of the SLAM system is its fast reconstruction speed. With the improvement of the robustness and accuracy of its algorithm, it makes real-time 3D reconstruction possible. However, due to the need for sensor participation, the reconstruction cost is high, and the process is relatively complicated.

2.2. Three-Dimensional Reconstruction Based on TSDF

The main idea of the 3D reconstruction method based on regression to a truncated signed distance function (TSDF) is to create a voxel grid, initialize the maximum truncated signed distance value for each voxel, fuse the depth map into the voxel through the TSDF [6], calculate the distance from each point to the center of each voxel in the voxel grid, and update the TSDF value of the voxel if the distance is within the truncation range. KinectFusion [7] uses the depth camera Kinect to scan and model indoor spaces and objects in real time, and uses a TSDF to manage spatial information. This method improves the efficiency of data processing and the accuracy of the reconstruction process. Atlas [8] provides an end-to-end reconstruction pipeline, using a 2D CNN to extract features from each image independently, and then using the intrinsic and extrinsic features of the camera to back-project and accumulate into voxel volumes. After accumulation, a 3D CNN refines the accumulated features and predicts the TSDF value. At the same time, the semantic segmentation goal is added to the model to accurately mark the generated surface, which improves the model's ability to handle occlusion and large room scenes. In order to reduce the computational burden, unlike Atlas, which processes the entire image sequence at once, NeuralRecon [9] proposed a coarse-to-fine framework that uses a recursive network to fuse features from previous fragments and reconstructs the entire scene by processing the local surface of each fragment sequence, making progress on datasets with high occlusion and scene complexity. However, due to its design idea of local estimation, TSDF-based 3D reconstruction methods have difficulty obtaining global reconstruction with fine details.

2.3. Three-Dimensional Reconstruction Based on MVS

The traditional multi-view stereo (MVS) method first estimates the depth map of each image based on multiple views, and then performs depth fusion to obtain the final reconstruction result. These methods can reconstruct relatively accurate 3D shapes and have been used in many downstream applications, such as new view synthesis. Schoenberger proposed a new general image 3D reconstruction method, which uses scale-invariant feature transform (SIFT) features and the Fast Library for Approximate Nearest Neighbors (FLANN) matching algorithm to improve the accuracy and efficiency of Structure from Motion (SFM), and open-sourced the project as COLMAP 3.10 [10] software for enthusiasts to use, which can perform dense point cloud reconstruction and surface reconstruction. Based on this, they further proposed the Pixelwise View Selection [11] method, which improved the poor reconstruction effect and efficiency caused by the previous input image specifications (such as different input image sizes, different lighting, etc.). This method regards the view selection problem as a binary classification problem, selects the best view for each pixel, and thus can use the information of all perspectives for better stereo matching. However, these methods perform poorly in large planar texture areas or areas with sparse textures because their optimization is highly dependent on the photometric process. In indoor scenes with planar texture areas, the inherent uniformity makes photometry ineffective, making it difficult to accurately estimate depth [12].

With the development of deep learning, learning-based MVS methods [13–15] have shown good performance in recent years. Mvsnet [16] uses convolutional neural networks to predict depth maps, integrates image information from multiple perspectives into a unified 3D space, and implements depth estimation by constructing a 3D matching cost-volume, which significantly improves the accuracy of depth estimation. Fast-Mvsnet [17] improves on Mvsnet by adopting a more efficient network structure and optimizing the depth map prediction process. By simplifying the representation of cost volume and adopting a lighter network architecture, the processing speed is improved, and the memory consumption is reduced. DeepV2D [18] combines the temporal information of video frames with visual depth estimation and adopts a two-stage network architecture. It first performs motion estimation on the video sequence, and then combines motion information and visual features to predict the depth map of each frame, further improving the quality of depth prediction. DeepMVS [19] uses a deep neural network to preprocess the input image, extract features, and predict the depth information of each pixel. It introduces a novel image warping and synthesis step to improve the consistency between views. UCS-Net [20] constructs the cost volume in a coarse-to-fine hierarchical manner to obtain higher resolution depth estimation. Although the MVS method based on deep learning has made great progress, it still faces some problems. Since the depth map is estimated separately for each view, there are often some geometric inconsistencies and scale ambiguities, resulting in holes and noise on the surface of the reconstructed result [21]. When encountering areas with relatively scarce textures, these models find it difficult to accurately predict depth information.

2.4. Three-Dimensional Reconstruction Based on Implicit Neural Networks

Coordinate-based implicit neural networks, which encode a field by regressing 3D coordinates to output values via an MLP, have become a popular approach for representing scenes due to their compactness and flexibility. The Dist [22] model proposed a method to learn 3D shapes from 2D images. IDR [23] models rely on view appearance and can be applied to non-Lambertian surface reconstruction. However, they require mask information to obtain reconstructions. NeRFs encode scene geometry via volume density and are suitable for the task of novel view synthesis for volume rendering. However, due to the lack of surface constraints, volume density cannot represent high-fidelity surfaces. Inspired by neural radiance fields, Neus [24] and VolSDF [25] attached volume rendering techniques to IDR and eliminated the need for mask information. Although these methods achieve stunning reconstruction results in small-scale, texture-rich scenes, they often perform poorly in large-scale indoor scenes with planar texture areas. Mip-NeRF360 [26] improves upon the original NeRF's problems of unbalanced details and proportions at near and far distances, as well as the limited nature of synthesized scenes, and can render unbounded scenes more realistically. To address the slow convergence of the original NeRF training process, NSVF [27] uses a sparse voxel octree to assist in spatial modeling, achieving significant improvements in training time. The traditional NeRF requires input from multiple views to estimate volume representations. If multi-view data are insufficient, the generated scene can easily collapse into a plane. To address this problem, Google researchers proposed LOLNeRF [28], which can train a NeRF model from a single viewpoint for the same type of object, without adversarial supervision, thereby enabling a single 2D image to generate a 3D model. All of the above are advances made by NeRFs in synthesizing new viewpoints. In recent years, researchers have gradually shifted their focus to using NeRFs for 3D reconstruction. Guy [29] et al. applied a NeRF to facial reconstruction, achieving the generation of high-quality 3D facial models from a single RGB image. By introducing the time dimension, this method can model and reconstruct the dynamic changes in facial shape and expression. However, all of the above NeRF studies are limited to object-level reconstruction. When the reconstructed objects are expanded to indoor scenes, the generated 3D models often contain a lot of noise and topological errors.

3. Methodology

This paper proposes a method for indoor scene 3D reconstruction that combines NeRF and SDF implicit expression. The volume density of the NeRF is used to provide geometric information for the SDF field, and the learning process is optimized by adding normal priors to strengthen the learning of geometric shapes and surfaces. The overall framework of the method proposed in this paper is shown in Figure 2. Multiple 2D images of indoor scenes are used as input, and the explicit 3D model mesh of the corresponding scene is output. This method mainly includes the following three modules:

1. Normal estimation module: This module uses a spatial rectifier-based method to generate the corresponding normal map for a single RGB image, and prepares data for the prior part of neural implicit reconstruction.
2. NeRF module: The appearance decomposition and feature processing of the scene image are performed through the neural radiant field, and the volume density and

color are obtained. The image under the corresponding perspective is obtained by volume rendering, and the MLP parameters are optimized inversely with the input image loss.

3. SDF field module: The purpose of this module is to learn a high-quality SDF from the network, and at the same time, strengthen the network's understanding of the geometric structure through the normal prior. The implicit-3D-expression SDF is converted into an explicit triangular mesh through the Marching Cubes algorithm.

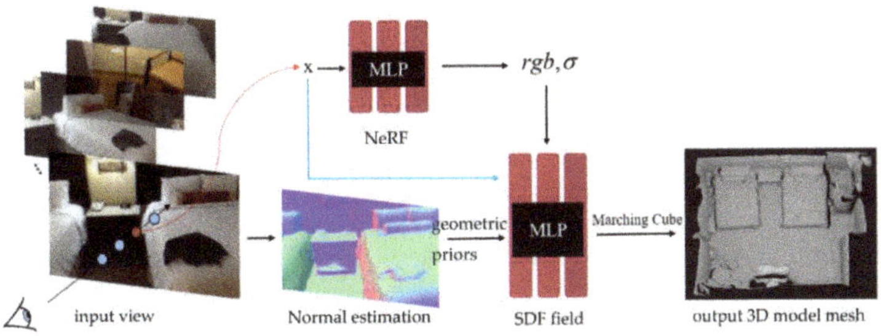

Figure 2. Overall framework of the method.

3.1. Optimization of Indoor 3D Reconstruction Based on Adaptive Normal Prior

A normal map is an important type of image, which represents the surface normal direction of each pixel in the image through the color of the pixel. In 3D graphics and computer vision, the normal is a vector perpendicular to the surface of an object, which can be used to describe the direction and shape of the surface. In the normal map, RGB color channels are generally used to represent the X, Y, and Z components of the normal vector, respectively. Normal information helps to correct errors in the reconstruction process and improve the reconstruction quality.

Currently, many monocular normal estimation methods have achieved high accuracy under a clear image input. However, considering that indoor scene images often have small blur and tilt, this paper selects TiltedSN [30] as the normal estimation module. Because the estimated normal map is usually over-smoothed, there are problems of inaccurate estimation on some fine structures, such as the chair legs in Figure 3a. Therefore, we adopt an adaptive method to use normal priors, using a mechanism based on the consistency of multiple views of the input image to evaluate the reliability of the normal prior. As shown in Figure 3b, this process is also called geometric checking. For areas that do not meet the consistency of multiple images, the normal prior is not applied. Instead, the appearance information is used for optimization to avoid the negative effects of incorrect normal maps that lead to misjudgment in reconstruction.

Given a reference image I_i, evaluate the consistency of the surface observed from pixel q. Define a local 3D plane $\{p | p^T n = dv^T n\}$ in the camera space associated with q, where v is the viewing direction, d is the distance to pixel q, and n is the normal estimate. Next, find a set of adjacent images, assuming that one of the adjacent images is I_j. The homography change from I_i to I_j can be calculated by the following formula:

$$H_{n,d} = K_j \left(R_j R_i^{-1} - \frac{(t_i - t_j) n^T}{dv^T n} \right) K_i^{-1} \quad (1)$$

where $\{K_*, R_*, t_*\}$ is the intrinsic parameter matrix, the rotation and translation camera parameters. For pixel q in I_i, find a square block P centered on it as the neighborhood, and warp the block to its adjacent view I_j using the calculated homography matrix. The block matching method (patchmatch) can be used to find similar image blocks on adjacent views, and the normalized cross-correlation (NCC) method is used to evaluate the visual consistency of (n, d). NCC is a method for measuring the similarity between two images. It evaluates the similarity between the two images by calculating the degree of correlation between them. Compared with simple cross-correlation, normalized cross-correlation is insensitive to changes in brightness and contrast, so it is more reliable in practical applications. The applied NCC formula is as follows:

$$NCC_j(P,n) = \frac{\sum_{q\in P} \hat{I}_i(q)\hat{I}_j(H_{n,d}(q))}{\sqrt{\sum_{q\in P} \hat{I}_i(q)^2 \sum_{q\in P} \hat{I}_j(H_{n,d}(q))^2}} \quad (2)$$

where $\hat{I}_*(q) = I_*(q) - \overline{I}_*(q)$, $I_i(q)$ and $I_j(q)$ represent two image regions to be compared, $\overline{I}_*(q)$ is the average value of $I_*(q)$, $\hat{I}_*(q)$ is the difference between them; the numerator calculates the sum of the products of the differences between the two image blocks, and the denominator calculates the square root of the product of the sum of the squares of the differences between the two image blocks. This process ensures the normalization of the results, making the NCC range within $(-1, 1)$. The closer the NCC value is to 1, the more similar the two image regions are; the closer the NCC value is to -1, the less similar they are; the closer the NCC value is to 0, the less obvious linear relationship there is between them.

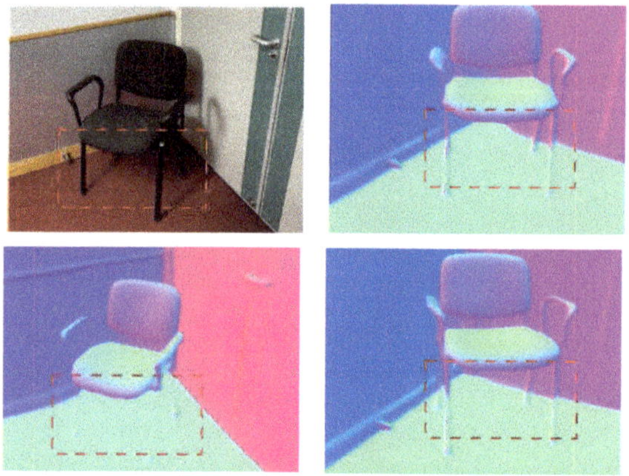

(a) (b)

Figure 3. (**a**) The normal estimation is inaccurate in some fine structures enclosed by red dashed box, such as chair legs, based on TiltedSN normal estimation module; (**b**) we use an adaptive normal prior method to derive accurate normals based on the consistency of adjacent images. In the red dashed box, the fine structures are accurately reconstructed.

If the reconstructed geometry is not accurate at the sampling pixel, it cannot meet the multi-view photometric consistency, which means that its related normal prior cannot provide help for the overall reconstruction. Therefore, a threshold ϵ is set, and by comparing the NCC at the sampling block with ϵ, the following indicator function can be used to adaptively determine the training weight of the normal prior.

$$\Omega_q(\hat{n}) = \begin{cases} 1 & if \sum_j NCC_j(P,\hat{n}) \geq \epsilon \\ 0 & if \sum_j NCC_j(P,\hat{n}) < \epsilon \end{cases} \quad (3)$$

The normal prior is used for supervision only when $\Omega_q(\hat{n}) = 1$. If $\Omega_q(\hat{n}) = 0$, the normal prior of the region is considered unreliable and will not be used in subsequent optimization processes.

3.2. Neural Implicit Reconstruction

Since the NeRF cannot express the surface of the object well, we are looking for a new implicit expression. The SDF can represent the surface information of objects and scenes and achieve better surface reconstruction. Therefore, we choose the SDF as the implicit scene representation.

3.2.1. Scene Representation

We represent the scene geometry as an SDF. An SDF is a continuous function f that, for a given 3D point, returns the distance from that point to the nearest surface:

$$f : \mathbb{R}^3 \to \mathbb{R} \qquad x \mapsto s = f(x) \tag{4}$$

Here, x represents a 3D point and s is the corresponding SDF value, thus completing the mapping from a 3D point to a signed distance. We define the surface S as the zero-level set of the SDF, expressed as follows:

$$S = \{x | f(x) = 0\} \tag{5}$$

By using the Marching Cubes algorithm on the zero horizontal plane of the SDF, that is, the surface of the object or scene, we can obtain a 3D mesh with a relatively smooth surface.

Using the SDF to get the mesh has the following advantages:

(1) Clear surface definition: The SDF provides the distance to the nearest surface for each spatial point, where the surface is defined as the location where the SDF value is zero. This representation is well suited for extracting clear and precise surfaces, making the conversion from SDF to mesh relatively direct and efficient.
(2) Geometric accuracy: The SDF can accurately represent sharp edges and complex topological structures, which can be maintained when converted to meshes, thereby generating high-quality 3D models.
(3) Optimization-friendly: The information provided by the SDF can be directly used to perform geometry optimization and mesh smoothing operations, which helps to further improve the quality of the model when generating the mesh.

3.2.2. Implicit Indoor 3D Reconstruction Based on Normal Prior

The reconstruction process based on the normal prior is shown in Figure 4. The input mainly consists of three parts.

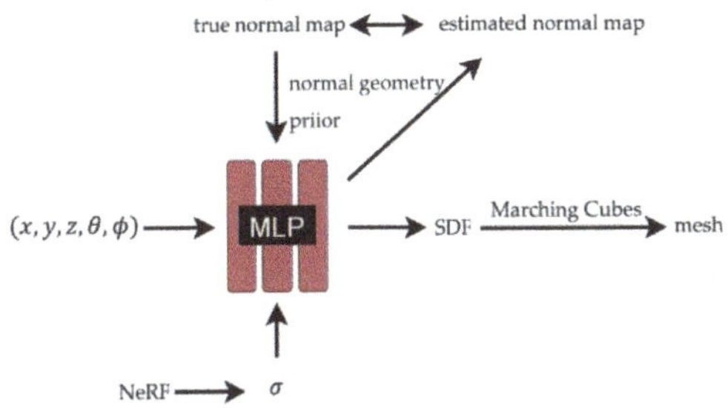

Figure 4. Neural implicit reconstruction process.

The first part is a five-dimensional vector representing the information of the sampling point (x, y, z, θ, ϕ). The second part is the volume density we obtained previously through the NeRF σ. Given that we use the SDF as a surface expression, the volume density obtained by the NeRF can represent more comprehensive scene geometry information. Because the scene contains multiple objects and air, it is difficult to represent a complex scene only through surface information. Traditional SDF-based methods are often limited to object reconstruction. The constraints of NeRF volume density combined with SDF reconstruction can reconstruct richer scene geometry. The third part is the normal geometry prior. The normal map here is obtained by the monocular normal estimation method mentioned earlier. The normal map provides the orientation information of the object surface, which is conducive to enhancing detail reconstruction.

The network used here is an improved NeRF, which also contains 2 MLPs. Like the NeRF, there is a color network, f_c, and the other grid has become an SDF network, f_{θ_g}, which can get the SDF value of the point through the 3D coordinates of the point.

Here, we will use volume rendering technology to get the predicted image, and optimize it with the input real image through the loss function. Specifically, for each pixel, we sample a set of points along the corresponding emission light, denoted as $p_i = o + d_i v$, where p_i is the sampling point, o is the camera center, and v is the direction of the light. The color value can be accumulated by the following volume rendering formula.

$$\hat{c} = \sum_{i=1}^{n} T_i \alpha_i c(p_i, v) \tag{6}$$

where $T_i = \prod_{j=1}^{i-1}(1-\alpha_j)$ is the cumulative transmittance, i.e., the probability that with no object occlusion, c is the color value, $\alpha_i = 1 - \exp\left(-\int_{t_i}^{t_{i+1}} \rho(t)dt\right)$ is the discrete opacity, and $\rho(t)$ is the opacity corresponding to the volume density σ in the original NeRF. Since the rendering process is fully differentiable, we learn the weights of f_c and f_{θ_g} by minimizing the difference between the rendering result and the reference image.

In addition to generating the appearance, the above volume rendering scheme can also obtain the normal vector. We can approximate the surface normal observed from a viewpoint by volume accumulation along this ray:

$$\hat{n} = \sum_{i=1}^{n} T_i \alpha_i n_i \tag{7}$$

where $n_i = \nabla f(p_i)$ is the gradient at point p_i.

At this time, we compare the true normal map obtained by the monocular method with the estimated normal map obtained by the volume rendering process, and further optimize the parameters of the MLP by calculating the loss to obtain a more accurate normal map and geometric structure.

3.3. Floating Debris Removal

The NeRF initially acts on the generation of new perspectives on objects, maintaining a high degree of clarity and realism. This is mainly due to volume rendering. However, when the NeRF is used for 3D reconstruction, some floating debris in the air often appears. This floating debris refers to small disconnected areas in the volume space and translucent substances floating in the air. In view synthesis work, this floating debris is often not easy to detect. However, if an explicit 3D model needs to be generated, this floating debris will seriously affect the quality and appearance of the 3D model. Therefore, it is very necessary to remove the floating debris in these incorrectly reconstructed places.

This floating debris often does not appear in object reconstruction. However, in scene-level reconstruction, due to the significant increase in environmental complexity and the lack of relevant constraints on the NeRF, there is a phenomenon of inaccurate local area density prediction. Therefore, this paper proposes a new regularization term to constrain the weight distribution of the NeRF.

First, simulate the sampling process of the NeRF. In a scene, assume that there are only rigid objects, excluding the existence of translucent objects. After a ray is shot out, there will be countless sampling points on this ray. The weight value of the sampling point before the ray encounters the object should be extremely low (close to 0). When the ray contacts the rigid object, the weight value here should soar, much higher than other values. After the object, the weight value returns to a lower range. This is the desired weight distribution in an ideal state, which is a relatively compact unimodal distribution. This method defines a regularization term, which is a step function defined by a set of standardized ray distances s and the weight w after parameterizing each ray:

$$L_{dist}(s,w) = \iint_{-\infty}^{\infty} w_s(u)w_s(v)|u-v|d_u d_v \quad (8)$$

Here, u and v refer to points on the sampling ray, that is, points on the x-axis in the weight distribution diagram, $|u-v|$ is the distance between the two points, and $w_s(u)$ and $w_s(v)$ are the weight values at point u and point v, respectively. Since all particle combinations from negative infinity to positive infinity need to be exhausted, integration is performed in the front. If you want to make the loss function as small as possible, there are mainly two situations:

(1) If the distance between point u and point v is relatively far, that is, the value of $|u-v|$ is large, if you want to ensure that the value of $L_{dist}(s,w)$ is as small as possible, then either $w_s(u)$ or $w_s(v)$ needs to be small and close to zero. That is, as shown in Figure 5, (A, B), (B, D), (A, D), etc. all satisfy that $|u-v|$ is large and the weight value of at least one point is extremely small (close to zero);

(2) If the values of $w_s(u)$ and $w_s(v)$ are both large, if you want to ensure that the value of $L_{dist}(s,w)$ is as small as possible, then the value of $|u-v|$ needs to be small, that is, the distance between points u and v is very close. As shown in Figure 5, only the combination of (B, C) satisfies the condition that the values of $w_s(u)$ and $w_s(v)$ are both large, and at this time, points u and v just meet the condition that the distance is very close.

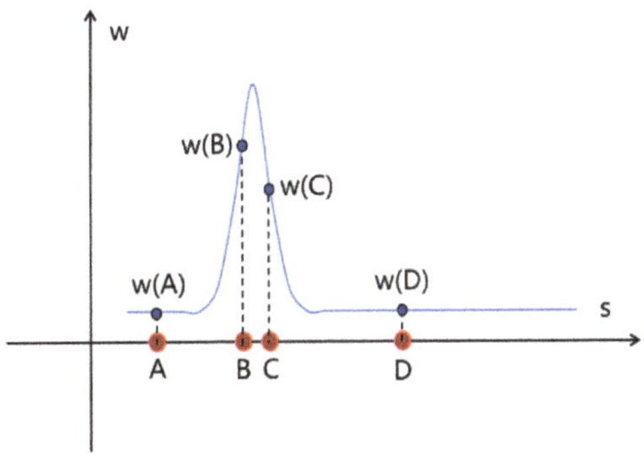

Figure 5. Distribution diagram of distance and weight values between sampling points.

Therefore, through the analysis of the above two cases, it can be found that the properties of the regularization term can constrain the density distribution to the ideal single-peak distribution with the good central tendency proposed before. The purpose of this regularization term is to minimize the sum of the normalized weighted absolute values of all samples along the ray, and encourage each ray to be as compact as possible, which is specifically reflected in the following steps:

1. Minimize the width of each interval;
2. Bring the intervals that are far apart closer to each other;
3. Make the weight distribution more concentrated.

The regularization term above cannot be used directly for calculation because it is in integral form. In order to facilitate calculation and use, it is discretized as the following:

$$L_{dist}(s,w) = \sum_{i,j} w_i w_j \left| \frac{s_i + s_{i+1}}{2} - \frac{s_j + s_{j+1}}{2} \right| + \frac{1}{3} \sum_i w_i^2 (s_{i+1} - s_i)$$

This discretized form also provides a more intuitive understanding of the behavior represented by this regularization, where the first term minimizes the weighted distance between points in all intervals, and the second term minimizes the weighted size of each individual interval.

3.4. Training and Loss Function

During the training phase, we sample a batch of pixels and adaptively minimize the difference between the color and normal estimates and the true normal map. We sample m pixels $\{q_k\}$ and their corresponding reference colors $\{I(q_k)\}$ and normals $\{N(q_k)\}$ in each iteration. For each pixel, we sample n points in the world coordinate system along the corresponding ray, and the total loss is defined as

$$L = \lambda_c L_c + \lambda_n L_n + \lambda_e L_{eik} + \lambda_d L_{dist} \quad (9)$$

Among them, λ_c, λ_n, λ_e, and λ_d are the hyperparameters of color loss, normal loss, Eikonal loss, and distortion loss, respectively.

Color loss, L_c, is used to measure the difference in color between the reconstructed image and the real image:

$$L_c = \frac{1}{m} \sum_k \| I(q_k) - \hat{c}(q_k) \| \quad (10)$$

In the training phase, a batch of pixels need to be sampled. Each iteration samples m pixels $\{q_k\}$ and the corresponding reference color $\{I(q_k)\}$, where $\hat{c}(q_k)$ represents the pixel color predicted by volume rendering.

The normal prior loss L_n is to render the reconstructed 3D mesh of the indoor scene as a normal map, and compare it with the real normal map generated by the monocular method to obtain a loss:

$$L_n = \sum_k \| N(q_k) - \hat{n}(q_k) \|_1 \cdot \Omega_{q_k}(\hat{n}(q_k)) \quad (11)$$

where $N(q_k)$ is the true normal information, and $\hat{n}(q_k)$ is the normal information predicted by the gradient. $\Omega_{q_k}(\hat{n}(q_k))$ is an indicator function used to judge the accuracy of the normal prior. Here, some data with inaccurate normal estimation are eliminated. The normal loss is mainly calculated by cosine similarity, because the direction of the normal vector is more important than its length. Cosine similarity is a good measure of the similarity of the two vectors in direction.

The Eikonal loss L_{eik} [31] of the regularized SDF is

$$L_{eik} = \sum_{x \in X} \left(\| \nabla f_\theta(x) \|_2 - 1 \right)^2 \quad (12)$$

where X is a set of sampling points in the 3D space and the area near the surface, and x represents one of the sampling points. The reason why the gradient $\nabla f_\theta(x)$ needs to be close to 1 is that the ideal SDF represents the shortest distance from the current point to the surface, so the gradient direction is the direction in which the distance field change is the steepest. Assuming that x moves toward the surface along this direction by Δd, the SDF should also change by Δd. Therefore, in the ideal state, $\nabla f_\theta(x) = \partial D(x) = \Delta D(x)/\Delta x = \Delta d/\Delta d = 1$, where $D(x)$ is the original Eikonal equation. By introducing the Eikonal loss, the properties of the SDF can be well constrained, thereby ensuring the smoothness and continuity of the reconstructed surface.

4. Experimentation

4.1. Dataset

This paper conducts experimental analysis on the ScanNet [31], Hypersim [32], and Replica [33] datasets, as shown in Table 1. The ScanNet dataset is the main dataset used for indoor scene 3D reconstruction tasks. Our method selects 10 scenes from the ScanNet dataset. For each scene, a set of equally spaced images (about 150–600 images) are sampled from the corresponding video and the images are adjusted to a resolution of 640 × 480. In addition, a scene is selected from the Hypersim dataset and the Replica dataset to test the generalization of this method in large-scale scenes. The results verify that this method has good reconstruction effects on other datasets in addition to the large public dataset ScanNet.

Table 1. Selected datasets to test our method.

Dataset	Scene Number	Scenes Selected in This Paper
ScanNet	1500+	10
Hypersim	461	10
Replica	18	10

4.2. Comparative Experiment

Five evaluation indicators are used: accuracy (Acc), completeness (Comp), precision (Prec), recall (Recall), and F1 score (F-score).

Accuracy is an indicator to measure the degree of consistency between the reconstructed mesh and the real scene mesh:

$$Acc = mean_{p \in P}\left(\min_{p^* \in P^*} \parallel p - p^* \parallel\right) \tag{13}$$

where P represents the set of points in the reconstructed grid, P^* is the set of points in the grid of the real scene, p is a point in the set P, and p^* is a point in the set P^*. $\parallel p - p^* \parallel$ represents the Euclidean distance between P and P^*. For each point p in P, find the point p^* in P^* that is closest to it, calculate the distance between them, and average all the distances.

Completeness is used to measure the extent to which the reconstructed model covers the original model or scene:

$$Comp = mean_{p^* \in P^*}\left(\min_{p \in P} \parallel p - p^* \parallel\right) \tag{14}$$

This metric measures completeness by calculating the average distance from each point in the true scene mesh P^* to the nearest point in the reconstructed mesh P.

Precision is a measure of the proportion of the correctly reconstructed part of the reconstructed model to the entire model. The calculation formula for precision is

$$Prec = mean_{p \in P}\left(\min_{p^* \in P^*} \parallel p - p^* \parallel < 0.05\right) \tag{15}$$

For each point p in the reconstructed mesh P, find the point p^* in the GT mesh P^* that is closest to it, calculate the distance between them, and compare it with the set threshold 0.05 to calculate the proportion less than 0.05.

Recall measures the proportion of points in the GT model that are covered and correctly reconstructed by the reconstruction model:

$$Recall = mean_{p^* \in P^*}\left(\min_{p \in P} \parallel p - p^* \parallel < 0.05\right) \tag{16}$$

For each point p^* in the reconstructed mesh P^*, find the point p in the GT mesh P that is closest to it, calculate the distance between them, and compare it with the previously set threshold of 0.05 to calculate the proportion less than 0.05.

The F1 score (F-score) is the harmonic mean of precision and recall, which can better measure some unbalanced datasets and provide a more comprehensive evaluation of the overall model:

$$F - score = \frac{2 \times Prec \times Recall}{Prec + Recall} \tag{17}$$

The architecture of the MLP encoding the SDF in our method consists of eight hidden layers of 256 channels. The training process and related parameters of the model are set as follows: the number of iterations is set to 160,000, the learning rate is set to 4×10^{-4}, the number of rays sampled in each batch of training (batchsize) is set to 1024, and each ray contains 64 coarse sampling points and 64 fine sampling points. In the patchmatching process, the patch size is set to 11×11, the step size is set to 2 for the block matching process, and the NCC threshold ϵ is 0.66. The weights of each loss function, λ_c, λ_n, λ_e and λ_d, are set to 1.0, 1.0, 0.1, and 0.5, respectively.

As shown in Table 2, the method in this paper is significantly better than the state-of-the art methods, and performs well in most indicators, far exceeding the traditional MVS method and the TSDF-based reconstruction method.

Table 2. Quantitative comparison with other methods.

Method	**Acc** ↓	**Comp** ↓	**Prec** ↑	**Recall** ↑	**F-Score** ↑
COLMAP [10]	0.047	0.235	0.711	0.441	0.537
Atlas [8]	0.211	0.070	0.500	0.659	0.564
NeuralRacon [11]	0.056	0.081	0.545	0.604	0.572
Neus [24]	0.179	0.208	0.313	0.275	0.291
VolSDF [25]	0.414	0.120	0.321	0.394	0.346
NeRF [34]	0.735	0.177	0.131	0.290	0.176
Manhattan-SDF [35]	0.072	0.068	0.621	0.586	0.602
MonoSDF [36]	**0.035**	**0.048**	<u>0.799</u>	<u>0.681</u>	<u>0.733</u>
I2-SDF [37]	0.066	0.070	0.605	0.575	0.590
Ours	<u>0.037</u>	**0.048**	**0.801**	**0.702**	**0.748**

Bold text indicates the best results and the underlined text indicates the second best results. ↓ indicates that the smaller the value of this indicator, the better; ↑ indicates that the larger the value of this indicator, the better.

Among the neural implicit reconstruction methods, Neus and VolSDF mainly focus on object-level reconstruction, so they perform poorly on the ScanNet dataset; the first-generation NeRF is mainly used for new perspective synthesis, and no algorithm for extracting grids is proposed. Therefore, direct use of Marching Cubes to extract grids will result in large-scale collapse. This article only draws on many modules of the NeRF, so it is compared here; the I2-SDF method has good results in the synthetic dataset proposed in this article, but it does not have generalization ability and does not work well on the ScanNet dataset with more noise and motion blur.

MonoSDF and Manhattan-SDF are second only to this method in terms of quantitative results. Manhattan-SDF is based on the Manhattan World hypothesis, which assumes that the world is mainly composed of planes aligned with the coordinate axes. It is easily affected by the complexity of indoor scenes. Strict reliance on planes aligned with the coordinate axes for reconstruction may lead to missing or inaccurate details. Therefore, the overall accuracy is lower than this method. MonoSDF has achieved good performance in various indicators, especially accuracy. This is because MonoSDF weakens the reconstruction of complex structures that are difficult to handle during the reconstruction process. Therefore, the reconstructed part is highly overlapped with the scene itself, but some parts of the reconstruction will be lost. Therefore, compared with other evaluation indicators, there is some gap between its recall rate and this method, because the recall rate measures the ratio of the real model that is reconstructed.

In addition to quantitative analysis, this chapter visualizes the reconstruction results and makes qualitative comparisons with the most advanced MonoSDF and Manhattan-SDF. As shown in Figure 6a, compared with GT, this method fills some holes that were not scanned at the time. Compared with Manhattan-SDF, this method has higher accuracy and a smoother reconstruction effect on the surface of objects. Compared with MonoSDF, this method has more accurate reconstruction in many structures (shown by the red dashed line). The specific details in the red dashed box in Figure 6a are shown in Figure 6b. Manhattan-SDF has low reconstruction accuracy. After zooming in on the detail image, larger triangular facets can be seen. From the visualization point of view, the details are far inferior to those of this method. MonoSDF has better overall detail reconstruction—especially, the wall area is relatively smooth—but there are many missing objects. In the right picture of Figure 6b, MonoSDF did not correctly reconstruct the lamp on the table, and only reconstructed a small part of the outline of the objects placed on the bookshelf, while this method restored these details of the real scene well. In the left picture of Figure 6b, MonoSDF and Manhattan-SDF are largely missing chair legs and armrests, while the method in this paper restores the chair structure in the real scene to a large extent.

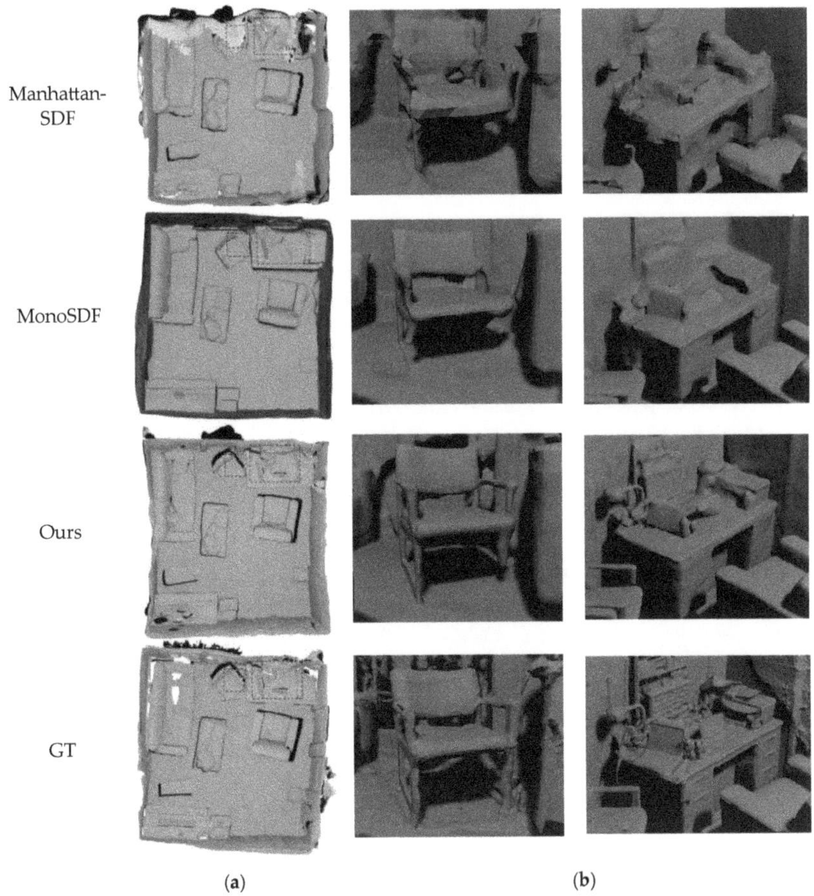

Figure 6. Three-dimensional model reconstructed from scenes in the ScanNet dataset. (**a**) Comparison of 3D models; (**b**) comparison of the specific details in the red dashed box.

4.3. Ablation Experiment

4.3.1. Normal Geometry Prior Ablation Experiment

This section will demonstrate the effectiveness of the adaptive normal prior scheme added in this paper through quantitative and qualitative ablation experimental analysis. There are three settings in this ablation experiment: (1) no normal prior is added, denoted as w/o N; (2) a normal prior is added, but the adaptive scheme is not used, denoted as w/N, w/o A; (3) the adaptive scheme is used with the normal prior, denoted as Ours. All settings are tested on the ScanNet dataset, Hypersim dataset, and Replica dataset. The evaluation indicators of each setting are shown in Table 3. It can be seen that the reconstruction quality can be significantly improved by adding the geometric prior, because it provides additional geometric constraints and alleviates many ambiguity problems caused by the lack of texture information in the original image. On this basis, the adaptive scheme can remove the incorrectly estimated normal map and further improve the reconstruction quality.

Table 3. Normal geometry prior ablation experiment.

Method	Acc ↓	Comp ↓	Prec ↑	Recall ↑	F-Score ↑
w/o N	0.183	0.152	0.286	0.290	0.284
w/N, w/o A	0.050	0.053	0.759	0.699	0.727
Ours	**0.037**	**0.048**	**0.805**	**0.709**	**0.753**

Bold text indicates the best results. ↓ indicates that the smaller the value of this indicator, the better; ↑ indicates that the larger the value of this indicator, the better.

The following will verify the effectiveness of the adaptive normal prior module used in this section through qualitative visualization results. This module provides a certain improvement in the reconstruction of all indoor scenes, but the improvement in thin structure areas and reflective areas is particularly obvious, so these two areas are selected as visualization displays. As shown in Figure 7, a qualitative analysis of the chair leg structure is performed. Figure 7a is the input RGB image, i.e., the reference image, and Figure 7b is the model reconstructed without using the normal prior. The overall reconstruction accuracy is not high, which is reflected in the fact that the reconstruction effect of the surrounding floor and walls is not flat enough, and the surface of the chair is not smooth enough. Figure 7c adds the normal prior, but does not use the adaptive scheme. Although the reconstruction effect of the wall and chair surface is improved, the chair leg part is not reconstructed due to the wrong normal estimation. Figure 7d is the method used in this paper, that is, the adaptive scheme uses the normal prior, which not only greatly improves the accuracy, but also completely reconstructs the thin structure areas such as the chair leg.

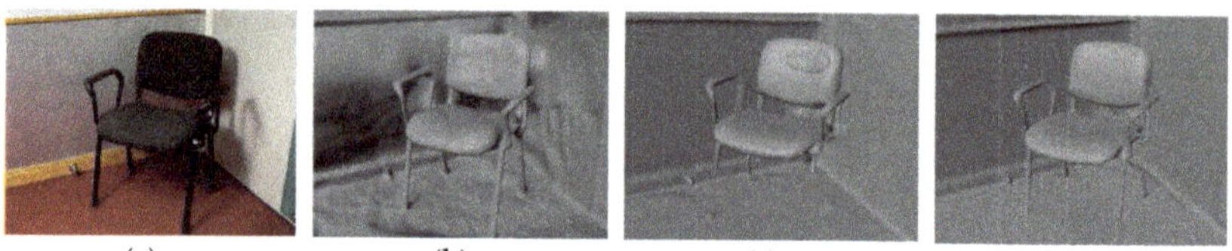

(a) (b) (c) (d)

Figure 7. Qualitive comparison for thin structure areas using ScanNet dataset: (**a**) reference image; (**b**) model reconstructed without using normal prior; (**c**) model reconstructed with normal prior and without adaptive scheme; (**d**) model reconstructed with normal prior and adaptive scheme.

In addition to having a good visual improvement effect in thin structure areas, the adaptive normal prior scheme is also effective in some areas due to lighting or specular reflection. The image shown in Figure 8a is a reference image. The sunlight projected from the window forms regular white light and shadows on the ground. These light and shadows usually affect the reconstruction of the flatness of the entire ground because the network will understand them as independent geometric structures. Figure 8b is a model reconstructed without using the normal prior. A more obvious step structure is reconstructed in the light and shadow area, which is not the result of correct reconstruction. The reconstruction granularity in other areas is also obviously insufficient. Figure 8c adds the normal prior but does not use the adaptive scheme. Since most of the normal priors have weakened the erroneous impact of this area on the reconstruction, the erroneous outline of this area is already vague and not as obvious as in Figure 8b. However, there are also a few normal maps closer to the light and shadow area that estimate this area as a geometric structure different from the floor. Therefore, after adding the adaptive scheme, these few normal estimates can be removed, and the reconstruction effect is as smooth as the ground area, as shown in Figure 8d.

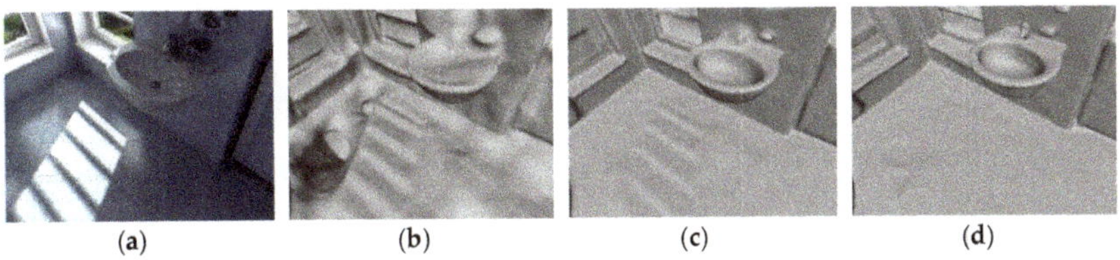

Figure 8. Qualitive comparison for reflective areas using Hypersim dataset: (**a**) reference image; (**b**) model reconstructed without using normal prior; (**c**) model reconstructed with normal prior and without adaptive scheme; (**d**) model reconstructed with normal prior and adaptive scheme.

Through the above quantitative and qualitative results, it is not difficult to find that the normal prior can significantly improve the reconstruction effect and geometric details in the indoor scene reconstruction task. The use of the adaptive normal prior can reduce the erroneous reconstruction of many geometric structures and regions, and has a good correction effect on thin structures and partially reflective areas, making the reconstruction of these parts more robust and reasonable and close to the real scene, further proving the effectiveness and usability of the adaptive normal prior module in this section.

4.3.2. Ablation Experiment of Distortion Loss Function

The effectiveness of the new regularization term, the distortion loss function, is proposed in this paper through quantitative and qualitative ablation experimental analysis. Since the distortion loss function L_{dist} is only applicable to scenes with floating debris, the ablation experiment is also conducted on these scene datasets. There are two different settings in this ablation experiment: (1) without adding the distortion loss function, denoted as w/o L_{dist}; (2) after adding the distortion loss function, denoted as Ours. By ablating it in five scenes with floating debris noise in ScanNet, the quantitative indicators shown in Table 4 can be obtained.

Table 4. Distortion loss function ablation experiment.

Method	Acc ↓	Comp ↓	Prec ↑	Recall ↑	F-Score ↑
w/o L_{dist}	0.055	**0.052**	0.742	0.701	0.721
Ours	**0.047**	**0.052**	**0.795**	**0.704**	**0.746**

Bold text indicates the best results. ↓ indicates that the smaller the value of this indicator, the better; ↑ indicates that the larger the value of this indicator, the better.

As can be seen from Table 4, Acc and Prec have been significantly improved, which means that the distortion loss function removes some erroneous reconstruction parts and achieves more accurate reconstruction, while Comp and Recall have not changed much, because the method in this paper does not produce too many additional areas for supplementary reconstruction.

Figure 9 shows a visual comparison of a scene with a large amount of floating debris. There are is a large amount of incorrectly reconstructed floating debris in Figure 9a. After adding the distortion loss function, most of this floating debris in Figure 9b is eliminated.

In addition, it also has a good removal effect on single floating debris areas in some scenes. As shown in Figure 10, there are single floating debris areas in both scenes in Figure 10a, which are successfully removed in Figure 10b.

The ablation experiment results on the distortion loss function show that, both qualitatively and quantitatively, the distortion loss function proposed in Chapter 3 of this paper can effectively remove floating object noise in the 3D model and improve the overall quality of the 3D model.

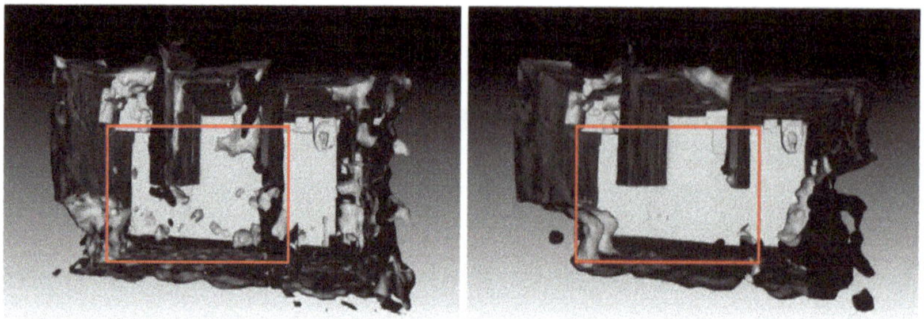

(**a**) W/o L_{dist} (**b**) Ours

Figure 9. Visual comparison for a scene with a large amount of floating debris using the ScanNet dataset; (**a**) reconstruction result without adding a distortion loss function; (**b**) reconstruction result with a distortion loss function.

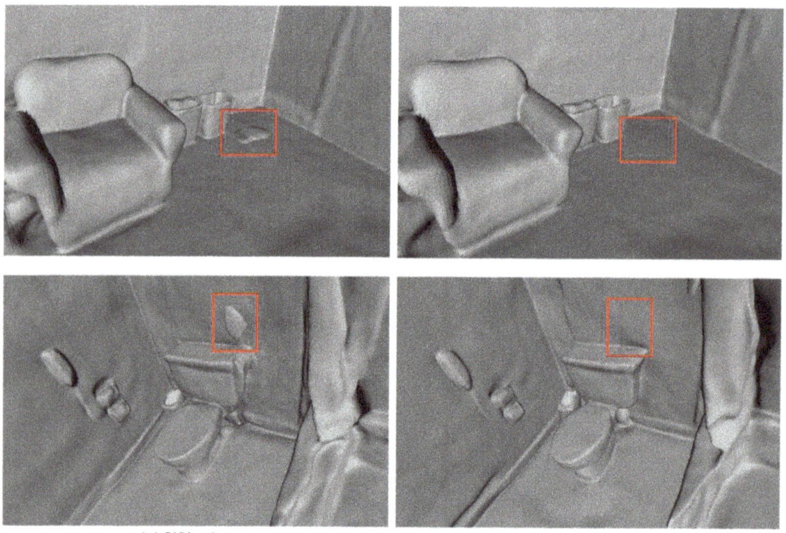

(**a**) W/o L_{dist} (**b**) Ours

Figure 10. Visual comparison for a scene with single floating debris areas enclosed by red dashed box using the ScanNet dataset; (**a**) reconstruction result without adding distortion loss function; (**b**) reconstruction result with distortion loss function.

4.4. Limitations

The comparative experiment part illustrates that the proposed method has advantages over the benchmark model in both quantitative indicators and visualization results, which fully proves the superiority of the proposed method. The ablation experiment part proves the effectiveness of each module of the proposed method. Nevertheless, the proposed method still has some limitations in some specific scenarios.

For scenes with messy objects, the reconstruction effect of this method is not perfect. This is because in the 3D reconstruction task of indoor scenes, the arrangement and combination of objects will greatly affect the reconstruction process. As shown in Figure 11, there are many irregularly shaped objects on the table in the input image, and they are placed in overlapping and mutually occluding situations. For soft and transparent objects, such as plastic bags, it is difficult to accurately estimate their surface details because their appearance features are not easy to capture. Although this method can reconstruct its general outline, it is difficult to judge that this is a plastic bag based on the outline. The reconstruction of such soft and non-solid objects in the scene has always been a difficult problem, and even the GT model obtained by scanning has not been able to restore this part well.

Figure 11. The limitations of this method in the 3D reconstruction of scenes with clutter, occlusion, soft non-solid objects, and blurred images, using the ScanNet dataset.

In addition, the fuzziness of the input image is also crucial to the reconstruction result. If the input image has a certain amount of camera blur, as shown in Figures 5–8, many details of the object will be lost in the image, and the loss of these details increases the difficulty of reconstruction. This blur can also lead to inaccurate feature point detection, affecting the feature extraction and feature matching process. This blur can also affect the reconstruction of some areas with insufficient texture features, such as the dividing line and gap between the drawers in Figure 11. Due to the image blur, the area is incorrectly reconstructed.

5. Conclusion and Future Work

Based on the implicit expression of the NeRF and SDF, this paper proposes an indoor scene reconstruction method based on adaptive normal priors, and optimizes the geometric learning process through adaptive normal priors. The method proposed in this paper can significantly improve the reconstruction quality of large plane textures and uneven lighting areas in indoor scenes, and can also remove floating debris in the reconstructed 3D model. However, there are still some problems that need to be further optimized and improved in the future:

1. When there are many objects and elements in the scene and they are irregular, this method can only reconstruct the general outline, and the object category cannot be directly determined by these outlines. At the same time, the reconstructed results at the connections between objects and between objects and backgrounds are discontinuous. One solution is to try to introduce more priors to allow the neural network to obtain more useful information to accurately learn and understand the elements in the scene. Another feasible solution is to find a way to distinguish between object areas and non-object areas, and then learn them separately, which is conducive to further capturing more complex details.
2. This method takes several hours to more than ten hours to train and optimize a single indoor scene, which limits the application of this method in reconstruction over a relatively large range and in real-time reconstruction. One possible solution to improve training efficiency is to use hashed multi-resolution encoding to use a smaller network without sacrificing quality, thereby significantly reducing the number of floating-point and memory access operations, allowing the neural network to be trained at a smaller computational cost while maintaining reconstruction quality, greatly reducing training time.

Author Contributions: Conceptualization, Z.L. and Y.H.; methodology, Z.L. and Y.H.; validation, Z.L. and Y.H.; formal analysis, Z.L., Y.H. and L.Y.; investigation, Z.L., Y.H. and L.Y.; resources, Z.L. and Y.H.; data curation, Z.L. and Y.H.; writing—original draft preparation, Z.L. and Y.H.; writing—review and editing, Z.L. and L.Y.; visualization, Z.L. and Y.H.; supervision, L.Y.; project administration, L.Y. All authors have read and agreed to the published version of the manuscript.

Funding: This research received no external funding.

Institutional Review Board Statement: Not applicable.

Informed Consent Statement: Not applicable.

Data Availability Statement: The dataset used in this study is open access and can be found here: ScanNet dataset: http://www.scan-net.org/, accessed on 4 July 2024; Hypersim dataset: https://github.com/apple/ml-hypersim, accessed on 4 July 2024; Replica dataset: https://github.com/facebookresearch/Replica-Dataset/releases, accessed on 4 July 2024.

Conflicts of Interest: The authors declare no conflicts of interest.

References

1. Kang, Z.; Yang, J.; Yang, Z.; Cheng, S. A review of techniques for 3d reconstruction of indoor environments. *ISPRS Int. J. Geo-Inf.* **2020**, *9*, 330. [CrossRef]
2. Li, J.; Gao, W.; Wu, Y.; Liu, Y.; Shen, Y. High-quality indoor scene 3d reconstruction with rgb-d cameras: A brief review. *Comput. Vis. Media* **2022**, *8*, 369–393. [CrossRef]
3. Hess, W.; Kohler, D.; Rapp, H.; Andor, D. Real-Time Loop Closure in 2d Lidar Slam. In Proceedings of the2016 IEEE International Conference on Robotics and Automation (ICRA), Stockholm, Sweden, 16–21 May 2016; pp. 1271–1278.
4. Zhang, J.; Singh, S. Loam: Lidar odometry and mapping in real-time. In *Robotics: Science and Systems*; University of California: Berkeley, CA, USA, 2014; Volume 2, pp. 1–9.
5. Shan, T.; Englot, B. Lego-loam: Lightweight and Ground-Optimized Lidar Odometry and Mapping on Variable Terrain. In Proceedings of the 2018 IEEE/RSJ International Conference on Intelligent Robots and Systems (IROS), Madrid, Spain, 1–5 October 2018; pp. 4758–4765.
6. Curless, B.; Levoy, M. A Volumetric Method for Building Complex Models from Range Images. In Proceedings of the 23rd Annual Conference on Computer Graphics and Interactive Techniques, New Orleans, LA, USA, 4–9 August 1996; pp. 303–312.
7. Newcombe, R.A.; Izadi, S.; Hilliges, O.; Molyneaux, D.; Kim, D.; Davison, A.J.; Kohi, P.; Shotton, J.; Hodges, S.; Fitzgibbon, A. Kinectfusion: Real-Time Dense Surface Mapping and Tracking. In Proceedings of the 2011 10th IEEE International Symposium on Mixed and Augmented Reality, Basel, Switzerland, 26–29 October 2011; pp. 127–136.
8. Murez, Z.; Van As, T.; Bartolozzi, J.; Sinha, A.; Badrinarayanan, V.; Rabinovich, A. Atlas: End-to-end 3d scene reconstruction from posed images. In *Computer Vision–ECCV 2020: 16th European Conference, Glasgow, UK, August 23–28, 2020, Proceedings, Part VII 16*; Springer: Berlin/Heidelberg, Germany, 2020; pp. 414–431.
9. Sun, J.; Xie, Y.; Chen, L.; Zhou, X.; Bao, H. Neuralrecon: Real-Time Coherent 3d Reconstruction from Monocular Video. In Proceedings of the IEEE/CVF Conference on Computer Vision and Pattern Recognition, Nashville, TN, USA, 20–25 June 2021; pp. 15598–15607.
10. Schonberger, J.L.; Frahm, J.M. Structure-from-Motion Revisited. In Proceedings of the IEEE Conference on Computer Vision and Pattern Recognition, Las Vegas, NV, USA, 26 June–1 July 2016; pp. 4104–4113.
11. Schönberger, J.L.; Zheng, E.; Frahm, J.M.; Pollefeys, M. Pixelwise View Selection for Unstructured Multi-View Stereo. In *Computer Vision–ECCV 2016: 14th European Conference, Amsterdam, The Netherlands, October 11–14,* ***2016****, Proceedings, Part III 14*; Springer: Berlin/Heidelberg, Germany, 2016; pp. 501–518.
12. Xu, Q.; Tao, W. Planar prior assisted patchmatch multi-view stereo. *Proc. AAAI Conf. Artif. Intell.* **2020**, *34*, 12516–12523. [CrossRef]
13. Im, S.; Jeon, H.G.; Lin, S.; Kweon, I.S. Dpsnet: End-to-end deep plane sweep stereo. *arXiv* **2019**, arXiv:1905.00538.
14. Wang, F.; Galliani, S.; Vogel, C.; Speciale, P.; Pollefeys, M. Patchmatchnet: Learned Multi-View Patchmatch Stereo. In Proceedings of the IEEE/CVF Conference on Computer Vision and Pattern Recognition, Nashville, TN, USA, 20–25 June 2021; pp. 14194–14203.
15. Xu, Q.; Tao, W. Pvsnet: Pixelwise visibility-aware multi-view stereo network. *arXiv* **2020**, arXiv:2007.07714.
16. Yao, Y.; Luo, Z.; Li, S.; Fang, T.; Quan, L. Mvsnet: Depth Inference for Unstructured Multi-View Stereo. In Proceedings of the European Conference on Computer Vision (ECCV), Munich, Germany, 8–14 September 2018; pp. 767–783.
17. Yu, Z.; Gao, S. Fast-Mvsnet: Sparse-to-Dense Multi-View Stereo with Learned Propagation and Gauss-Newton Refinement. In Proceedings of the IEEE/CVF Conference on Computer Vision and Pattern Recognition, Seattle, WA, USA, 13–19 June 2020; pp. 1949–1958.
18. Teed, Z.; Deng, J. Deepv2d: Video to depth with differentiable structure from motion. *arXiv* **2018**, arXiv:1812.04605.
19. Huang, P.H.; Matzen, K.; Kopf, J.; Ahuja, N.; Huang, J.B. Deepmvs: Learning Multi-View Stereopsis. In Proceedings of the IEEE Conference on Computer Vision and Pattern Recognition, Salt Lake City, UT, USA, 18–22 June 2018; pp. 2821–2830.
20. Cheng, S.; Xu, Z.; Zhu, S.; Li, Z.; Li, L.E.; Ramamoorthi, R.; Su, H. Deep Stereo using Adaptive thin Volume Representation with Uncertainty Awareness. In Proceedings of the IEEE/CVF Conference on Computer Vision and Pattern Recognition, Seattle, WA, USA, 13–19 June 2020; pp. 2524–2534.
21. Yang, H.; Chen, R.; An, S.P.; Wei, H.; Zhang, H. The growth of image-related three dimensional reconstruction techniquesin deep learning-driven era: A critical summary. *J. Image Graph.* **2023**, *28*, 2396–2409.

22. Liu, S.; Zhang, Y.; Peng, S.; Shi, B.; Pollefeys, M.; Cui, Z. Dist: Rendering deep implicit signed distance function with differentiable sphere tracing. In Proceedings of the IEEE/CVF Conference on Computer Vision and Pattern Recognition, Seattle, WA, USA, 13–19 June 2020; pp. 2019–2028.
23. Yariv, L.; Kasten, Y.; Moran, D.; Galun, M.; Atzmon, M.; Ronen, B.; Lipman, Y. Multiview neural surface reconstruction by disentangling geometry and appearance. *Adv. Neural Inf. Process. Syst.* **2020**, *33*, 2492–2502.
24. Wang, P.; Liu, L.; Liu, Y.; Theobalt, C.; Komura, T.; Wang, W. Neus: Learning neural implicit surfaces by volume rendering for multi-view reconstruction. *arXiv* **2021**, arXiv:2106.10689.
25. Yariv, L.; Gu, J.; Kasten, Y.; Lipman, Y. Volume rendering of neural implicit surfaces. *Adv. Neural Inf. Process. Syst.* **2021**, *34*, 4805–4815.
26. Barron, J.T.; Mildenhall, B.; Verbin, D.; Srinivasan, P.P.; Hedman, P. Mip-nerf 360: Unbounded Anti-Aliased Neural Radiance Fields. In Proceedings of the IEEE/CVF Conference on Computer Vision and Pattern Recognition, New Orleans, LA, USA, 19–24 June 2022.
27. Liu, L.; Gu, J.; Zaw Lin, K.; Chua, T.S.; Theobalt, C. Neural sparse voxel fields. *Adv. Neural Inf. Process. Syst.* **2020**, *33*, 15651–15663.
28. Rebain, D.; Matthews, M.; Yi, K.M.; Lagun, D.; Tagliasacchi, A. Lolnerf: Learn from one look. In Proceedings of the IEEE/CVF Conference on Computer Vision and Pattern Recognition, New Orleans, LA, USA, 19–24 June 2022.
29. Gafni, G.; Thies, J.; Zollhofer, M.; Nießner, M. Dynamic Neural Radiance Fields for Monocular 4d Facial Avatar Reconstruction. In Proceedings of the IEEE/CVF Conference on Computer Vision and Pattern Recognition, Nashville, TN, USA, 20–25 June 2021.
30. Do, T.; Vuong, K.; Roumeliotis, S.I.; Park, H.S. Surface Normal Estimation of Tilted Images Via Spatial Rectifier. In *Computer Vision–ECCV 2020: 16th European Conference, Glasgow, UK, August 23–28, 2020, Proceedings, Part IV 16*; Springer International Publishing: New York, NY, USA, 2020; pp. 265–280.
31. Dai, A.; Chang, A.X.; Savva, M.; Halber, M.; Funkhouser, T.; Nießner, M. Scannet: Richly-Annotated 3d Reconstructions of Indoor Scenes. In Proceedings of the IEEE Conference on Computer vision and Pattern Recognition, Honolulu, HI, USA, 21–26 July 2017; pp. 5828–5839.
32. Roberts, M.; Ramapuram, J.; Ranjan, A.; Kumar, A.; Bautista, M.A.; Paczan, N.; Webb, R.; Susskind, J.M. Hypersim: A Photorealistic Synthetic Dataset for Holistic Indoor Scene Understanding. In Proceedings of the IEEE/CVF International Conference on Computer Vision, Montreal, QC, Canada, 10–17 October 2021; pp. 10912–10922.
33. Straub, J.; Whelan, T.; Ma, L.; Chen, Y.; Wijmans, E.; Green, S.; Engel, J.J.; Mur-Artal, R.; Ren, C.; Verma, S.; et al. The replica dataset: A digital replica of indoor spaces. *arXiv* **2019**, arXiv:1906.05797.
34. Mildenhall, B.; Srinivasan, P.P.; Tancik, M.; Barron, J.T.; Ramamoorthi, R.; Ng, R. Nerf: Representing scenes as neural radiance fields for view synthesis. *Commun. ACM* **2021**, *65*, 99–106. [CrossRef]
35. Guo, H.; Peng, S.; Lin, H.; Wang, Q.; Zhang, G.; Bao, H.; Zhou, X. Neural 3d scene reconstruction with the Manhattan-world assumption. In Proceedings of the IEEE/CVF Conference on Computer Vision and Pattern Recognition, New Orleans, LA, USA, 19–24 June 2022; pp. 5511–5520.
36. Yu, Z.; Peng, S.; Niemeyer, M.; Sattler, T.; Geiger, A. Monosdf: Exploring monocular geometric cues for neural implicit surface reconstruction. *Adv. Neural Inf. Process. Syst.* **2022**, *35*, 25018–25032.
37. Zhu, J.; Huo, Y.; Ye, Q.; Luan, F.; Li, J.; Xi, D.; Wang, L.; Tang, R.; Hua, W.; Bao, H.; et al. I2-SDF: Intrinsic Indoor Scene Reconstruction and Editing via Raytracing in Neural SDFs. In Proceedings of the IEEE/CVF Conference on Computer Vision and Pattern Recognition, Vancouver, BC, Canada, 18–22 June 2023; pp. 12489–12498.

Journal of
Imaging

MDPI

Article

Single-Image-Based 3D Reconstruction of Endoscopic Images

Bilal Ahmad *, Pål Anders Floor, Ivar Farup and Casper Find Andersen

Department of Computer Science, Norwegian University of Science & Technology, 2815 Gjøvik, Norway; paal.anders.floor@ntnu.no (P.A.F.); ivar.farup@ntnu.no (I.F.); casper.andersen@ntnu.no (C.F.A.)
* Correspondence: bilal.ahmad@ntnu.no

Abstract: A wireless capsule endoscope (WCE) is a medical device designed for the examination of the human gastrointestinal (GI) tract. Three-dimensional models based on WCE images can assist in diagnostics by effectively detecting pathology. These 3D models provide gastroenterologists with improved visualization, particularly in areas of specific interest. However, the constraints of WCE, such as lack of controllability, and requiring expensive equipment for operation, which is often unavailable, pose significant challenges when it comes to conducting comprehensive experiments aimed at evaluating the quality of 3D reconstruction from WCE images. In this paper, we employ a single-image-based 3D reconstruction method on an artificial colon captured with an endoscope that behaves like WCE. The shape from shading (SFS) algorithm can reconstruct the 3D shape using a single image. Therefore, it has been employed to reconstruct the 3D shapes of the colon images. The camera of the endoscope has also been subjected to comprehensive geometric and radiometric calibration. Experiments are conducted on well-defined primitive objects to assess the method's robustness and accuracy. This evaluation involves comparing the reconstructed 3D shapes of primitives with ground truth data, quantified through measurements of root-mean-square error and maximum error. Afterward, the same methodology is applied to recover the geometry of the colon. The results demonstrate that our approach is capable of reconstructing the geometry of the colon captured with a camera with an unknown imaging pipeline and significant noise in the images. The same procedure is applied on WCE images for the purpose of 3D reconstruction. Preliminary results are subsequently generated to illustrate the applicability of our method for reconstructing 3D models from WCE images.

Keywords: 3D reconstruction; image enhancement; endoscopy; medical imaging

1. Introduction

Wireless capsule endoscopy (WCE) was pioneered by Given Imaging in the year 2000 [1]. It offers numerous advantages over traditional endoscopic procedures. It is less invasive, requires no sedation, and offers a painless and comfortable experience for patients. It is used to visually inspect the entire gastrointestinal (GI) tract, from the esophagus to the large intestine, using a small swallowable capsule equipped with a miniature camera. It is used to diagnose inflammatory bowel disease, GI bleeding, and polyps [2]. Despite its many advantages, WCE images also entail several challenges. These include issues related to uneven and low illumination, low resolution, and noise [3]. Moreover, the lack of control over the capsule's movement within the GI tract restricts the thorough examination of areas of particular interest.

Three-dimensionally (3D)-reconstructed models of WCE images can be effective for conducting a comprehensive analysis of specific areas of interest. By employing 3D reconstruction algorithms, it becomes feasible to transform the 2D images captured by the capsule camera into a 3D representation of the GI tract. Three-dimensional models along with their images can allow gastroenterologists to visualize internal organs from different angles and perspectives, aiding in the identification of abnormalities and facilitating more precise planning for interventions and surgeries. Results in [4] have shown that gastroenterologists find 3D models useful to an extent that they sometimes prefer them over original images.

Within the realm of computer vision, 3D reconstruction poses an intriguing challenge, which requires the utilization of different techniques to image data [5]. Vision-based depth estimation techniques can be classified into different categories. A range of techniques for monocular image-based depth estimation have been developed, including texture gradient analysis [6], image focus analysis [7], and photometric methods [8]. Other approaches leverage multiple images, relying on camera motion or variations in relative camera positions [9]. The integration of 3D reconstruction techniques finds extensive applications across diverse fields, spanning cultural heritage, robotics, medical diagnostics, video surveillance, and more [10,11].

In numerous real-world applications, capturing multiple images of a scene or object from various angles can be challenging. Consequently, single-image-based methods prove effective and suitable in such situations. This is particularly evident in the case of WCE, where the capsule relies on the natural peristaltic contractions to traverse through the human GI tract. Given its low frame rate, it happens that the scene within the GI tract is captured only once. In such circumstances, single-image-based 3D reconstruction techniques are the only viable option.

Shape from shading (SFS) is a method that requires only one image for 3D reconstruction, and therefore, it is a potential candidate for WCE application. Horn and Brooks [12] were among the first to recover the 3D shape of the surface using the SFS method. They obtained surface gradients through an iterative approach relying on a nonlinear first-order partial differential equation (PDE), establishing a relationship between 3D shape and intensity variations within an image. By applying integrability constraints, Frankot and Chellappa [13] demonstrated superior accuracy and efficiency in estimating the depth variable compared with the approach by Horn and Brooks. Kimmel and Sathian [14] employed the numerical scheme based on the fast marching method to recover depth, yielding a numerically consistent, computationally optimal, and practically fast algorithm for the classical SFS problem. Tankus et al. [15] remodeled the SFS method under the framework of perspective projection, expanding its range of potential applications. Similarly, Wu et al. [16] also solved the SFS problem under perspective projection without assuming the light source at the camera center, with a specific focus on medical endoscopy.

The method proposed by Wu et al. [16] closely aligns with the WCE setting, featuring a near-light model with multiple light sources positioned around the camera center. Consequently, we selected their method as a starting point for further experimentation. The methodology follows a two-step process for shape reconstruction. Initially, it involves deriving a reflectance function by considering the relative positions of the light sources, camera, and surface reflectance properties. Following this, the error between the reflectance function and image irradiance is minimized by formulating an image irradiance equation (IIE). While a typical solution to IIE involves an L2 regularizer as a smoothness constraint, we opted for anisotropic diffusion (AD) due to its superior accuracy compared with the L2 regularizer [17].

WCE presents considerable challenges in the domain of 3D reconstruction due to its inherent limitations. The device lacks controllability over its light settings, requiring expensive equipment for operation, which is often unavailable. These practical constraints pose significant challenges when attempting to conduct extensive experiments regarding the assessment of 3D reconstruction quality for WCE images. To address these challenges, we successfully conducted a comprehensive investigation on the 3D reconstruction of synthetic colon images captured with a camera in a virtual environment [4]. In the following experiments, we initially employ images of an artificial colon captured under a controlled environment using an industrial endoscope for the purpose of 3D reconstruction, before transitioning to the analysis of images obtained from WCE. The imaging system of the endoscope behaves like that of WCE, though it introduces significantly less lens distortion. Moreover, it offers higher resolution than a typical WCE image, and the light strength can be manually controlled. The endoscope has six rectangular-shaped light-emitting diodes (LEDs) surrounding the camera behind a protective glass covering. The known dimensions of the artificial colon provide a reference for assessing the correctness of the reconstructed 3D colon model.

This article utilizes a single-image-based method to reconstruct the 3D shape of the artificial colon. The camera is corrected for lens distortion, and the light source intensity of the endoscope has also been measured. The camera response function (CRF) is estimated to convert the device's output grayscale image to image irradiance. The method proposed by Andersen et al. [18] is employed, which uses a single image of a ColorChecker to compute the CRF of a camera with an unknown imaging pipeline. Wu et al. [16] assume an ideal multiple-point light model in their PSFS approach. Given that the endoscope is equipped with six light sources, it should closely align with the characteristics of the ideal six-point light model. However, the endoscope light sources produce a different pattern due to their rectangular shape and the presence of a glass covering, which can lead to scattering and interference effects. Therefore, corrections are applied to the captured image to account for this deviation. Thereafter, the near-light perspective SFS (PSFS) algorithm that integrates AD as a smoothness constraint is applied to reconstruct the 3D shapes of the endoscopic images. The PSFS algorithm utilizes grayscale images. Therefore, the albedo is simply a reflection factor between 0 and 1. Initially, well-defined primitive objects are tested to assess the method's robustness and accuracy. Afterward, the same methodology is applied to recover the geometry of the colon. The known dimensions of the artificial colon also provide a reference for assessing the correctness of the reconstructed 3D colon model. In the end, we present preliminary results of 3D reconstruction using PillCam images, illustrating the potential applicability of our method across various endoscopic devices. The core contributions of the paper are as follows:

- We present a comprehensive pipeline for step-by-step 3D reconstruction using an AD-based PSFS algorithm, as demonstrated in Figure 1. This pipeline is generic and applicable to any endoscopic device, provided that we have access to the required image data for 3D reconstruction, as well as data for geometric and radiometric calibration.
- We utilized JPG images and opted for an endoscope where access to RAW image data was unavailable, reflecting real-world scenarios where RAW data may not be accessible. This choice underscores the practical applicability of our approach, as in many real-world applications, access to RAW data is limited.
- We validated the AD-based PSFS method in real-world scenarios by conducting 3D reconstruction on simple primitives and comparing the results with ground truth—a practice seldom addressed in the literature. This rigorous validation process enhances the credibility and reliability of our approach.
- We present simple methods for estimating the spatial irradiance and light source intensity of the endoscope, designed for scenarios where relying on multiple images for radiometric calibration is not feasible. Further details on these methods are provided in Section 2.4 of the article.

The rest of the article is organized as follows: Section 2 provides an overview of various methodologies for 3D reconstruction, encompassing the PSFS model with anisotropic diffusion, geometric and radiometric calibration of the endoscope, albedo measurement, image rescaling, and denoising. Section 3 details the entire experimental setup, beginning with the creation of ground truth models, followed by image capture, and concluding with the reconstruction of 3D surfaces for primitives and an artificial colon. Additionally, preliminary results for WCE images are presented. Lastly, Section 4 concludes the article.

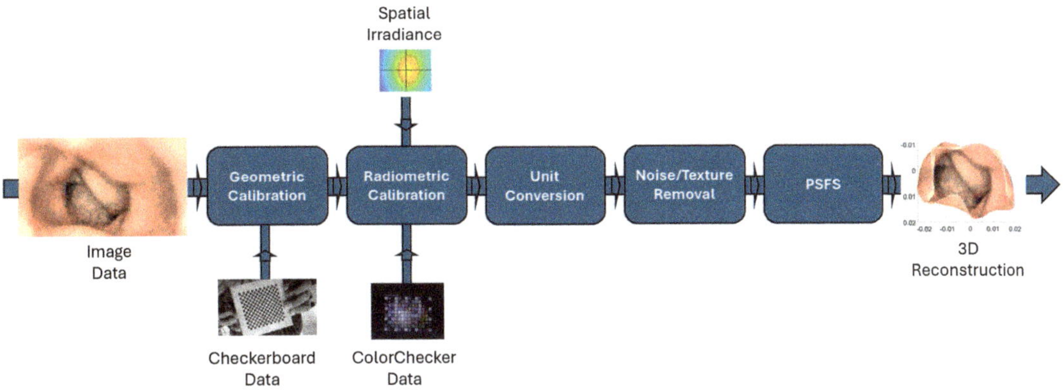

Figure 1. Comprehensive pipeline for 3D reconstruction using PSFS algorithm.

2. Methods Overview

This section covers various methods involved in 3D reconstruction using the PSFS method with an output image from an endoscope. We begin by introducing the PSFS method with AD (Section 2.1). Following that, we discuss the different calibration and preprocessing steps necessary before inputting the image into the PSFS algorithm. Initially, geometric calibration of the endoscope is conducted by capturing images of a checkerboard to correct distortion and determine camera intrinsic parameters, such as focal length (Section 2.2). Subsequently, the captured endoscopic image intended for 3D reconstruction undergoes radiometric calibration, involving the computation of CRF and spatial irradiance (Section 2.4). The radiometrically corrected image is then rescaled (Section 2.5) and denoised (Section 2.6). The comprehensive pipeline of the 3D reconstruction algorithm using the PSFS method is illustrated in Figure 1.

2.1. PSFS Model

In this section, we cover the PSFS method, where six-point light sources are placed around a camera and the camera is directed towards the negative z-axis, as shown in Figure 2. Under perspective projection, the relationship between image coordinates $(\widetilde{x}, \widetilde{y})$ and the camera coordinates (x, y, z) is given as follows:

$$x = \widetilde{x}\frac{z}{f} \qquad y = \widetilde{y}\frac{z}{f}, \tag{1}$$

where f denotes the camera's focal length. Assuming a diffuse surface, the reflected light from the point $\mathbf{P}$ can be determined using Lambert's cosine law and inverse square fall-off law from multiple light sources as follows [16]:

$$R(\widetilde{x}, \widetilde{y}, z, p, q) = I_o \rho \sum_{i=1}^{6} \left(\frac{\mathbf{n}(\widetilde{x}, \widetilde{y}, z, p, q) \cdot \mathbf{l}_i(\widetilde{x}, \widetilde{y}, z)}{r_i(\widetilde{x}, \widetilde{y}, z)^2} \right), \tag{2}$$

where I_o represents the intensity of the light source(s), ρ denotes the albedo of the surface, and p and q are the surface gradient components along the x and y directions, respectively. Furthermore, $r_i(\widetilde{x}, \widetilde{y}, z)^2$ accounts for the inverse square fall-off distance from each point light source, $\mathbf{l}_i$ is a unit vector aligned along the ith light ray, and $\mathbf{n}$ refers to the surface unit normal, which is computed as follows [12]:

$$\mathbf{n} = \frac{[-\frac{\partial z}{\partial x}, -\frac{\partial z}{\partial y}, 1]}{\sqrt{(\frac{\partial z}{\partial x})^2 + (\frac{\partial z}{\partial y})^2 + 1}}. \tag{3}$$

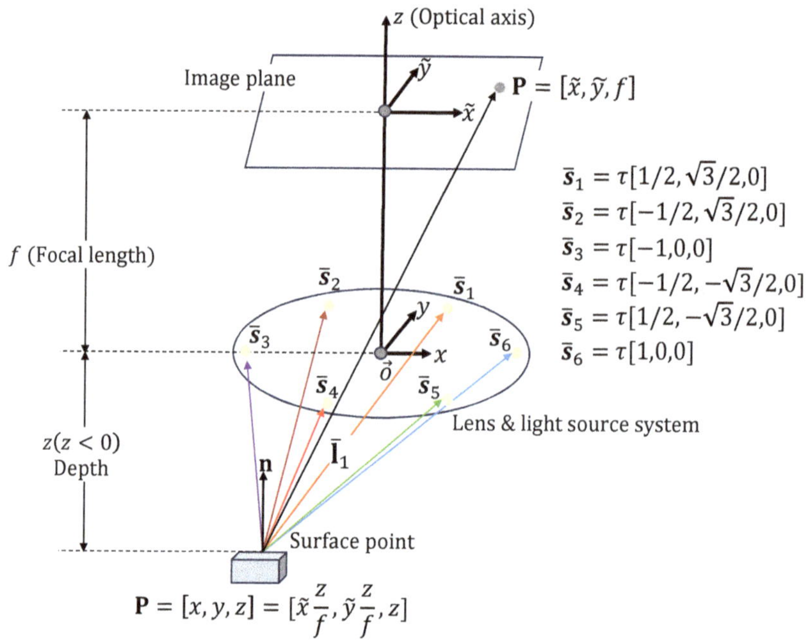

Figure 2. PSFS model with light source at the camera center **O**. (x, y, z) represents the camera coordinate system, which is centered at **O**. The z-axis is the optical axis, pointing towards the image plane.

Given the distance from the camera center to a light source, we can explicitly write the light source vector from the point **P** as follows:

$$\bar{\mathbf{l}}_i = \left[\tau\cos\theta_i - \tilde{x}\frac{z}{f}, \tau\sin\theta_i - \tilde{y}\frac{z}{f}, -z\right], \tag{4}$$

where τ is the distance from the camera center to a light source, $\theta_i = 2\pi i/6$ for $i \in [1, 6]$. The unit vector $\mathbf{l}_i$ can be expressed as $\mathbf{l}_i = \bar{\mathbf{l}}_i / \parallel \bar{\mathbf{l}}_i \parallel$.

According to Horn and Brooks [12], IIE can be written as follows:

$$R(\tilde{x}, \tilde{y}, z, p, q) = I(\tilde{x}, \tilde{y}), \tag{5}$$

where $I(\tilde{x}, \tilde{y})$ is the image irradiance. Equation (5) is solved to determine the optimal depth value z by minimizing the difference between $I(\tilde{x}, \tilde{y})$ and $R(\tilde{x}, \tilde{y}, z, p, q)$. The optimization equation is established for z, while the values of p and q are updated through the gradients of the modified z [17]. The error $E(z)$ is minimized as follows:

$$E(z) = \lambda e_i(z) + (1 - \lambda)e_s(z), \tag{6}$$

where e_i and e_s represent irradiance error and smoothness constraint, respectively. λ is a weighting factor and controls the scaling between e_i and e_s. $e_i(z)$ can be computed over the image domain $\Omega \subset \mathbb{R}^2$ as follows:

$$e_i(z) = \int_{\Omega} (I(\tilde{x}, \tilde{y}) - R(\tilde{x}, \tilde{y}, z, p, q))^2 d\Omega. \tag{7}$$

$e_s(z)$ is typically a L2 regularizer. However, we have employed AD as a smoothness constraint because it not only enhances the accuracy of the depth map by suppressing noise but also demonstrates effectiveness in preserving structural details of the reconstructed scene, outperforming the L2 regularizer [17,19].

AD is introduced as a smoothness constraint by first calculating a 2×2 structure tensor ($S_{i,j}$) based on the gradient of the depth z [20]. $S_{i,j}$ is given as [20] as follows:

$$S_{i,j} = \frac{\partial z}{\partial x^i}\frac{\partial z}{\partial y^j}. \tag{8}$$

Subsequently, we compute the corresponding eigenvalues (λ_+, λ_-) and eigenvectors (θ_+, θ_-) following a similar approach to [21]. Utilizing (λ_+, λ_-) and (θ_+, θ_-), the diffusion tensor $\mathbf{D}$ is then derived as follows:

$$\mathbf{D} = \frac{\partial \psi}{\partial \lambda_+}\theta_+\theta_+^T + \frac{\partial \psi}{\partial \lambda_-}\theta_-\theta_-^T. \tag{9}$$

In terms of (λ_+, λ_-), Lagrangian density ψ can be written as follows [22]:

$$e_s(z) = \int_\Omega \psi(\lambda_+, \lambda_-)d\Omega. \tag{10}$$

Equations (7) and (10) are combined in Equation (6) and can be formulated as follows:

$$E(z) = \int_\Omega (\lambda(I - R)^2 + (1 - \lambda)\psi(\lambda_+, \lambda_-))d\Omega. \tag{11}$$

The solution to Equation (11) is given by Euler–Lagrange PDE:

$$\lambda(I - R)\frac{\partial R}{\partial z} + (1 - \lambda)\nabla \cdot (\mathbf{D}\nabla z) = 0, \tag{12}$$

which we numerically solve by gradient descent:

$$\frac{\partial z}{\partial t} = \nabla \cdot (\mathbf{D}\nabla z) + \frac{\lambda}{1 - \lambda}(I - R)\frac{\partial R}{\partial z}. \tag{13}$$

Similar to [17], $I(\tilde{x}, \tilde{y})$ is utilized to derive the structure tensor. Through this single-step computation of the structure tensor, the process becomes efficient, making the computation task simpler and more linear.

2.2. Geometric Calibration

Geometric calibration is needed to estimate the camera's intrinsic parameters as well as its lens distortion. It has been observed that the endoscope exhibits minimal lens distortion towards its periphery. However, the necessity arises to rectify this distortion for the sake of precise depth estimation, as the SFS algorithm assumes a pinhole model.

For geometric calibration, we employed a standard checkerboard measuring 10×10 cm, with each individual square on the board measuring 4 mm. The images are taken at a 10 cm distance from the tip of the camera at different angles. The MATLAB camera calibration toolbox is used for the geometric calibration of the endoscope [23]. The intrinsic parameters are computed using Heikkila's method [24] with two extra distortion coefficients corresponding to tangential distortion.

The MATLAB camera calibration toolbox basically requires between 10 and 20 images of the checkerboard from different viewing angles. A total of 15 images of the checkerboard are used in our case. An image of the checkerboard is shown in Figure 3a. The camera model is set to standard, and radial distortion is set to 2 coefficients as it is observed that the endoscope camera has little distortions towards the periphery. Figure 3b shows a sample image of the colon corrected for lens distortion.

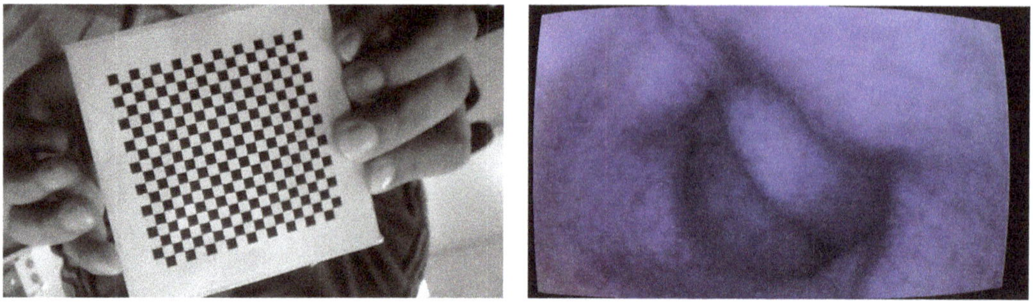

(a) (b)

Figure 3. Geometric calibration: (**a**) checkerboard and (**b**) geometric calibration.

It is important to mention here that the procedure is repeated three times with three different sets of checkerboard images to confirm the consistency in the results. The estimated focal length is around 2.4 ± 0.1 mm for all three sets, and there is no skew observed.

2.3. Albedo Measurement

Albedo is the fraction of incident light that a surface reflects. It has a value between 0 and 1, where 0 corresponds to all the incident light being absorbed by the surface and 1 corresponds to a body that reflects all incident light. The primitives have diffused white surfaces. Therefore, the albedo is assumed to be $\rho = 1$ for all the primitive objects.

The artificial colon consists of a soft rubber material with a nearly uniform pinkish color. Therefore, it is necessary to measure the albedo of the surface. The albedo of the colon is measured by taking the image of the colon and a diffuse spectralon tile placed side by side. Both the spectralon and the colon are kept at an equal distance from the camera, and an image is taken outside so that both surfaces have a uniform distribution of light, as shown in Figure 4a. The albedo of the surface is measured by taking the ratio between the colon and the spectralon pixel value at any given location. The estimated albedo value of the artificial colon is $\rho = 0.60$.

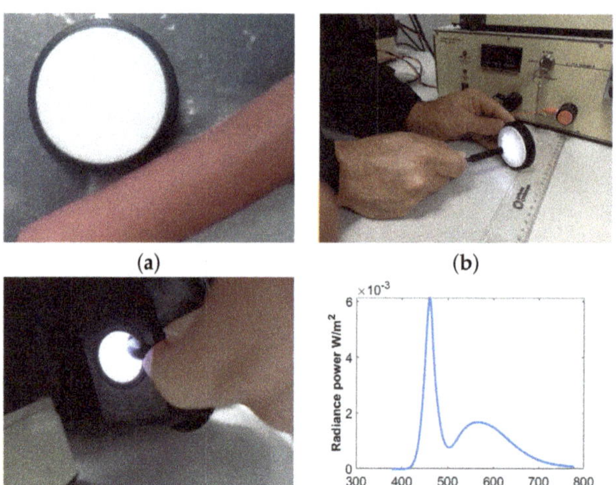

(c) (d)

Figure 4. Radiance intensity and albedo measurement: (**a**) albedo, (**b**) nonisotropic light, (**c**) uniform light, and (**d**) radiance power.

2.4. Radiometric Calibration

Radiometric calibration has been performed to measure the light intensity, CRF, and spatial distribution of the light intensity on the image. The PSFS algorithm assumes a pinhole model with ideal multiple-point light sources. Therefore, it is crucial to convert from a grayscale image to image irradiance via CRF and correct for the anisotropy of the light source [16], as discussed in Section 2.4.2. Sections 2.4.2 and 2.4.3 provide detailed discussions on the CRF estimation and anisotropy correction, respectively. Measuring light source intensity is also important, as it is a crucial parameter for computing the reflection function given in Equation (2).

2.4.1. Light Source Intensity Measurement

The light intensity of the endoscope is measured by using a CS2000 spectroradiometer [25]. An integrating sphere (IS) must be used to measure intensity because of the nonisotropic behavior of the light source. The IS is a hollow spherical cavity with its interior coated with diffused white reflective material. The aim of the integrating sphere is to provide a stable and uniform illumination condition. An endoscope is placed inside the IS, and radiance power P is measured over the visible spectrum. After measuring the solid angle ω of the endoscope light, I_o is calculated as follows: $I_o = P/(4\pi) \times \omega$. The nonuniformity of the light source, the uniformity of the endoscope light inside the IS, and spectra of the light are shown in Figure 4b–d, respectively.

2.4.2. Camera Response Function

CRF is essential to convert the device's output grayscale image to image irradiance [16]:

$$I(\tilde{x},\tilde{y}) = \frac{r^{-1}[v(\tilde{x},\tilde{y})]}{M(\tilde{x},\tilde{y})}, \tag{14}$$

where $I(\tilde{x},\tilde{y})$ is the image irradiance, $v(\tilde{x},\tilde{y})$ is the grayscale image, and $r(\cdot)$ is the CRF. $M(\tilde{x},\tilde{y})$ incorporates the deviation from the ideal point-light source assumed by PSFS.

The endoscope used in this work has an unknown imaging processing chain, and there are no means of controlling the exposure time. This decision was intentional, reflecting the common limitation among WCE devices available in the market, which generally do not offer any control over the exposure time. By selecting an endoscope that mimics the behavior of typical WCE devices, our approach demonstrates applicability to a broader range of endoscopic devices.

Through experimental observations with the endoscope, it has been observed that the camera performs automatic exposure adjustments. During the image capture process of the SG ColorChecker [26], we have further noted the camera's automatic color adjustment and white balancing mechanisms in operation. It is worth noting that this is similar to the functionality of a standard WCE. These complicating factors have compelled us to abstain from methods that utilize multiple images for the estimation of the CRF.

The method by Andersen et al. [18] is applied to measure the CRF. The method requires only a single image of a ColorChecker to estimate volumetric, spatial, and per-channel nonlinearities. These nonlinearities involve compensating for both physical scene and camera properties through a series of successive signal transformations, bridging the gap between the estimated linear and recorded responses. The estimation process relies on a novel principle of additivity, computed using the spectral reflectances of the colored patches on the ColorChecker. The SG ColorChecker [26] is used to estimate the CRF. An image of the ColorChecker from endoscope and camera response curves is shown in Figure 5.

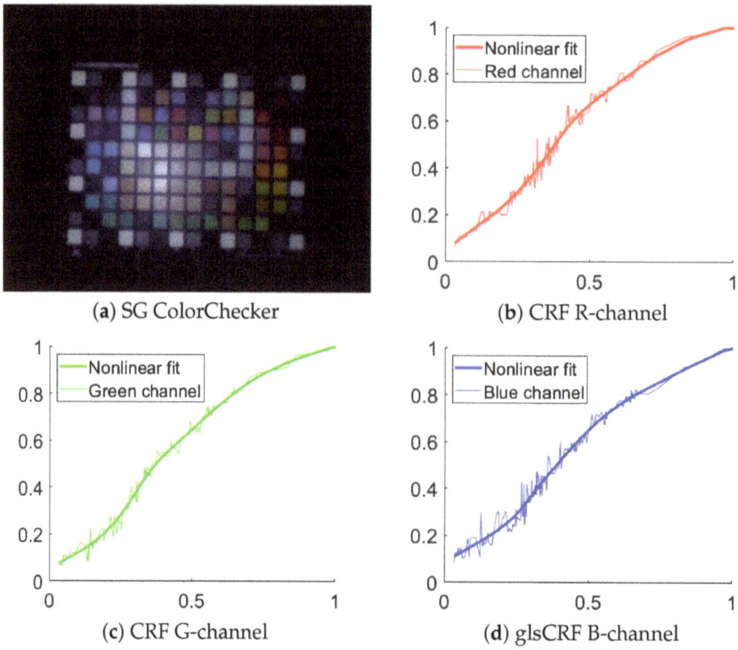

Figure 5. Results of camera response function. (**a**) Image of SG chart captured with an endoscope and used for estimating the CRF and the light distribution. (**b**) CRF in red channel. The red dotted line represents the data point. The red line represents the nonlinear fit. The horizontal axis represents the normalized image intensity, and the vertical axis represents the normalized image irradiance, the same in (**c**,**d**). (**c**) The green dotted red line represents the data point. The green line represents the nonlinear fit. (**d**) The blue dotted line represents the data point. The blue line represents the nonlinear fit.

2.4.3. Spatial Irradiance

The reflection model mentioned in the PSFS method is based on six-point light sources and demands an ideal six-point light distribution in the image to correctly determine the 3D geometry. The endoscope light deviates from an ideal six-point light distribution model due to the rectangular shape of the light sources and the scattering and interference effect caused by the glass on top of the endoscope lens. An inclination in the light sources has been detected, and also, due to the presence of six noncentral light sources, we observe a deviation where the maximum intensity does not align precisely with the image center. Therefore, it is important to quantify these additional effects and compensate for them so that the resulting reflection model satisfies the conditions of six-point light sources. According to [16],

$$\widetilde{M}(\widetilde{x},\widetilde{y}) = M(\widetilde{x},\widetilde{y}) \cdot \sum_{i=1}^{6} \frac{\mathbf{n} \cdot \mathbf{l}_i}{r_i^2}, \tag{15}$$

where the second term on the right side in Equation (15) represents the light distribution from the six-point light sources.

An image of a white diffuse paper, considered as $\widetilde{M}(\widetilde{x},\widetilde{y})$ in our context, was captured and is displayed in Figure 6a. It is noticeable from the image that the endoscopic lighting deviates from the ideal six-point light configuration, exhibiting an oval pattern with an offset from the image center. The ideal six-point light distribution model is constructed by physically measuring the distance from the diffused paper to the tip of the endoscope, as shown in Figure 6b. Finally, $M(\widetilde{x},\widetilde{y})$ is recovered using Equation (15) and then compensated for in the image. $M(\widetilde{x},\widetilde{y})$ is shown in Figure 6c.

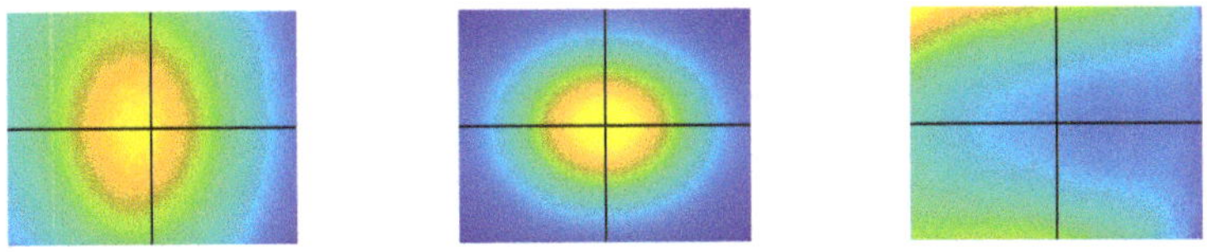

Figure 6. Correction of light distribution. The point where horizontal and vertical lines intersect denotes the image center: (**a**) $\widetilde{M}(\widetilde{x},\widetilde{y})$, (**b**) $\sum_{i=1}^{6}\frac{\mathbf{n}\cdot\mathbf{l}_i}{r_i^2}$, and (**c**) $M(\widetilde{x},\widetilde{y}) = \widetilde{M}(\widetilde{x},\widetilde{y}) / \sum_{i=1}^{6}\frac{\mathbf{n}\cdot\mathbf{l}_i}{r_i^2}$.

2.5. Unit Conversion

The parameters computed thus far are in physical units, leading to the estimation of R in physical coordinates. To establish a consistency between $I(\widetilde{x},\widetilde{y})$ and R, as outlined in Equation (13), $I(\widetilde{x},\widetilde{y})$ is transformed from pixel units to physical units. This conversion is achieved as follows:

$$I_p(\widetilde{x},\widetilde{y}) = \frac{I(\widetilde{x},\widetilde{y}) - \min I(\widetilde{x},\widetilde{y})}{\max I(\widetilde{x},\widetilde{y}) - \min I(\widetilde{x},\widetilde{y})} \times \left(I_o\rho\frac{\cos\theta}{r^2}\right), \tag{16}$$

where $I_p(\widetilde{x},\widetilde{y})$ denotes the physical value of the image irradiance. θ is the angle between the surface normal and the light ray at the point on the surface where illumination is maximized. r is the distance from the light source to the point on the surface where illumination is maximized. In the case of the primitives, the points are measured, whereas, in the case of the colon, the estimation of the parameters r and θ relies on factors such as the field of view (FOV) of the camera, the total length of the colon, and the position of the camera within the colon.

2.6. Image Denoising

In endoscope images, significant noise is observed, mainly due to JPEG compression artifacts. These artifacts include blocky patterns and color distortions. A noisy image when fed into the SFS algorithms can destabilize the differential equations due to inaccuracies and ambiguities in shading information, which can lead to inaccuracies in the estimation of surface normals and object shape.

In order to reduce the noise, the method by Xu et al. is utilized [27]. The method essentially separates the visual information related to the surface texture of an object from its underlying structural components within an image. We employ this method to remove noise from the image while retaining its structural details. The method is based on the relative total variation scheme, which captures the essential difference between texture and structure by utilizing their different properties. Later, they employed an optimization method that leverages novel variation measures, including inherent variation and relative total variation, to identify significant structures while disregarding the underlying texture patterns.

2.7. Assessment Criteria

The reconstructed 3D shapes of the different primitives are compared with ground truth models by measuring relative root-mean-square error (r_{RMSE}) and relative max-depth error (r_{MDE}). These metrics are chosen to quantify depth errors with respect to a reference depth, making the results easily interpretable. r_{RMSE} quantifies the overall geometric deformation present in the reconstructed 3D model, while r_{MDE} highlights the maximum deviation observed between the 3D-reconstructed model and the ground truth.

r_{RMSE} allows for the evaluation of geometric distortion in the 3D-reconstructed model. A perfect 3D reconstruction is indicated by an error value of 0, whereas a highly distorted 3D reconstruction corresponds to a value of 1. r_{RMSE} is computed as follows:

$$r_{\text{RMSE}} = \frac{1}{d_{\max}} \sqrt{\frac{1}{n} \sum_{i=1}^{n} \mid \widehat{D_i} - D_i \mid^2}, \tag{17}$$

where D, $d_{\max}$, $\widehat{D}$, and n represent ground truth depth, maximum ground truth depth point, depth of the recovered 3D shape, and total number of depth points considered for error estimation, respectively.

r_{MDE} indicates the relative maximum deviation between the estimated depth values produced by a 3D reconstruction algorithm and the ground truth depth values. A low r_{MDE} suggests that the majority of depth estimates are close to their ground truth counterparts, indicating high accuracy in the 3D reconstruction. Conversely, a high r_{MDE} implies significant discrepancies between the estimated and ground truth depth values, indicating poorer accuracy in some places in the reconstruction. r_{MDE} is computed as follows:

$$r_{\text{MDE}} = \frac{1}{d_{\max}} \max \mid \widehat{D_i} - D_i \mid . \tag{18}$$

3. Experiments and Results

3.1. Ground Truth Models

The primary goal of performing depth estimation on simple primitives is to gauge the accuracy and effectiveness of the PSFS method on endoscopic images. This evaluation was conducted in the context of a camera system that contains an unknown imaging pipeline, and where the captured images exhibit significant noise. This approach, involving the reconstruction of fundamental geometric shapes and their subsequent comparison with ground truth models, will prove effective in achieving the desired evaluation.

The experiments are conducted on geometric primitives with known dimensions, including a sphere, a cube, and a pyramid. These primitives have a diffuse surface with an albedo $\rho \approx 1$. Given that these primitives have well-defined geometry and known dimensions, ground truth models of these three primitives are generated in MATLAB to compare them with reconstructed surfaces.

3.2. Image Acquisition

The PSFS algorithm is subjected to comprehensive testing using a variety of images, including synthetic colon images and different geometric primitives of known dimensions. The images of the primitives are captured by placing a diffuse paper beneath them to ensure a uniform albedo throughout the scene. A series of images capturing these different primitives are presented in Figure 7.

(a) (b) (c)

Figure 7. Images of primitives captured with an endoscope: (**a**) sphere, (**b**) cube, and (**c**) pyramid.

Additionally, the images of the synthetic colon are acquired to assess the method's applicability for potential 3D reconstruction applications within the context of WCE. A synthetic colon [28] is an artificial phantom of a colon without deformation and has a smooth wall with a diameter and length of 0.028 m and 0.3 m, respectively. Therefore, a deformed support [29] is used to hold this colon and produce deformations similar to a real colon.

The colon is placed in its support, and one of the ends is closed with a clip. The endoscope is inserted from the other end, and a series of images of the deformed colon are captured, as shown in Figure 8. The endoscope used in the experiment is an *Oiiwak WiFi endoscope* [30], which is an industrial endoscope that wirelessly transmits the acquired images or videos to an android device via a software named MoView. Images of the colon, deformed support, endoscope, and setup for capturing images of the artificial colon are shown in Figure 9.

(a) (b) (c)

Figure 8. Images of artificial colon captured with an endoscope: (**a**) ROI-1, (**b**) ROI-2, and (**c**) ROI-3.

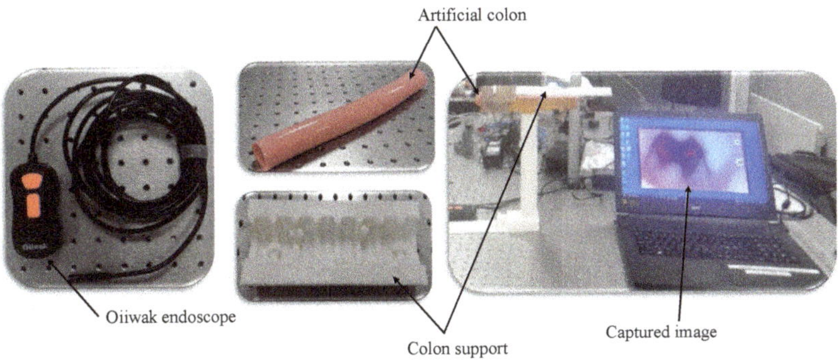

Figure 9. Setup and equipment used for image acquisition of synthetic colon.

3.3. 3D Reconstruction

In the first step, the image captured with the endoscope is corrected for lens distortions. Subsequently, the image undergoes correction by utilizing the CRF and addressing the anisotropy of the six-point light using Equation (14). The image is then converted to physical units using Equation (16). Thereafter, the image is denoised and then input into the PSFS algorithm. A reflectance map is derived using Equation (2) with a flat surface as initial depth z. Subsequently, the z's are updated using Equation (13), where the gradients p and q are computed as $\frac{\partial z}{\partial x}$ and $\frac{\partial z}{\partial y}$, respectively. Notably, the parameter λ assumes different values in distinct cases and is determined empirically within our experimental setup.

In the case of primitives, the images are cropped to 500×500 pixels because we are interested in recovering the shape of the primitives rather than the whole image. The ground truth models of the primitives are shown in Figure 10, whereas the recovered shape of all the primitives is shown in Figure 11. The reconstructed primitives are compared with ground truth models by computing r_{RMSE} and r_{MDE} using Equations (17) and (18), respectively. We achieve r_{RMSE} at around 0.04 and r_{MDE} at around 0.10 with respect to ground truth for different primitives, as shown in Table 1.

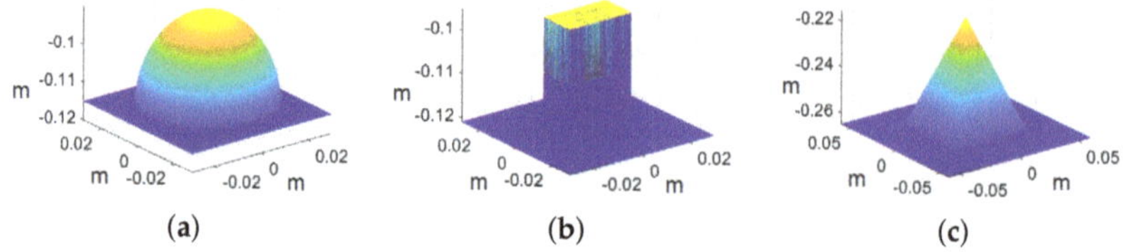

Figure 10. Ground truth models of primitives. The axis represents the values in meters: (**a**) sphere, (**b**) cube, and (**c**) pyramid.

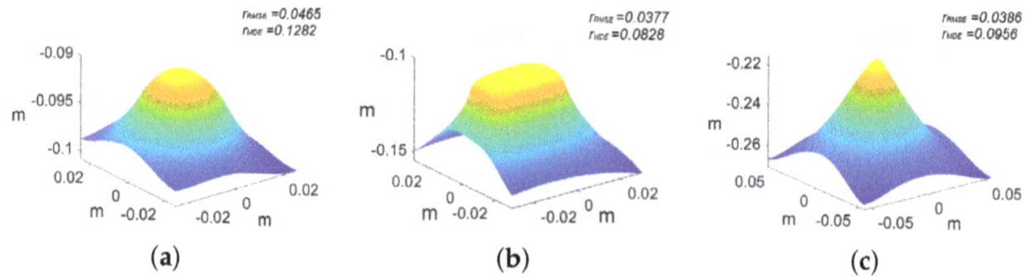

Figure 11. Recovered 3D primitives. The axis represents the values in meters: (**a**) sphere, (**b**) cube, and (**c**) pyramid.

Table 1 indicates that the sphere exhibits higher errors compared with the pyramid and the cube. This disparity can be attributed to the presence of an inclination in one of the endoscope's light sources. This manufacturing error poses a significant challenge in accurately modeling the light distribution within our PSFS model. As the sphere covers a larger part of the captured view, in comparison with the other shapes evaluated, the impact of the inclination becomes more pronounced. Consequently, these factors collectively contribute to greater errors in the case of the sphere model.

Table 1. Quantitative evaluation for primitives.

Primitives	**Cube**	**Sphere**	**Pyramid**
r_{RMSE}	0.0377	0.0465	0.0386
r_{MDE}	0.0828	0.1282	0.0956

Full-sized images are used for the reconstruction of the colon model. Color correction is applied to the colon images since their original hue is somewhat pinkish, which appears purplish due to the bluish nature of the endoscope's lighting. The colon color depicted in Figure 4a serves as the reference color. The difference in hue between the original and the endoscope-captured image of the colon is identified using a chromaticity diagram. Subsequently, the color of the synthetic colon was adjusted to align with its original shade.

The endoscope is equipped with LEDs, which behave similar to point light sources. This behavior causes dim illumination in deeper regions of the captured images due to the inverse square fall-off law. To address this problem, we have adopted the approach proposed in [4] to enhance contrast, especially in images capturing larger depths. The method involves illuminating the deeper regions by transitioning the lighting in the image from point light to directional light. This transformation is achieved by utilizing surface normals derived from reconstructed 3D models. Following color correction and contrast enhancement on the images of the artificial colon, a notable noise is observed due to the inherent noise in the original images. To address this issue, the enhanced images are further denoised. The enhanced images are converted into their luma and chroma components, with a focus on addressing significant noise present in the luma component. Subsequently, the luma of all the images is subjected to denoising using anisotropic diffusion. The diffusion tensor is derived like in Equation (9), after applying a Gaussian filter to the luma component of the enhanced images, ensuring the preservation of edges in the resulting denoised images. The final geometrically corrected enhanced images of the colon are presented in Figure 12, and the subsequent 3D models, wrapped with enhanced images, are illustrated in Figure 13.

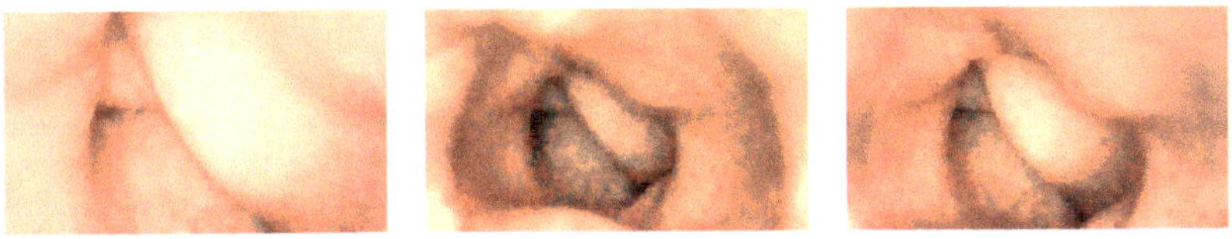

(a) (b) (c)

Figure 12. Color-corrected and directionally lit colon images: (**a**) ROI-1, (**b**) ROI-2, and (**c**) ROI-3.

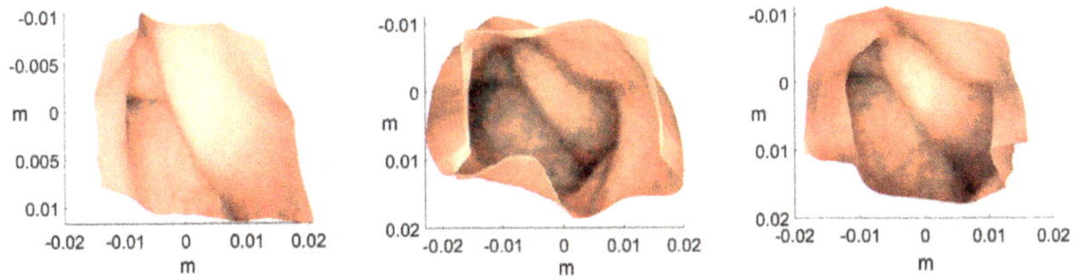

(a) (b) (c)

Figure 13. Recovered 3D colon models. The axis represents the values in meters: (**a**) ROI-1, (**b**) ROI-2, and (**c**) ROI-3.

3.4. Discussion

The PSFS algorithm demonstrates robustness in handling noisy endoscope-captured images. Initially, the method is tested on simple primitives to assess accuracy by comparing the reconstructions with ground truth models. The results in Table 1 indicate a notable level of accuracy, which, in turn, served as an indicator of the method's potential for accurately reconstructing the 3D geometry of the colon.

While reconstructing 3D shapes, the number of iterations in the PSFS algorithm varies across experiments. We terminate the process when successive iterations show no significant change in irradiance error $e_i(z)$ according to Equation (7). Throughout the experimentation with the PSFS algorithm, $e_i(z)$ is continuously reduced, indicating, as referenced in [17], an improvement in the quality of depth estimation.

The known dimensions of the artificial colon played a crucial role in assessing the accuracy of the reconstructed 3D colon model during laboratory experimentation. As previously stated, the artificial colon has a diameter of 0.028 m, a value closely matched by all the reconstructed colon models shown in Figure 13. Another significant advantage of employing the endoscope is the extensive laboratory experiments that are challenging to replicate with a WCE in a controlled environment, mainly due to the unavailability of high-cost equipment required for WCE operation. However, after successfully reconstructing the colon shapes using an endoscope that closely mimics the behavior of WCE, confidence is established in the feasibility of applying the same procedure to reconstruct 3D shapes from WCE images.

3.5. Preliminary Results of WCE

After experimenting with an endoscope, we subsequently test images of the GI system captured with WCE. These images were acquired during clinical trials involving the examinations of ten patients. The pilot study was conducted in collaboration between Innlandet Hospital Trust and NTNU, Gjøvik, Norway, in 2021, under the consultation of the professor and gastroenterologist Øistein Hovde. The capsule modality used in the examinations was PillCam COLON 2.

Three distinct images from the colon region of the GI tract are selected, as illustrated in Figure 14. We choose images without any artifacts or deficiencies to demonstrate the applicability of our method on PillCam images. We apply a similar procedure as used with the endoscope, with a notable difference in radiometric calibration due to unavailability of the SG ColorChecker images captured with PillCam. The geometric calibration is performed using images of a checkerboard acquired during the pilot study. For physical unit conversion, r and θ are empirically estimated by leveraging the optical properties of WCE, such as effective visibility distance, as provided in the PillCam information manual [31]. Due to the absence of radiometric results, certain assumptions are made for the CRF and the spatial irradiance. It is assumed that the four light sources of PillCam COLON 2 are similar to an ideal four-point light distribution model. The conversion of image intensity values to image irradiance utilizes standard sRGB curves. The mucosal texture of the GI tract is removed using the method proposed by Xu et al. [27] to approximate a uniform albedo throughout the scene. Finally, the PSFS method is employed to reconstruct 3D shapes of the PillCam images.

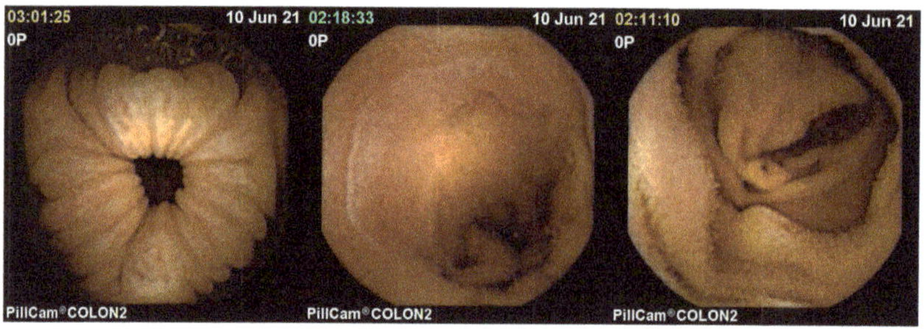

Figure 14. Images of the human colon captured with PillCam COLON 2.

Image are cropped to a size of 275 × 275, as the MATLAB camera calibration toolbox is unable to handle regions towards the periphery quite well. Cropped images utilized for 3D reconstruction are shown in Figure 15, and the corresponding reconstructed 3D models of all three images are shown in Figure 16. In Figure 17, the side view of all the 3D models are shown, which clearly illustrates that our method successfully reconstructs a significant portion of the structure, even in these preliminary results, despite the absence of radiometric data from the PillCam camera.

(a) (b) (c)

Figure 15. Geometrically corrected cropped images utilized for 3D reconstruction: (**a**) PC-1, (**b**) PC-2, and (**c**) PC-3.

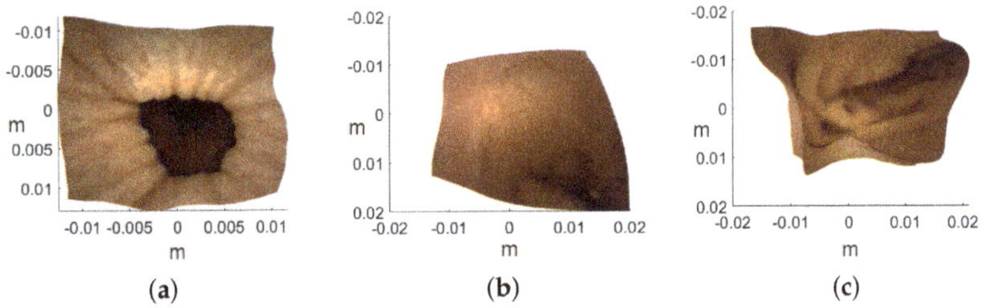

Figure 16. Top view of the recovered 3D human GI regions. The axis represents the values in meters: (**a**) PC-1, (**b**) PC-2, and (**c**) PC-3.

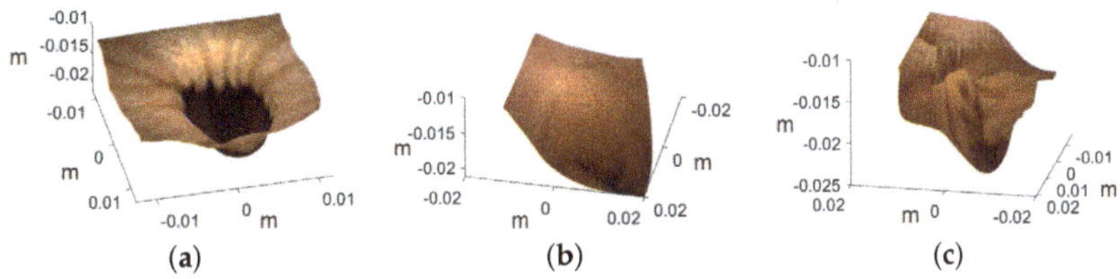

Figure 17. Side view of the recovered 3D human GI regions. The axis represents the values in meters: (**a**) PC-1, (**b**) PC-2, and (**c**) PC-3.

The 3D reconstruction results can be further enhanced by conducting radiometric calibration on PillCam, as it is an important parameter to convert the device's output grayscale image to image irradiance, as per Equation (14). The geometric calibration can be further improved by utilizing other methods that deal with fish-eye lenses [32]. Nevertheless, these preliminary results are quite convincing, demonstrating the capability of our method to handle images with significant lens distortion, even in the absence of radiometric calibration results and albedo values. Further investigation is encouraged to enhance the accuracy of 3D models, as they are recognized as valuable tools during diagnostic assessment in gastroenterology, as highlighted in [4].

4. Conclusions

This article investigates the possibility of reconstructing endoscopic images using the PSFS algorithm employed with anisotropic diffusion. Images of simple primitives are initially tested to evaluate the accuracy of the method on endoscopic images by comparing the reconstructed geometries with ground truth models. Afterward, single images of the endoscopes are used to reconstruct the colon surface. Results show that our systematic approach can handle a camera with an unknown imaging pipeline and noisy images and can accurately reconstruct the geometry of the colon.

Additionally, we have implemented a technique utilizing surface normals derived from the 3D-reconstructed models to improve illumination and thereby enhance contrast in images capturing larger depths. This is achieved by changing the illumination in such images from point light to directional light. Various other techniques have also been discussed for the geometric and radiometric calibration of an endoscope camera. This calibration is essential for accurately reconstructing 3D shapes using the PSFS algorithm. In the end, preliminary 3D reconstruction results using PillCam images are provided, demonstrating the potential applicability of our method to different endoscopic devices. In future works, efforts will be made to fully calibrate the PillCam COLON 2 camera to further enhance the 3D reconstruction results.

Author Contributions: Conceptualization, B.A., P.A.F., and I.F.; methodology, B.A., P.A.F., I.F., and C.F.A.; software, B.A. and C.F.A.; validation, B.A., P.A.F., I.F., and C.F.A.; data curation, B.A.; writing—original draft preparation, B.A.; writing—review and editing, B.A., P.A.F., I.F., and C.F.A.; revision, B.A., P.A.F., I.F., and C.F.A.; supervision, P.A.F. and I.F. All authors have read and agreed to the published version of the manuscript.

Funding: Funding was provided by the Research Council of Norway under the projects CAPSULE, grant no. 300031, and Individual Colour Vision-Vased Image Optimisation, grant no. 287209.

Institutional Review Board Statement: Not applicable.

Informed Consent Statement: Informed consent has been obtained from the patients for this study.

Data Availability Statement: No new data were created during this study.

Acknowledgments: The authors express their gratitude to Fereshteh Mirjalili, Majid Ansari-Asl, and Peter Nussbaum for their valuable assistance in the radiometric calibration of the camera. Special thanks are extended to Charlet Valentine for designing the support structure for the artificial colon. The authors also acknowledge Øistein Hovde and Anuja Vats for their contribution to the collection of the dataset during clinical trials. Furthermore, the authors would like to thank Marius Pedersen for designing and 3D-printing the stand for PillCam COLON 2.

Conflicts of Interest: The authors declare no conflicts of interest.

References

1. Iddan, G.; Meron, G.; Glukhovsky, A.; Swain, P. Wireless capsule endoscopy. *Nature* **2000**, *405*, 417.
2. Silva, J.; Histace, A.; Romain, O.; Dray, X.; Granado, B. Toward embedded detection of polyps in wce images for early diagnosis of colorectal cancer. *Int. J. Comput. Assist. Radiol. Surg.* **2014**, *9*, 283–293.
3. Pan, G.; Wang, L. Swallowable wireless capsule endoscopy: Progress and technical challenges. *Gastroenterol. Res. Pract.* **2011**, *2012*, 841691.
4. Ahmad, B.; Floor, P.A.; Farup, I.; Hovde, Ø. 3D Reconstruction of Gastrointestinal Regions Using Single-View Methods. *IEEE Access* **2023**, *11*, 61103–61117.
5. Aharchi, M.; Ait Kbir, M. A review on 3D reconstruction techniques from 2D images. In Proceedings of the Innovations in Smart Cities Applications, Paris, France, 4–6 October 2020; pp. 510–522.
6. Verbin, D.; Zickler, T. Toward a universal model for shape from texture. In Proceedings of the IEEE/CVF Conference on Computer Vision and Pattern Recognition, Seattle, WA, USA, 14–19 June 2020; pp. 422–430.
7. Nayar, S.K.; Nakagawa, Y. Shape from focus. *IEEE Trans. Pattern Anal. Mach. Intell.* **1994**, *16*, 824–831.
8. Zhang, R.; Tsai, P.S.; Cryer, J.E.; Shah, M. Shape-from-shading: A survey. *IEEE Trans. Pattern Anal. Mach. Intell.* **1999**, *21*, 690–706.
9. Schonberger, J.L.; Frahm, J.M. Structure-from-motion revisited. In Proceedings of the IEEE conference on computer vision and pattern recognition, Las Vegas, NV, USA, 27–30 June 2016; pp. 4104–4113.
10. Palade, D.O.; Cobzeanu, B.M.; Zaharia, P.; Dabija, M. 3D Reconstruction role in surgical treatment of sinonasal tumours. *Rev. Chim.* **2018**, *69*, 1455–1457.
11. Münster, S.; Köhler, T. 3D reconstruction of cultural heritage artifacts. In *Virtual Palaces, Part II: Lost Palaces and Their Afterlife. Virtual Reconstruction between Science and Media*; ART-Books: Heidelberg, Germany, 2016; pp. 87–102.
12. Horn, B.K.; Brooks, M.J. The variational approach to shape from shading. *Comput. Vision Graph. Image Process.* **1986**, 33, 174–208.
13. Frankot, R.T.; Chellappa, R. A method for enforcing integrability in shape from shading algorithms. *IEEE Trans. Pattern Anal. Mach. Intell.* **1988**, *10*, 439–451.
14. Kimmel, R.; Sethian, J.A. Optimal algorithm for shape from shading and path planning. *J. Math. Imaging Vis.* **2001**, *14*, 237–244.
15. Tankus, A.; Sochen, N.; Yeshurun, Y. Shape-from-shading under perspective projection. *Int. J. Comput. Vis.* **2005**, *63*, 21–43.
16. Wu, C.; Narasimhan, S.G.; Jaramaz, B. A multi-image shape-from-shading framework for near-lighting perspective endoscopes. *Int. J. Comput. Vis.* **2010**, *86*, 211–228.
17. Ahmad, B.; Floor, P.A.; Farup, I. A Comparison of Regularization Methods for Near-Light-Source Perspective Shape-from-Shading. In Proceedings of the 2022 IEEE International Conference on Image Processing (ICIP), Bordeaux, France, 16–19 October 2022; pp. 3146–3150.
18. Andersen, C.F.; Farup, I.; Hardeberg, J.Y. Additivity Constrained Linearisation of Camera Calibration Data. *IEEE Trans. Image Process.* **2023**, *32*, 3774–3789.
19. Ahmad, B. Anisotropic Diffusion for Depth Estimation in Shape from Focus Systems. In Proceedings of the VISAPP 2024: 19th International Conference on Computer Vision Theory and Applications, Rome, Italy, 27–29 February 2024.
20. Di Zenzo, S. A note on the gradient of a multi-image. *Comput. Vision Graph. Image Process.* **1986**, *33*, 116–125.
21. Sapiro, G.; Ringach, D.L. Anisotropic diffusion of multivalued images with applications to color filtering. *IEEE Trans. Image Process.* **1996**, *5*, 1582–1586.
22. Tschumperlé, D.; Deriche, R. Vector-valued image regularization with PDEs: A common framework for different applications. *IEEE Trans. Pattern Anal. Mach. Intell.* **2005**, *27*, 506–517.

23. Bouguet, J.Y. Camera Calibration Toolbox for Matlab. 2008. Available online: http://www.vision.caltech.edu/bouguetj/calib_doc (accessed on 25 March 2024).
24. Heikkila, J.; Silvén, O. A four-step camera calibration procedure with implicit image correction. In Proceedings of the computer society conference on computer vision and pattern recognition, San Juan, PR, USA, 17–19 June 1997; pp. 1106–1112.
25. KonicaMinolta. CS2000 Spectroradiometer. Available online: https://sensing.konicaminolta.us/us/products/cs-2000-spectroradiometer/ (accessed on 25 March 2024).
26. X-Rite. Digital SG ColorChecker. Available online: https://www.xrite.com/categories/calibration-profiling/colorchecker-digital-sg (accessed on 25 March 2024).
27. Xu, L.; Yan, Q.; Xia, Y.; Jia, J. Structure extraction from texture via relative total variation. *ACM Trans. Graph.* **2012**, *31*, 139.
28. LifeLikeBioTissue. Single Layer Bowel. Available online: https://lifelikebiotissue.com/shop/general/single-layer-bowel (accessed on 25 March 2024).
29. Charlet, V. *Creation of a 3D Modular System for Colon Feformation*; Technical Report for Colorlab; NTNU: Torgarden, Norway, 2022.
30. Oiiwak. Dual Camera WiFi Endoscope. Available online: https://www.oiiwak.com/ (accessed on 25 March 2024).
31. Metronic. PillCam COLON 2. Available online: https://www.medtronic.com/covidien/en-us/products/capsule-endoscopy/pillcam-colon-2-system.html (accessed on 25 March 2024).
32. Kannala, J.; Brandt, S.S. A generic camera model and calibration method for conventional, wide-angle, and fish-eye lenses. *IEEE Trans. Pattern Anal. Mach. Intell.* **2006**, *28*, 1335–1340.

Journal of
Imaging

MDPI

Article

Neural Radiance Field-Inspired Depth Map Refinement for Accurate Multi-View Stereo †

Shintaro Ito *, Kanta Miura, Koichi Ito * and Takafumi Aoki

Graduate School of Information Sciences, Tohoku University, 6-6-05, Aramaki Aza Aoba, Sendai 9808579, Japan; kanta@aoki.ecei.tohoku.ac.jp (K.M.); aoki@ecei.tohoku.ac.jp (T.A.)

* Correspondence: shintaro@aoki.ecei.tohoku.ac.jp (S.I.); ito@aoki.ecei.tohoku.ac.jp (K.I.)

† This paper is an extended version of our paper published in Ito, S.; Miura, K.; Ito, K.; Aoki, T. Depth map estimation from multi-view images with nerf-based refinement. In Proceedings of the 2023 IEEE International Conference on Image Processing (ICIP), Kuala Lumpur, Malaysia, 8–11 October 2023; pp. 2955–2959.

Abstract: In this paper, we propose a method to refine the depth maps obtained by Multi-View Stereo (MVS) through iterative optimization of the Neural Radiance Field (NeRF). MVS accurately estimates the depths on object surfaces, and NeRF accurately estimates the depths at object boundaries. The key ideas of the proposed method are to combine MVS and NeRF to utilize the advantages of both in depth map estimation and to use NeRF for depth map refinement. We also introduce a Huber loss into the NeRF optimization to improve the accuracy of the depth map refinement, where the Huber loss reduces the estimation error in the radiance fields by placing constraints on errors larger than a threshold. Through a set of experiments using the Redwood-3dscan dataset and the DTU dataset, which are public datasets consisting of multi-view images, we demonstrate the effectiveness of the proposed method compared to conventional methods: COLMAP, NeRF, and DS-NeRF.

Keywords: multi-view stereo; neural radiance fields; depth map estimation; 3D reconstruction

1. Introduction

Multi-View Stereo (MVS) is a technique for acquiring 3D data from target objects or scenes from multiple images captured by a camera [1–3]. Since MVS requires only camera images, it is not restricted to the capturing environment, reduces the effort required for capturing images, and can more easily acquire 3D data compared to active scanners.

MVS estimates depth maps from images taken from different viewpoints and integrates them to reconstruct 3D data [2–7]. A depth map is an image in which the pixel values represent the distance, i.e., the depth, from the camera to the object. To estimate the depth map for each viewpoint, MVS performs image matching between multi-view images. One of the typical methods is plain sweeping [4,8]. In plain sweeping, the most optimal depth is searched for in each pixel of the input image based on the similarity of textures in the local region of the image while varying the depth from the camera to the object, where Normalized Cross-Correlation (NCC) or Zero-mean Normalized Cross-Correlation (ZNCC) between local regions is generally used as the similarity [3,8]. Although the optimal depth can be estimated by taking into account the geometric consistency among multi-view images, the number of image-matching operations becomes large since a full depth search is required for each pixel [7]. To reduce the number of image matching operations in MVS, efficient methods using PatchMatch [9,10] have been proposed [3,7,11,12]. Among them, COLMAP [3,13] has been proposed as a pipeline for 3D reconstruction using PatchMatch and is used as a de facto standard method for MVS. PatchMatch-based methods assign depth and normal as parameters to each pixel. For example, the initial value of the depth is a random number within the acceptable range of depth estimation obtained from the epipolar constraints between cameras, and the initial value of the normal is a random number within $\pm\pi/3$ for the angles of the X and Y axes [7]. Then, the parameters are optimized by matching corresponding pixels in different viewpoints according to these parameters. The depth map can be estimated with fewer matching operations than a full search by using random numbers as the initial values of the parameters. Since the parameters are optimized using image matching, the accuracy of depth estimation is degraded in poor-texture regions and at object boundaries, and occlusion prevents depth estimation. Recently, depth map estimation methods using deep learning have been proposed [14–17]. In this paper, "texture" refers to the spatial distribution of colors and their intensities in an image. "Rich texture" indicates that there is a large difference in intensity values between pixels and that the texture has a complex pattern, while a "poor texture" indicates that there is almost no difference in intensity values between pixels and that the texture is uniform. "Object boundary" indicates the boundary between the foreground and background in the image, where the pixels have significantly different depths. A typical method, namely MVSNet [15], projects feature maps extracted by a Convolutional Neural Network (CNN) [18] to another viewpoint based on plain sweeping, and estimates the depth of each pixel based on the similarity of the features. Depth map estimation with training is more accurate than that without training since CNN-based methods can use features considering the shape and positions of the neighboring regions of the pixel of interest as well as textures. On the other hand, even with deep learning, depth estimation is difficult in poor-texture regions and object boundaries. Thus, the depth map estimation in MVS can accurately estimate the depth on object surfaces with rich texture, while the estimation accuracy is degraded in poor-texture regions and at object boundaries.

Neural Radiance Fields (NeRFs) [6] have been proposed as another method for depth map estimation from multi-view images. NeRF represents a 3D space as a radiance field, which is parametrized with a Multi Layer Perceptron (MLP). The MLP is trained so as to estimate a volume density and view-dependent emitted radiance given the spatial location and view direction of the camera from multi-view images. The use of the trained MLP makes it possible to synthesize images from novel viewpoints based on the radiance field on the ray connecting the camera and the object. NeRF can not only generate novel view images from the radiance field, but can also generate depth maps. Depth can be synthesized pixel by pixel using the radiance field, even for poor-texture regions and object boundaries. On the other hand, it is not always possible to accurately estimate the depth on the surface of an object using NeRF compared with MVS.

As described above, MVS estimates depths based on image matching and thus can accurately estimate depths on object surfaces with rich texture, while the accuracy of depth estimation is degraded in poor-texture regions and at object boundaries. On the other hand, NeRF estimates the radiance field of a scene from multi-view images and estimates depth for each pixel from the radiance field, and thus can estimate depth for poor-texture regions and object boundaries, while the accuracy is not always high for object surfaces. In this paper, we propose a method to refine the depth maps obtained by MVS through the iterative optimization of NeRF. The standard NeRF trains an MLP to generate novel view images, while the proposed method refines the depth map by iteratively optimizing an MLP, so that the MLP can render the input image and the depth map obtained by MVS. Therefore, the proposed method only performs iterative optimization of the radiance field and does not require any training. Through a set of experiments using the Redwood-3dscan dataset [19] and the DTU dataset [20], which are public datasets consisting of multi-view images, we demonstrate the effectiveness of the proposed method compared to conventional methods. In the experiments, we employ an evaluation metric that is invariant to the depth scale [21], in addition to the widely used evaluation metrics for depth map estimation.

2. Related Work

This section summarizes the depth map estimation methods using MVS and NeRF that are related to this study.

2.1. MVS-Based Approaches

Here, we give an overview of COLMAP [3,13] using PatchMatch and MVSNet [15] using deep learning as MVS-based depth map estimation methods.

2.1.1. COLMAP

COLMAP is a pipeline for 3D reconstruction from multi-view images that consists of Structure from Motion (SfM) [13] and MVS [3]. SfM is a method for 3D reconstruction and camera parameter estimation by sequentially adding images using the principle of triangulation used in stereo vision [1]. Correspondence point pairs are obtained based on the similarity between feature points, 3D points are reconstructed using the correspondence point pairs and camera parameters based on the principle of triangulation, and camera parameters are optimized by minimizing reprojection errors. SfM estimates camera parameters and reconstructs sparse 3D point clouds from multi-view images. SfM in COLMAP is a de facto standard method among many MVS methods for estimating the camera parameters of multi-view images. MVS estimates the depth map of each viewpoint using the results of SfM and reconstructs dense 3D point clouds. MVS in COLMAP, similar to PatchMatch, assigns depth and normal to each pixel as parameters initialized with random numbers, and then iteratively performs image matching and parameter propagation among multi-view images to optimize the depth and normal. To improve the accuracy of depth map estimation using PatchMatch, MVS in COLMAP utilizes the following ideas: (i) propagates parameters taking into account the geometry by selecting the pixel of interest and the corresponding view for each pixel based on the camera rotation, occlusion obstructing the view, and image resolution, (ii) employs NCC with bilateral weights for image matching in local regions, (iii) improves the accuracy of depth estimation by maximizing the photometric consistency and minimizing the geometric consistency based on reprojection errors between viewpoints, and (iv) removes outliers according to the confidence value calculated by the photometric consistency and the geometric consistency. COLMAP also has problems with low depth estimation accuracy in poor-texture regions and at object boundaries.

2.1.2. Deep Learning

Recently, a number of depth map estimation methods using deep learning have been developed [14–17,22]. Here, we describe one of the typical methods, MVSNet [15]. MVSNet estimates a depth map for each viewpoint through three steps: feature extraction from multi-view images, the creation of a cost volume, and depth map estimation. Let the image for which the depth map is to be obtained be the reference image, and the images in the neighborhood of the reference image be the neighboring images. Feature maps are extracted from both the reference image and the neighboring images using a 2D CNN. A virtual plane is assumed in the depth direction of the camera for the reference image, the feature maps of the neighboring images are projected onto the virtual plane by homography transformation, a feature volume at each viewpoint is created, and a scene cost volume is created by aggregating the feature volumes of the reference image and the neighboring images. The cost volume is used to determine the existence probability of object surfaces in the depth direction, and the depth of each pixel is estimated from its expected value.

As described above, MVS estimates the depth map using image matching based on the texture in the images and the features extracted by CNN. Because of the use of image matching, the depth map can be estimated with high accuracy in rich-texture regions, while the estimation accuracy degrades in poor-texture regions and at object boundaries. In addition, MVS is difficult to estimate the depth in regions containing occlusions even though the deformation between images is normalized using a homography transformation to improve the accuracy of image matching.

2.2. NeRF-Based Approaches

We describe a novel view synthesis method, i.e., NeRF [6] and depth map estimation using NeRF. We also describe Depth-Supervised NeRF (DS-NeRF) [23], which utilizes sparse 3D point clouds reconstructed by SfM, as a depth map estimation method using NeRF.

2.2.1. NeRF

NeRF estimates the radiance field of a 3D scene from multi-view images and camera parameters using an MLP, and synthesizes a novel view by volume rendering [24] the radiance field [6]. The MLP takes the coordinates $\boldsymbol{x} = (x, y, z)$ of a 3D point in its direction (θ, ϕ) as the input and the RGB value $\boldsymbol{c} = (r, g, b)$ of the 3D point and the density σ representing the opacity of the 3D point as the output. The ray $\boldsymbol{r}_i$ from the camera center $\boldsymbol{o}$ in the camera image I through the pixel i in the camera image I and the 3D point $\boldsymbol{x}_i$ corresponding to the pixel $i \in I$ is defined by

$$\boldsymbol{r}_i(t) = \boldsymbol{o} + t\boldsymbol{d}_i, \tag{1}$$

where t is the position on the ray and $\boldsymbol{d}_i$ is its direction (θ_i, ϕ_i) which observes the 3D point $\boldsymbol{x}$. From the RGB value $\boldsymbol{c}(\boldsymbol{r}_i(t), \boldsymbol{d}_i)$ of a 3D point on the ray and the density $\sigma(\boldsymbol{r}_i(t))$ of 3D points, the pixel value $\boldsymbol{C}_i$ at pixel i is calculated by

$$\boldsymbol{C}_i = \int_{t_{\text{near}}}^{t_{\text{far}}} T_i(t)\sigma(\boldsymbol{r}_i(t))\boldsymbol{c}(\boldsymbol{r}_i(t), \boldsymbol{d}_i)\, dt, \tag{2}$$

where t_{near} and t_{far} indicate the range of volume rendering and $T_i(t)$ is an accumulated transmittance function, which describes the phenomenon that the brightness of rays is attenuated by objects, and is defined by

$$T_i(t) = \exp\left(- \int_{t_{\text{near}}}^{t} \sigma(\boldsymbol{r}_i(s))\, ds \right). \tag{3}$$

In practice, since N 3D points on the sampled rays $\hat{\boldsymbol{r}}$ are used, Equation (2) can be rewritten as

$$\hat{\boldsymbol{C}}(\hat{\boldsymbol{r}}) = \sum_{j=1}^{N} T_j(1 - \exp(-\sigma_j\delta_j))\boldsymbol{c}_j, \tag{4}$$

where $\delta_j = t_{j+1} - t_j$ denotes the distance between adjacent 3D points located on the ray and T_j is given by

$$T_j = \exp\left(- \sum_{k=1}^{j-1} \sigma_k\delta_k \right). \tag{5}$$

The MLP is trained with the loss function $\mathcal{L}$ between the pixel values $\hat{\boldsymbol{C}}(\hat{\boldsymbol{r}})$ of the image synthesized by volume rendering and $\boldsymbol{C}^{gt}(\hat{\boldsymbol{r}})$ of the camera image, which is defined by

$$\mathcal{L} = \sum_{\hat{\boldsymbol{r}} \in \boldsymbol{R}} ||\hat{\boldsymbol{C}}(\hat{\boldsymbol{r}}) - \boldsymbol{C}^{gt}(\hat{\boldsymbol{r}})||^2, \tag{6}$$

where $\boldsymbol{R}$ is a set of rays passing through each pixel. In NeRF, the depth $D(\hat{\boldsymbol{r}})$ is calculated by using the density σ of sampled 3D points on the ray $\hat{\boldsymbol{r}}$ and T_i obtained from the density [25–28] as follows:

$$D(\hat{\boldsymbol{r}}) = \sum_{j=1}^{N} T_j\{1 - \exp(-\sigma_j\delta_j)\}t_j. \tag{7}$$

A depth map for each viewpoint can be obtained by calculating the depth for all the pixels. NeRF does not use image matching for local regions, and therefore can estimate depth maps with high accuracy in poor-texture regions and at object boundaries.

2.2.2. DS-NeRF

Here, we describe Depth-Supervised NeRF (DS-NeRF) [23], which combines NeRF and SfM in COLMAP as a method to improve the performance of NeRF. As mentioned above, NeRF trains an MLP using the color reconstruction loss between the synthesized image and the camera image. In addition to the color reconstruction loss, DS-NeRF uses the depth loss between the depth obtained by volume rendering and the depth obtained from the sparse 3D point cloud in SfM. DS-NeRF can train an MLP more efficiently than NeRF and can synthesize novel views from a small number of images. The depth loss $\mathcal{L}_{Depth}$ used in DS-NeRF is calculated based on KL divergence as follows:

$$\mathcal{L}_{Depth} \approx \mathbb{E}_{x_i \in \boldsymbol{X}_j} \sum_k \log h_k \exp\left\{ -\frac{\left(t_k - \boldsymbol{D}_{ij}\right)^2}{2\hat{\sigma}_i^2} \right\} \Delta t_k, \tag{8}$$

where X_j indicates a set of feature points visible from camera j, x_i indicates the i-th feature point, h_k indicates the existence probability of the object surface at the k-th sampling point on the ray, $\hat{\sigma}_i$ indicates the reprojection error at the i-th feature point x_i, and $\boldsymbol{D}_{ij}$ indicates the distance from camera j to the feature point i. The larger the reprojection error of the feature points, the weaker the loss constraint is to take into account the estimation error of the 3D points by SfM. Although depth maps can be estimated from a small number of images, sparse depth maps have to be used for training in DS-NeRF. Therefore, it is not always possible to synthesize a highly accurate depth map by volume rendering using the radiance field.

Recently, RC-MVSNet [17] has been proposed, which combines CasMVSNet [22] and NeRF to train CasMVSNet by unsupervised learning. Although unsupervised learning reduces the limitation on the amount of training data, the depth map cannot always be estimated with high accuracy since NeRF is estimated based on the depth map generated by CasMVSNet.

3. NeRF-Inspired Depth Map Refienment

As mentioned above, the depth map estimated by MVS does not obtain depth in poor-texture regions, occlusions, and at object boundaries. We propose a depth map estimation method multi-view images with NeRF-inspired depth map refinement. The proposed method differs from general NeRF in that it iteratively optimizes the MLP to synthesize the input image and the depth map estimated by MVS, rather than training the MLP to synthesize novel view images. NeRF trains the radiance field of the scene using the input multi-view images and uses it to synthesize novel view images. On the other hand, the proposed method refines the depth map by optimizing the radiance field of the scene so that the input multi-view images and the dense depth map can be synthesized. The proposed method corresponds to overfitting the training data from the viewpoint of NeRF. Since NeRF aims to synthesize novel view images, while the proposed method aims to refine the input depth maps, the proposed method can achieve its objective even by overfitting the training data in NeRF. In the following, we refer to "optimize" as overfitting the MLP to the training data to estimate a depth map from the same viewpoint as the training data. We also refer to "train" as synthesizing a novel view by training the MLP with the training data, i.e., normal NeRF. We describe an overview of the proposed method, the network architecture of the MLP used in the proposed method, and the objective functions for optimization in the following.

3.1. Overview

The proposed method consists of camera parameter estimation by SfM, depth map estimation by MVS, and depth map refinement by NeRF optimization, taking multi-view images as the input. The framework of the proposed method is shown in Figure 1, which is inspired by DS-NeRF [23]. DS-NeRF uses sparse 3D point clouds obtained by SfM to train the MLP so as to synthesize novel views using NeRF. On the other hand, the proposed method refines the depth map obtained by MVS through the optimization of an MLP, which is different to DS-NeRF. The proposed method uses COLMAP to estimate the camera parameters [13] and depth maps [3] to compare the performance of the proposed method with that of DS-NeRF. Therefore, it should be noted that the COLMAP process can be replaced by other SfM and/or MVS methods in the proposed method. The proposed method iteratively optimizes the MLP representing the radiance field using the dense depth map estimated by MVS. We optimize the MLP so that the depth map is synthesized by volume rendering to be close to the depth map estimated by MVS, and so that the image from the same viewpoint as the input image is synthesized. As a result, it is possible to estimate the depth in poor-texture regions and at object boundaries that cannot be estimated by MVS. We obtain a depth map that is more accurate than MVS by volume rendering the depth map using the optimized MLP.

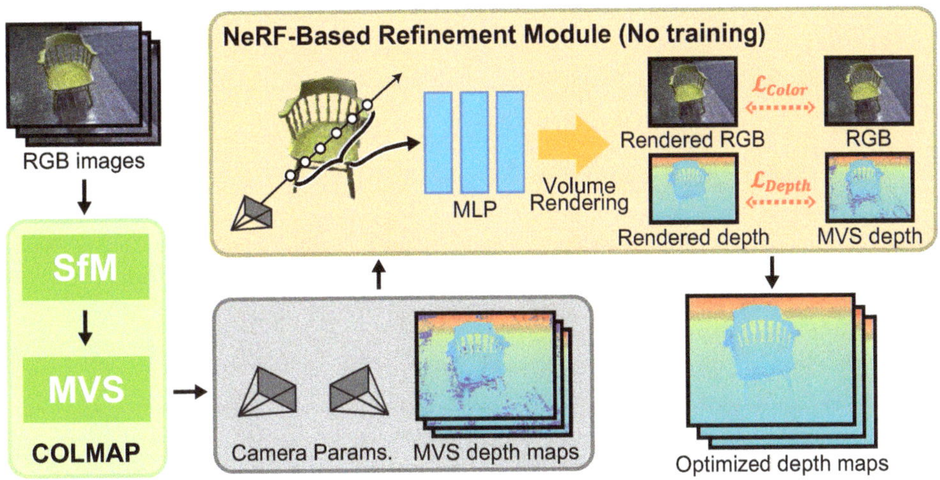

Figure 1. Overview of the proposed method (SfM: Structure from Motion, MVS: Multi-View Stereo).

3.2. Network Architecture of an MLP

An MLP, which refines the depth maps obtained by MVS, consists of the network architecture as shown in Figure 2. This network architecture is designed based on DS-NeRF [23]. A 3D point $\boldsymbol{x} = (x, y, z)$ and its direction $\boldsymbol{d} = (\phi, \theta)$ are inputs, and RGB values $\boldsymbol{c}$ and the density σ of $\boldsymbol{x}$ are outputs. Three-dimensional points $\boldsymbol{x}$ and view direction $\boldsymbol{d}$ are applied during positional encoding [6] to create higher dimensional vectors $\gamma(\boldsymbol{x})$ and $\gamma(\boldsymbol{d})$, which are input to the MLP. We generate 256-dimensional feature vectors passing $\gamma(\boldsymbol{x})$ through eight fully-connected layers with the ReLU activation function. The output of the fifth layer is concatenated with $\gamma(\boldsymbol{x})$ using skip connection. Then, the 3D point density σ and 256-dimensional feature vectors are obtained by passing them through a fully-connected layer. The output feature vector is then concatenated with the feature vector $\gamma(\boldsymbol{d})$, and the RGB values of the 3D points are output through a fully connected layer.

3.3. Objective Functions

We describe the objective functions that are required in the optimization of the MLP to refine the depth maps obtained by MVS. Note that we use the term "loss" in the following since the only differences between the loss function used in training the MLP and the objective function used in MLP optimization are the expressions "loss" and "error". The proposed method employs the color reconstruction loss $\mathcal{L}_{Color}$ [6] as the objective function for color reconstruction and the depth loss $\mathcal{L}_{Depth}$ based on Huber loss [29] as the objective function for depth reconstruction.

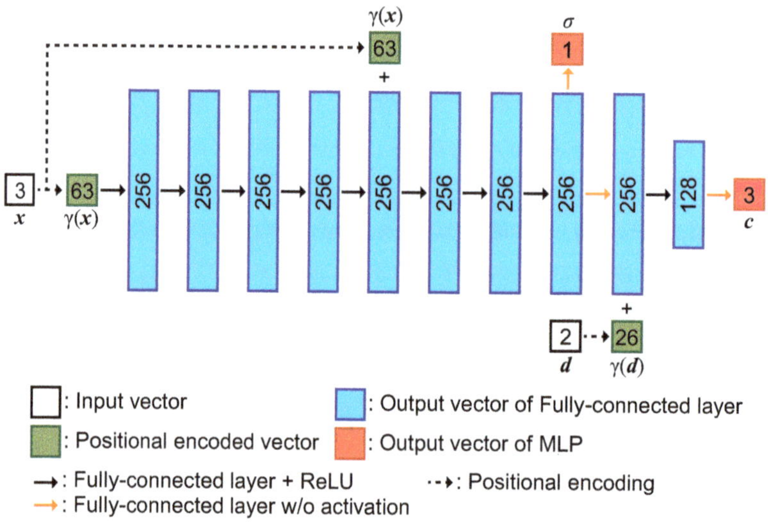

Figure 2. The network architecture of the MLP used in the proposed method, where the number inside the boxes indicates the dimension of each feature vector.

3.3.1. Color Reconstruction

The color reconstruction loss, $\mathcal{L}_{Color}$, is the mean squared error loss between the pixel values estimated by volume rendering using Equation (4) and the pixel values of the same pixel in the input image and is defined by

$$\mathcal{L}_{Color} = \sum_{j \in J} ||C_j - C_j^{gt}||^2, \tag{9}$$

where J indicates a set of pixels in the input image, C_j indicates pixel values synthesized by volume rendering at pixel j, and C_j^{gt} indicates pixel values of the same pixel in the input image.

3.3.2. Depth Reconstruction

We propose a new loss function based on Huber loss [29] for depth reconstruction that is robust against outliers. We consider that it is important to have robustness against outliers since the depth maps obtained by MVS in COLMAP contain many outliers. Huber loss is a loss function that combines L1 loss and L2 loss. Using the idea of Huber loss, the proposed method uses the mean squared error loss, i.e., L2 loss, when the error between the depth obtained by volume rendering and the depth obtained by MVS in COLMAP is smaller than a threshold ϵ, and the absolute error loss, i.e., L1 loss, when the error is larger than ϵ. The term $H(D_k, D_k^{mvs})$ based on Huber loss used in the depth loss $\mathcal{L}_{Depth}$ is defined by

$$H(D_k, D_k^{mvs}) = \begin{cases} \frac{a^2}{2} & |a| \leq \epsilon \\ \epsilon(|a| - \frac{\epsilon}{2}) & \text{otherwise} \end{cases}, \tag{10}$$

where D_k indicates the depth at pixel k obtained by volume rendering, D_k^{mvs} indicates the depth at pixel k in the depth map estimated by MVS in COLMAP, $a = D_k - D_k^{mvs}$, and $\epsilon = \frac{t_{far} - t_{near}}{N_{coarse} - 1}$. As mentioned above, the depth is obtained by accumulating the densities of 3D points on the rays in volume rendering. The error between the depth obtained by volume rendering and the depth obtained by MVS in COLMAP should be smaller than the distance between adjacent 3D points. Therefore, we use the number of sampling points N_{coarse} used for coarse sampling in hierarchical volume sampling as the threshold ϵ. Then, the depth loss $\mathcal{L}_{Depth}$ used in the proposed method is defined by

$$\mathcal{L}_{Depth} = \frac{1}{K} \sum_{k \in K} H(D_k, D_k^{mvs}), \tag{11}$$

where K indicates a set of pixels in the input image whose depth D_k^{mvs} is obtained by MVS in COLMAP. Note that the depth loss $\mathcal{L}_{Depth}$ is calculated only for pixels with depth obtained by MVS in COLMAP.

The iterative optimization of the MLP used in the proposed method employs an objective function that combines the color reconstruction loss and depth loss described above, which is given by

$$\mathcal{L} = \mathcal{L}_{Color} + \lambda_D \mathcal{L}_{Depth}, \tag{12}$$

where λ_D indicates a hyper parameter.

4. Experiments and Discussion

This section describes experiments to evaluate the accuracy of the proposed method using public datasets of multi-view images. We describe the dataset used in the experiments, the experimental conditions, evaluation metrics, ablation study of depth loss, accuracy comparison with conventional methods, and 3D reconstruction in the following.

4.1. Dataset

We describe two multi-view image datasets, i.e., the Redwood-3d scan dataset (https://redwood-data.org/3dscan/index.html (accessed on 7 February 2024)) [19] and the DTU dataset (https://roboimagedata.compute.dtu.dk/?page_id=36 (accessed on 7 February 2024)) [20], which are used in the experiments.

4.1.1. Redwood-3d Scan Dataset (Redwood)

Redwood consists of 10,933 RGB-D video images taken in a variety of scenes and 441 3D mesh models. There are 44 different categories of scenes, such as chairs, tables, sculptures, and plants. The RGB-D video images were taken by non-experts in computer vision, and many of them contain low-quality frames and poor-texture regions. Therefore, it is difficult to reconstruct 3D shapes from the multi-view images in Redwood using MVS due to external factors such as motion blur, noise, poor-textured objects, and illumination changes. In our experiments, we use 12 scenes: "amp#05668", "chair#04786", "chair#05119", "childseat#04134", "garden#02161", "mischardware#05645", "radio#09655", "sculpture#06287", "table#02169", "telephone#06133", "travelingbag#01991", and "trash-container#07226" as shown in Figure 3. We extract 11 frames from the RGB-D video image of each scene, and use the RGB image with 640 × 480 pixels of each frame as the input and the depth map as the ground truth for accuracy evaluation. The camera parameters for each viewpoint used in all the depth map estimation methods are estimated by SfM in COLMAP [13].

Figure 3. Example of images from Redwood used in the experiments, where images are extracted from the RGB-D video.

4.1.2. DTU Dataset (DTU)

DTU consists of multi-view images of a variety of objects, a 3D point cloud measured by a laser scanner, and the camera parameters. The multi-view images consist of a set of images with 1600×1200 pixels, which are taken of each object from 49 or 64 viewpoints. Multi-view images in DTU are acquired under the controlled environment. Therefore, we can evaluate the potential performance of MVS methods themselves since there are few external factors using DTU. There are 124 types of objects, such as building models, animal figurines, plants, and vegetables. We use the "scan9", "scan33", and "scan118" as shown in Figure 4. Due to the processing time, we resize the images to 800×600 pixels and use them as input images. Since the images in DTU were taken under seven different lighting conditions, we use the multi-view image taken under one of the seven lighting conditions. The camera parameters for each view used in all the depth map estimation methods are estimated by SfM in COLMAP [13]. Since DTU does not have the ground truth for evaluating the accuracy of the depth map estimation, we use the depth maps created by Yao et al. [5,15].

Figure 4. Example of images from DTU used in the experiments.

4.2. Experimental Condition

In our experiments, we compare the accuracy of depth map estimation among COLMAP [3], NeRF [6], DS-NeRF [23], RC-MVSNet [17], and the proposed method to demonstrate the effectiveness of the proposed method. NeRF and DS-NeRF train an MLP that represents the radiance field using multi-view images so that novel view images can be synthesized. By inputting a novel view direction to the trained MLP, the image and depth map of that view can be synthesized. NeRF and DS-NeRF need to train an MLP using training data and evaluate it on test data. On the other hand, the proposed method optimizes an MLP that represents the radiance field so that the input images and depth maps can be synthesized. In this experiment, we estimate depth maps for the input known viewpoints. To evaluate the accuracy under the same conditions as the proposed method, NeRF and DS-NeRF trained an MLP using the input multi-view images and use the trained MLP to synthesize depth maps for the input multi-view images. Therefore, we trained NeRF and DS-NeRF a certain number of times as in the proposed method. Table 1 shows the hyper parameters used in the experiments. The number of training or optimization iterations was set to 15,000 for Redwood and 100,000 for DTU, since the number of images and the number of pixels are different for each dataset. The batch size, which represents the number of rays in each iteration, was set to 5120. DS-NeRF and the proposed method, which require the depth map loss to be calculated, have a parameter λ_D that controls the ratio of depth rays used to calculate the depth loss within the batch size and the weights of the loss function. The ratio of depth rays used in DS-NeRF and the proposed method are set to 0.5 and 0.2, respectively, and λ_D is set to 0.1 for both methods. NeRF, DS-NeRF, and the proposed method use hierarchical volume sampling [6] as a sampling method based on the density of points on a ray. Hierarchical volume sampling first produces the color and density of N_{coarse} 3D points in a coarse network, and then produces the color and density of N_{fine} 3D points belonging to high-density regions in a fine network. We set $N_{coarse} = 64$ and $N_{fine} = N_{coarse} + 128$ for all the methods in the experiments. In the experiments, Adam [30] is used as the optimizer. The learning rate begins at 5.0×10^{-4} and decays exponentially to 5.0×10^{-5} during the optimization process. For RC-MVSNet, we use the trained model and evaluation code available in the official GitHub repository (https://github.com/Boese0601/RC-MVSNet (accessed on 26 Feburary 2024)). The threshold for the reprojection error used in depth map filtering is set to 0.5 pixels.

Table 1. A set of hyper parameters used in the experiments.

Method	# of Iterations		Batch Size	Ratio of Depth Rays	λ_D
	Redwood [19] [Times]	DTU [20] [Times]	[Rays]	[Rays/Batch Size]	
COLMAP [3]	–	–	–	–	–
NeRF [6]	15,000	100,000	5120	–	–
DS-NeRF [23]	15,000	100,000	5120	0.5	0.1
Proposed	15,000	100,000	5120	0.2	0.1

4.3. Evaluation Metrics

We evaluate the accuracy of depth map estimation using the following five evaluation metrics. In the following, y_i denotes the depth of the pixel i in the estimated depth map, y_i^* denotes the depth of pixel i in the ground-truth depth map, and T denotes a set of pixels for evaluation.

The first metric is the scale invariant logarithmic error (SILog) [21], which is defined by

$$\text{SILog}: \frac{1}{2\|T\|} \sum_{i \in T} \left(\log \frac{y_i}{y_i^*} + \frac{1}{\|T\|} \sum_{i \in T} \log \frac{y_i^*}{y_i} \right)^2. \quad (13)$$

This is a metric that evaluates the scale-independent error between the ground truth and estimated depths, where lower values indicate that the estimated depths are correct. For example, in Redwood, the depth map estimated by COLMAP is scale-independent, while the ground truth is millimeter-scale. In our experiments, the scale between the ground truth and the estimated depth map is estimated by the least-squares algorithm and adjusted to the millimeter scale for a fair evaluation. If the estimated depths contain outliers, the scale estimation has errors. SILog evaluates scale-invariant errors and is therefore less sensitive to errors in scale fitting.

The second metric is the Absolute Relative Difference (AbsRel), which is defined by

$$\text{AbsRel}: \frac{1}{\|T\|} \sum_{i \in T} \|y_i - y_i^*\| / y^*. \quad (14)$$

This is a metric that evaluates the absolute relative error between the ground truth and the estimated depths, where lower values indicate that the estimated depths are correct.

The third metric is the Squared Relative Difference (SqRel), which is defined by

$$\text{SqRel} : \frac{1}{\|T\|} \sum_{i \in T} \|y_i - y_i^*\|^2 / y_i^*. \tag{15}$$

This is a metric that evaluates the squared relative error between the ground truth and the estimated depths, where lower values indicate that the estimated depths are correct. SqRel is sensitive to outliers since the larger the error in the estimated value, the larger the evaluated value.

The fourth metric is Root Mean Squared Error (RMSE(log)) on a logarithmic scale, which is defined by

$$\text{RMSE (log)} : \sqrt{\frac{1}{\|T\|} \sum_{i \in T} \| \log y_i - \log y_i^* \|^2}. \tag{16}$$

This is a metric that evaluates the root mean square error between the ground truth and estimated depths, where lower values indicate that the estimated depths are correct.

The fifth metric evaluates the ratio between the ground truth and the estimated depths that is less than the threshold, which is given by

$$\delta < \text{threshold} : \% \text{ of } y_i \ s.t. \max_i (y_i / y_i^*, y_i^* / y_i) = \delta < \text{threshold}. \tag{17}$$

This indicates that, the larger the value, the more accurate the estimated depth.

The first to fourth metrics evaluate the error between the ground truth and the estimated depths, and the fifth evaluates the accuracy of the estimated depths. As mentioned in the first metric, the depth maps estimated by the conventional and proposed methods are different in scale from the ground truth measured in millimeters. Therefore, except for SILog, the scale of the depth maps has to be aligned when evaluating accuracy. In our experiments, the scale is obtained using the least-squares algorithm so that the error between the sparse depth at each view created from the sparse 3D point cloud estimated by SfM and the corresponding ground truth is small. Using the obtained scale, we evaluate the estimation accuracy by converting the depth maps estimated by each method to the millimeter scale.

4.4. Ablation Study of Depth Loss

In this subsection, we describe an ablation study on the depth loss of the proposed method to confirm the dependence of the proposed method on the parameters. In this experiment, we use amp#05668 in Redwood.

4.4.1. Threshold of Huber Loss

The depth loss used in the proposed method is designed based on the Huber loss as described in Section 3.3.2. Huber loss uses L2 loss if the difference between the estimated depth and the true value is less than or equal to the threshold ϵ, otherwise L1 loss is used. Therefore, ϵ has an impact on the accuracy of the depth map estimation. Table 2 shows the accuracy of depth map estimation using the proposed method when Huber loss ϵ is multiplied by the scale factor s, where the numbers in bold and underlined indicate the best and second best in each evaluation metric, respectively. In the case of ϵ multiplied by 0.5, i.e., $s = 0.5$, AbsRel, SqRel, and RMSE, the accuracy of the depth estimation is the best, while SILog and $\delta < 1.25$ are the third most accurate. In the case of ϵ multiplied by 0.1 and 2, i.e., $s = 0.1, 2.0$, the accuracy of the depth estimation is degraded for most of the evaluation metrics. On the other hand, when ϵ is used, i.e., $s = 1.0$, the accuracy of depth estimation is within the top two across all of the evaluation metrics. From the above, the proposed method employs $s = 1.0$ as a scale factor for the threshold ϵ for depth loss.

Table 2. Experimental results of depth map estimation using the proposed method when ϵ of the Huber loss is multiplied by the scale factor s. The numbers in bold and underlined indicate the best and the second best in each evaluation metric, respectively. The up arrow indicates that higher values represent better results, while the down arrow indicates that lower values represent better results, respectively.

s	Error↓				Accuracy ↑
	SILog [log(mm) × 100]	**AbsRel** [%]	**SqRel** [%]	**RMSE (log)** [log(mm)]	**$\delta < 1.25$** [%]
0.1	**0.5626**	0.0804	3.135	0.1097	97.61
0.5	0.5787	**0.0786**	**2.990**	**0.1094**	98.51
1.0	0.5759	**0.0786**	2.992	0.1095	98.53
2.0	0.5970	0.0779	2.988	0.1099	**98.55**

4.4.2. Hyper Parameter of Objective Function

The objective function used in the proposed method has a hyperparameter λ_D that adjusts the balance between the color reconstruction loss $\mathcal{L}_{Color}$ and the depth loss $\mathcal{L}_{Depth}$. In this experiment, we perform an ablation study on λ_D. Table 3 shows the accuracy of the depth map estimation of the proposed method when λ_D is changed. The accuracy of the depth map estimation of the proposed method is the highest when $\lambda_D = 0.1$. Therefore, $\lambda_D = 0.1$ is used in the following experiments.

Table 3. Experimental results of depth map estimation using the proposed method when λ_D of the objective function is changed. The numbers in bold indicate the best in each evaluation metric. The up arrow indicates that higher values represent better results, while the down arrow indicates that lower values represent better results, respectively.

λ_D	Error↓				Accuracy ↑
	SILog [log(mm) × 100]	**AbsRel** [%]	**SqRel** [%]	**RMSE (log)** [log(mm)]	**$\delta < 1.25$** [%]
0.05	0.5822	0.0789	3.008	0.1099	98.45
0.1	**0.5759**	**0.0786**	**2.992**	**0.1095**	**98.53**
0.2	0.6086	0.0791	3.032	0.1115	98.44
0.5	0.6180	**0.0786**	3.040	0.1115	98.41

4.4.3. Difference between Other Depth Loss

In this experiment, we conducted the ablation study for the proposed method using MSE (L2 loss), MAE (L1 loss), and the proposed depth loss based on Huber loss as the depth loss $\mathcal{L}_{Depth}$ of the proposed method. We used "amp#05668" from Redwood as input images in this experiment. Table 4 shows the results of the ablation study. As for MSE, the accuracy of depth map estimation is high for AbsRel and RMSE(log), which is comparable to that using the proposed depth loss. As for the proposed depth loss, the accuracy of depth map estimation is high for SILog, SqRel, and $\delta < 1.25$. Since the SILog of the proposed depth loss is the highest, the use of the proposed depth loss makes it possible to estimate a smooth and highly accurate depth map. As mentioned above, the evaluation metrics other than SILog are sensitive to the scale between the estimated depth map and the ground truth. The high value of SILog indicates that the estimation accuracy of the depth map is high independent of the scale fitting error. Figure 5 shows the depth maps obtained by each method. In the case of MSE, the object boundary is smooth, although there are some missing areas on the surface of the amplifier. This is because MSE is sensitive to outliers, and the MLP was optimized to be close to the outlier of the depth map estimated by MVS in COLMAP. In the case of using MAE, there is no missing area on the object surface, although the object boundary is not smooth. In the case of the proposed depth loss, there is no missing area on the object's surface and the object boundary is sharp. As a result, the depth map can be estimated with the highest accuracy using the proposed depth loss.

Table 4. Summary of qualitative experimental results in the ablation study for the proposed methods with a variety of depth loss functions. The numbers in bold indicate the best in each evaluation metric. The up arrow indicates that higher values represent better results, while the down arrow indicates that lower values represent better results, respectively.

Depth Loss	Error↓				Accuracy ↑
	SILog [log(mm) × 100]	AbsRel [%]	SqRel [%]	RMSE (log) [log(mm)]	$\delta < 1.25$ [%]
MSE	0.6413	**0.0773**	3.025	**0.1097**	98.05
MAE	0.6623	0.0790	3.107	0.1139	98.24
Huber (Proposed)	**0.5765**	0.0791	**3.015**	0.1099	**98.47**

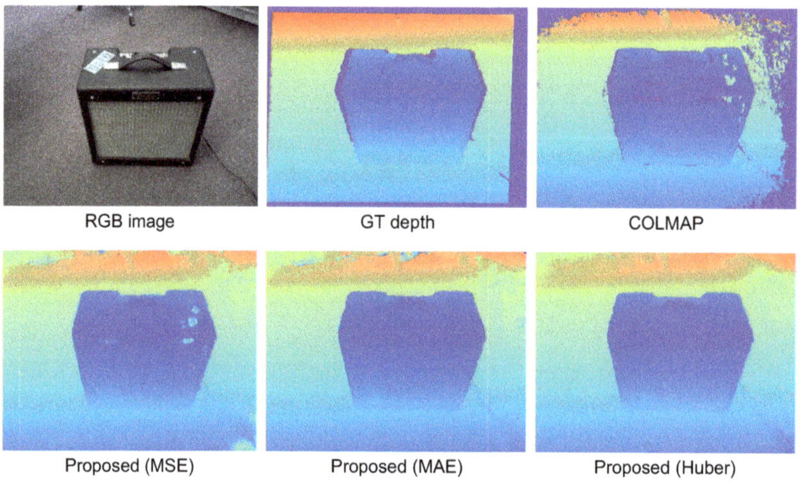

Figure 5. Depth maps estimated by COLMAP and the proposed method with a variety of depth loss functions, where blue in the depth map indicates close to the camera and red indicates far from the camera.

4.5. Comparison with Conventional Methods

This section demonstrates the effectiveness of the proposed method by comparing the accuracy of depth map estimation using the conventional and proposed methods using Redwood and DTU.

Tables 5 and 6 show the quantitative results for Redwood. COLMAP and NeRF have larger errors and lower accuracy than the other methods, indicating that the depths contain large errors. RC-MVSNet and the proposed method exhibit better results than other methods in most evaluation metrics. In particular, the SILog for the proposed method is smaller than that for COLMAP, NeRF, DS-NeRF, and RC-MVSNet in most cases. This result indicates that the depth map refined by the proposed method contains fewer errors. Figure 6 shows the depth maps estimated by each method. RC-MVSNet shows comparable results to the proposed method in the quantitative evaluation; however, it has more missing regions in the depth map compared to the other methods. The reason for this is that RC-MVSNet uses filtering of the depth map based on reprojection errors. Therefore, the estimated depths are highly accurate, while the depth maps include missing regions.

The proposed method estimates the depth map more smoothly than the conventional methods. For example, the proposed method can estimate accurate and smooth depths of flat surfaces such as the floor and the ground in "amp#05668" and "childseat#04134". This is because the proposed method optimizes the radiance field based on the depth map estimated by COLMAP, unlike NeRF and DS-NeRF. These results indicate that the depth map estimated by COLMAP can be refined through the iterative optimization of an MLP representing the radiance field since the proposed method has fewer missing regions than the depth map estimated by COLMAP. On the other hand, neither COLMAP nor the proposed method could estimate the depth of the surface of the trashcan with poor texture in "trashcontainer#07226". In "travelingbag#01991", COLMAP has missing depths for the surface of the traveling bag, while the proposed method smoothly estimated their depths. The difference between "trashcontainer#07226" and "travelingbag#01991" is the size of the missing region in the depth map estimated by COLMAP. If the missing regions in the input depth map are large, the proposed method cannot interpolate the depth map.

Table 5. Summary of qualitative experimental results in Redwood. The numbers in bold indicate the best in each evaluation metric. The up arrow indicates that higher values represent better results, while the down arrow indicates that lower values represent better results, respectively.

Datasets	Method	Error↓				Accuracy ↑
		SILog [log(mm) $\times$ 100]	**AbsRel** [%]	**SqRel** [%]	**RMSE (log)** [log(mm)]	$\delta < 1.25$ [%]
amp #05668	COLMAP [3]	6.533	0.0906	4.248	0.2711	97.64
	NeRF [6]	8.313	0.2087	9.784	0.3913	84.64
	DS-NeRF [23]	0.7120	0.0794	3.289	0.1146	**99.40**
	RC-MVSNet [17]	4.637	0.0857	3.175	0.1678	98.64
	Proposed	**0.5759**	**0.0786**	**2.992**	**0.1095**	99.18
chair #04786	COLMAP [3]	12.86	0.1745	8.686	0.3982	95.94
	NeRF [6]	27.22	0.5636	24.51	1.252	40.91
	DS-NeRF [23]	2.895	0.1795	8.638	0.2523	97.22
	RC-MVSNet [17]	**0.9583**	0.1595	**6.554**	**0.1974**	**98.81**
	Proposed	1.315	**0.1556**	6.785	0.1982	98.62
chair #05119	COLMAP [3]	17.45	0.1016	8.400	0.4359	95.28
	NeRF [6]	16.41	0.3130	19.53	0.5966	71.09
	DS-NeRF [23]	1.098	0.0754	5.089	0.1167	**99.51**
	RC-MVSNet [17]	**0.8361**	**0.0656**	**4.396**	0.1054	99.35
	Proposed	1.026	0.0663	4.507	0.1130	99.21
childseat #04134	COLMAP [3]	4.151	0.0539	2.674	0.2009	99.25
	NeRF [6]	3.328	0.1345	5.598	0.1900	99.99
	DS-NeRF [23]	0.1874	0.0527	1.890	0.0624	**100.0**
	RC-MVSNet [17]	**0.1023**	**0.0481**	1.692	**0.0554**	**100.0**
	Proposed	0.1280	0.0488	**1.670**	0.0563	**100.0**
garden #02161	COLMAP [3]	3.916	0.0928	5.647	0.2174	98.47
	NeRF [6]	9.752	0.2282	14.92	0.4288	83.19
	DS-NeRF [23]	0.8502	0.0892	5.165	0.1272	99.33
	RC-MVSNet [17]	**0.4220**	**0.0824**	**4.139**	**0.1032**	**99.73**
	Proposed	0.8336	0.0883	4.969	0.1269	99.09
mischardware #05645	COLMAP [3]	16.25	0.1213	14.77	0.4189	95.09
	NeRF [6]	7.251	0.2172	12.80	0.3726	90.79
	DS-NeRF [23]	1.913	0.1001	5.966	0.1700	99.73
	RC-MVSNet [17]	2.886	**0.0656**	**3.837**	0.1327	99.37
	Proposed	**0.8973**	0.0664	4.030	**0.1137**	**99.74**

Table 6. Summary of qualitative experimental results in Redwood (continued). The numbers in bold indicate the best in each evaluation metric. The up arrow indicates that higher values represent better results, while the down arrow indicates that lower values represent better results, respectively.

Datasets	Method	Error↓				Accuracy ↑
		SILog [log(mm) × 100]	AbsRel [%]	SqRel [%]	RMSE (log) [log(mm)]	$\delta < 1.25$ [%]
radio #09655	COLMAP [3]	7.308	0.0606	14.98	0.2666	98.07
	NeRF [6]	3.968	0.1430	6.854	0.2345	98.497
	DS-NeRF [23]	1.501	0.0596	4.134	0.1233	99.81
	RC-MVSNet [17]	7.078	0.0350	2.521	0.1702	98.25
	Proposed	**0.2654**	**0.0235**	**1.730**	**0.0520**	**99.99**
sculpture #06287	COLMAP [3]	31.79	0.3789	10.38	0.7903	86.08
	NeRF [6]	6.3333	0.5077	13.431	0.7880	44.12
	DS-NeRF [23]	1.136	0.3292	8.511	0.4179	97.58
	RC-MVSNet [17]	5.573	0.3340	**8.103**	0.4584	96.64
	Proposed	**0.8592**	**0.3230**	8.280	**0.4037**	**98.08**
table #02169	COLMAP [3]	5.456	0.1145	8.168	0.2391	97.53
	NeRF [6]	23.94	0.1973	20.15	0.5058	87.62
	DS-NeRF [23]	5.005	0.1320	10.358	0.2251	95.90
	RC-MVSNet [17]	2.106	**0.1091**	**5.761**	**0.1517**	**99.03**
	Proposed	**1.920**	**0.1091**	6.927	0.1572	98.50
telephone #06133	COLMAP [3]	13.98	0.1245	7.174	0.3890	94.29
	NeRF [6]	16.52	0.2938	13.25	0.5518	75.79
	DS-NeRF [23]	2.935	0.1196	5.848	0.1949	97.87
	RC-MVSNet [17]	7.957	0.0962	**4.651**	0.2602	97.07
	Proposed	**2.412**	**0.0915**	4.909	**0.1676**	**98.19**
trashcontainer #07226	COLMAP [3]	22.14	0.1117	6.877	0.4908	94.15
	NeRF [6]	2.085	0.1083	6.027	0.1874	99.03
	DS-NeRF [23]	0.2313	**0.0563**	2.481	0.0716	**99.99**
	RC-MVSNet [17]	**0.0332**	0.0566	**1.930**	**0.0611**	**99.99**
	Proposed	0.1365	0.0564	2.159	0.0672	99.98
travelingbag #01991	COLMAP [3]	12.18	0.0800	7.401	0.347	95.85
	NeRF [6]	1.401	0.0760	5.071	0.1231	98.86
	DS-NeRF [23]	0.9334	0.0540	3.691	0.090	**99.06**
	RC-MVSNet [17]	29.25	0.1075	7.537	0.4499	92.53
	Proposed	**0.9091**	**0.0487**	**3.497**	**0.087**	**99.06**

Table 7 shows the quantitative results for DTU. The proposed method exhibits better results than the conventional methods in most evaluation metrics. In particular, the SILog for the proposed method is smaller or equal to that for COLMAP, NeRF, DS-NeRF, and RC-MVSNet. The proposed method has few large outliers in the depths since the errors are small and the accuracy is high, as shown in Table 7. Figure 7 shows the depth maps estimated by each method. All of the methods estimated depth maps with high accuracy in DTU. As mentioned in the experimental results for Redwood, RC-MVSNet stands out as having missing regions compared to the other methods. NeRF, DS-NeRF, and the proposed method estimated accurate depth maps even for poor-texture regions compared to COLMAP since the depth maps are synthesized from the radiance field.

In "scan118", NeRF has small missing regions on the object surface, while DS-NeRF and the proposed method do not have such regions. Since the proposed method has less noise near the object boundaries than DS-NeRF, the complementarity between MVS and NeRF can be utilized to estimate the depth map. On the other hand, the proposed method did not significantly improve the accuracy of depth map estimation for DTU compared to Redwood. This is because the size and number of input images differ between DTU and Redwood. Redwood uses 11 images with 640 × 480 pixels, while DTU uses 49 images with

800 × 600 pixels. The multi-view images in DTU have a sufficient number of viewpoints for depth map estimation and are rich enough in object texture to allow the depth map to be estimated with high accuracy even with conventional methods.

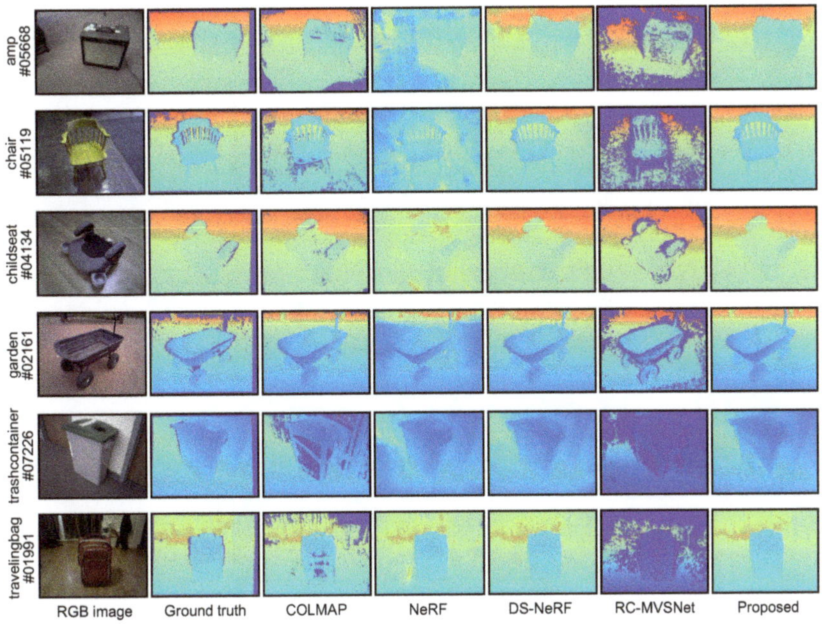

Figure 6. Estimated depth maps for each method in Redwood, where blue in the depth map indicates close to the camera and red indicates far from the camera.

Table 7. Summary of qualitative experimental results in DTU. The numbers in bold indicate the best in each evaluation metric. The up arrow indicates that higher values represent better results, while the down arrow indicates that lower values represent better results, respectively.

Scene	Method	Error↓				Accuracy ↑
		SILog [log(mm) × 100]	AbsRel [%]	SqRel [%]	RMSE (log) [log(mm)]	$\delta < 1.25$ [%]
scan9	COLMAP [3]	6.039	0.3602	9.059	0.5097	98.10
	NeRF [6]	0.8856	**0.3528**	8.793	0.4324	99.72
	DS-NeRF [23]	0.7815	0.3529	8.784	**0.4327**	**99.74**
	RC-MVSNet [17]	13.83	0.3785	9.6952	0.6160	93.40
	Proposed	**0.7280**	0.3530	**8.783**	0.4330	**99.74**
scan33	COLMAP [3]	14.68	0.1064	4.428	0.3971	96.76
	NeRF [6]	1.115	0.0840	3.230	0.1251	99.70
	DS-NeRF [23]	1.034	**0.0837**	3.093	0.1231	99.71
	RC-MVSNet [17]	7.376	0.0935	3.646	0.2886	97.71
	Proposed	**0.9809**	**0.0837**	**3.002**	**0.1219**	**99.72**
scan118	COLMAP [3]	6.932	0.0372	2.968	0.2629	98.61
	NeRF [6]	0.9723	0.0342	3.004	0.0984	99.38
	DS-NeRF [23]	0.7852	0.0302	2.490	0.0888	99.43
	RC-MVSNet [17]	7.091	0.0421	2.996	0.2476	97.69
	Proposed	**0.7282**	**0.0296**	**2.318**	**0.0855**	**99.45**

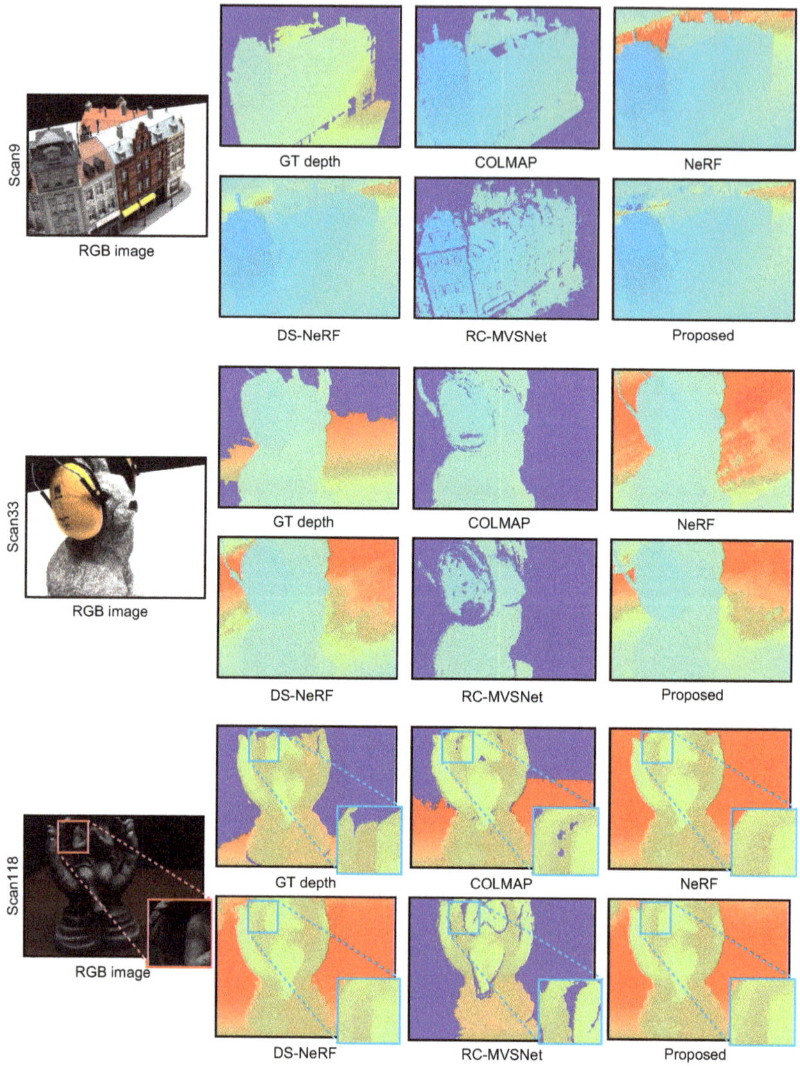

Figure 7. Estimated depth maps for each method in DTU, where blue in the depth map indicates close to the camera and red indicates far from the camera.

4.6. 3D Reconstruction

We reconstructed the 3D point clouds by applying depth map fusion [7,12,31] to the depth maps estimated by COLMAP and the proposed method. In this experiment, we used "Scan9", "Scan33", and "Scan118" from the DTU dataset, and multi-view images taken outdoors by the authors.

Figure 8 shows the reconstructed 3D point clouds for COLMAP and the proposed method. Note that the background regions are detected by image segmentation using SAM [32] and are masked in the depth maps to reconstruct the 3D point clouds for better visibility. In "scan 9", the proposed method has fewer missing regions on the roofs of building, and fewer outliers around chimneys and walls than COLMAP. In "scan33", COLMAP cannot reconstruct 3D points in the region with poor texture on the headset, while the proposed method can reconstruct 3D points even in such a region. In "scan118", the proposed method can reconstruct the 3D points in the region where COLMAP cannot. In particular, in "scan118", the proposed method has a wider reconstruction range than COLMAP. These results indicate that the proposed method can reconstruct regions that cannot be reconstructed by COLMAP by refining the depth map estimated by COLMAP.

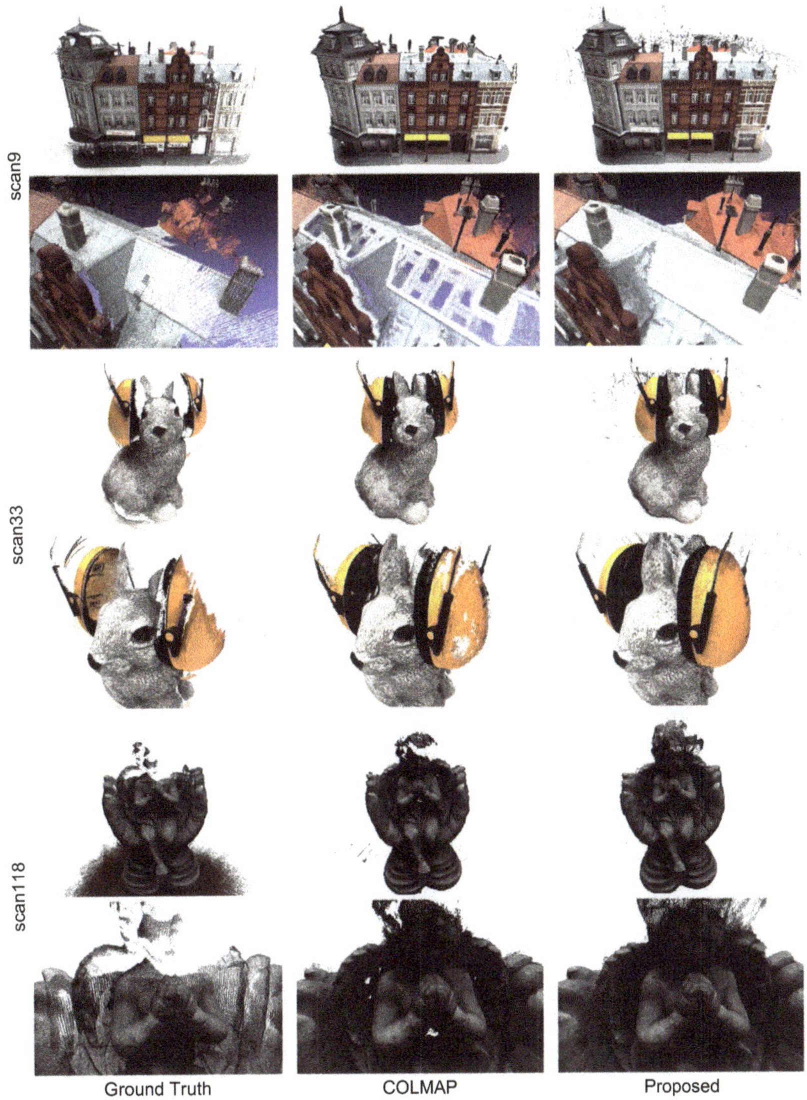

Figure 8. 3D point clouds reconstructed from estimated depth maps by COLMAP and the proposed method in DTU.

We evaluate the applicability of the proposed method by performing 3D reconstruction from multi-view images taken outdoors using an ordinary camera. The dataset consists of 35 RGB images of "Shore to Shore", which is a 14-foot bronze-cast sculpture located in Vancouver's Stanley Park, Canada, taken by the authors in June 2023. Figure 9 shows examples of images used in this experiment. It is a difficult situation to apply multi-view stereo and NeRF to since not only the sculpture but also dynamic objects such as tourists are in the image. Figure 10 shows the results of 3D reconstruction from multi-view images using COLMAP and the proposed method. COLMAP reconstructs the details of the sculpture, while there are many outliers on the object's surface and at the object boundaries. The proposed method reconstructs the sculpture with high accuracy due to there being few outliers on the object's surface. From the above, the proposed method can refine the depth maps estimated by COLMAP in real-world environments.

Figure 9. Examples of images of "Shore to Shore", which is a 14-foot bronze-cast sculpture located in Vancouver's Stanley Park, Canada, taken by the authors in June 2023.

Figure 10. Three-dimensional point clouds reconstructed from depth maps estimated by COLMAP and the proposed method in our dataset.

5. Conclusions

In this paper, we proposed a method to refine the depth maps obtained by MVS through the iterative optimization of an MLP in NeRF. We focused on the fact that MVS can accurately estimate depths in rich-texture regions and NeRF can accurately estimate depths in poor-texture regions and object boundaries, and exploited the complementarity between them. From the viewpoint of NeRF, this approach corresponds to overfitting the MLP with training data, while we conceived of optimizing the MLPs using input images to refine their depth maps. Through a set of experiments using the Redwood-3dscan dataset [19] and the DTU dataset [20], we clearly demonstrated the effectiveness of the proposed method compared to conventional methods. One of the challenging tasks in MVS is to reconstruct the 3D shapes of transparent and translucent objects [33]. The method described in this paper cannot reconstruct the 3D shapes of transparent and translucent objects since the depth map estimated by COLMAP is used. The 3D shapes of transparent and translucent objects can be reconstructed by using photometric stereo, which estimates surface normals from images taken by a camera under varying lighting [34]. NeRF can also consider the degree of transparency on the rays to take into account transparent and translucent objects. We expect that the combination of photometric stereo and the proposed method will be effective in addressing this task. Thus, we will consider refining the depth maps obtained by other MVS using the proposed method and also optimizing the camera parameters by NeRF in our framework.

Author Contributions: Funding acquisition, K.I. and T.A.; Methodology, S.I. and K.M.; Supervision, K.I. and T.A.; Writing—original, S.I.; Writing—review and editing, K.I. All authors have read and agreed to the published version of the manuscript.

Funding: This work was supported, in part, by JSPS KAKENHI Grant Numbers 21H03457 and 23H00463.

Institutional Review Board Statement: Not applicable.

Informed Consent Statement: Not applicable.

Data Availability Statement: Publicly available datasets were analyzed in this study. The Redwood-3dscan dataset can be found here: http://redwood-data.org/3dscan/ (accessed on 7 February 2024). The DTU dataset can be found here: https://roboimagedata.compute.dtu.dk/?page_id=36 (accessed on 7 February 2024).

Conflicts of Interest: The authors declare no conflicts of interest.

References

1. Szeliski, R. *Computer Vision: Algorithms and Applications*; Springer: New York, NY, USA, 2010.
2. Seitz, S.M.; Curless, B.; Diebe, J.; Scharstein, D.; Szeliski, R. A comparison and evaluation of Multi-View Stereo reconstruction algorithms. In Proceedings of the 2006 IEEE Computer Society Conference on Computer Vision and Pattern Recognition, New York, NY, USA, 17–22 June 2006; pp. 519–528.
3. Schönberger, J.L.; Zheng, E.; Pollefeys, M.; Frahm, J. Pixelwise view selection for unstructured Multi-View Stereo. In *Proceedings of the European Conference Computer Vision, Amsterdam, The Netherlands, 11–14 October 2016*; Springer: Cham, Switzerland, 2016; pp. 501–518.
4. Collins, R.T. A space-sweep approach to true multi-image matching. In Proceedings of the CVPR IEEE Computer Society Conference on Computer Vision and Pattern Recognition, San Francisco, CA, USA, 18–20 June 1996; pp. 358–363.
5. Yao, Y.; Luo, Z.; Li, S.; Shen, T.; Fang, T.; Quan, L. Recurrent MVSNet for high-resolution multi-view stereo depth inference. In Proceedings of the 2019 IEEE/CVF Conference on Computer Vision and Pattern Recognition, Long Beach, CA, USA, 6–20 June 2019; pp. 5525–5534.
6. Mildenhall, B.; Srinivasan, P.P.; Tancik, M.; Barron, J.T.; Ramamoorthi, R.; Ng, R. NeRF: Respresenting scenes as neural radiance fields for view synthesis. In Proceedings of the 16th European Conference Computer Vision, Glasgow, UK, 23–28 August 2020; pp. 405–421.
7. Ito, K.; Ito, T.; Aoki, T. PM-MVS: Patchmatch multi-view stereo. *Mach. Vis. Appl.* **2023**, *34*, 32–47. [CrossRef]
8. Gallup, D.; Frahm, J.M.; Mordohai, P.; Yang, Q.; Pollefeys, M. Real-time plane-sweeping stereo with multiple sweeping directions. In Proceedings of the 2007 IEEE Conference on Computer Vision and Pattern Recognition, Minneapolis, MN, USA, 17–22 June 2007; pp. 1–8.
9. Barnes, C.; Shechtman, E.; Flinkelstein, A.; Goldman, B.D. PatchMatch: A randomized correspondence algorithm for structural image editing. *ACM Trans. Graph.* **2009**, *28*, 24. [CrossRef]
10. Bleyer, M.; Rhemann, C.; Rother, C. PatchMatch Stereo-Stereo matching with slanted support windows. In Proceedings of the British Machine Vision Conference, Dundee, UK, 29 August–2 September 2011; pp. 1–11.
11. Zhang, E.; Dunn, E.; Joic, V.; Frahm, J.M. PatchMatch based joint view selection a nd depthmap estimation. In Proceedings of the 2014 IEEE Conference on Computer Vision and Pattern Recognition, Columbus, OH, USA, 23–28 June 2014; pp. 1510–1517.
12. Hiradate, K.; Ito, K.; Aoki, T.; Watanabe, T.; Unten, H. An extension of PatchMatch stereo for 3D reconstruction from multi-view images. In Proceedings of the 2015 3rd IAPR Asian Conference on Pattern Recognition (ACPR), Kuala Lumpur, Malaysia, 3–6 November 2015; pp. 61–65.
13. Schönberger, J.L.; Frahm, J. Structure-from-Motion revisited. In Proceedings of the IEEE Conference Computer Vision and Pattern Recognition, Las Vegas, NV, USA, 27–30 June 2016; pp. 4104–4113.
14. Yao, Y.; Luo, Z.; Li, S.; Fang, T.; Quan, L. Deepmvs: Learning multi-view stereopsis. In Proceedings of the IEEE Conference Computer Vision and Pattern Recognition, Salt Lake City, UT, USA, 18–22 June 2018; pp. 2821–2830.
15. Yao, Y.; Luo, Z.; Li, S.; Fang, T.; Quan, L. MVSNet: Depth inference for unstructured Multi-View Stereo. In Proceedings of the European Conference Computer Vision, Munich, Germany, 8–14 September 2018; pp. 767–783.

16. Wang, F.; Galliani, S.; Vogel, C.; Speciale, P.; Pollefeys, M. PatchmatchNet: Learned multi-view patchmatch stereo. In Proceedings of the IEEE Conference Computer Vision and Patter Recognition, Nashville, TN, USA, 20–25 June 2021; pp. 14194–14203.
17. Chang, D.; Božič, A.; Zhang, T.; Yan, Q.; Chen, Y.; Süsstrunk, S.; Nießner, M. RC-MVSNet: Unsupervised multi-view stereo with neural rendering. In Proceedings of the European Conference Computer Vision, Tel Aviv, Israel, 23–27 October 2022; pp. 665–680.
18. Goodfellow, I.; Bengio, Y.; Courville, A. *Deep Learning*; MIT Press: Cambridge, MA, USA, 2016.
19. Choi, S.; Zhou, Q.; Miller, S.; Koltun, V. A large dataset of object scans. *arXiv* **2016**, arXiv:1602.02481.
20. Jensen, R.; Dahl, A.; Vogiatzis, G.; Tola, E.; Aanæs, H. Large scale multi-view stereopsis evaluation. In Proceedings of the IEEE Conference Computer Vision and Pattern Recognition, Columbus, OH, USA, 23–28 June 2014; pp. 406–413.
21. Eigen, D.; Puhrsch, C.; Fergus, R. Depth map prediction from a single image using a multi-scale deep network. In Proceedings of the 27th International Conference on Neural Information Processing Systems, Montreal, QC, Canada, 8–13 December 2014; pp. 1–9.
22. Gu, X.; Fan, Z.; Zhu, S.; Dai, Z.; Tan, F.; Tan, P. Cascade cost volume for high-resolution multi-view stereo and stereo matching. In Proceedings of the IEEE/CVF Conference Computer Vision and Pattern Recognition, Seattle, WA, USA, 13–19 June 2020; pp. 2495–2504.
23. Deng, K.; Liu, A.; Zhu, J.Y.; Ramanan, D. Depth-supervised NeRF: Fewer views and faster training for free. In Proceedings of the IEEE/CVF Conference Computer Vision and Pattern Recognition, New Orleans, LA, USA, 18–24 June 2022; pp. 12882–12891.
24. Kajiya, J.T.; Herzen, B.P.V. Ray tracing volume densities. In Proceedings of the 11th Annual Conference on Computer Graphics and Interactive Techniques, Minneapolis, MN, USA, 23–27 July 1984; Volume 18, pp. 165–174.
25. Roessle, B.; Barron, J.T.; Mildenhall, B.; Srinivasan, P.P.; Nießner, M. Dense depth priors for neural radiance fields from sparse input views. In Proceedings of the IEEE/CVF Conference Computer Vision and Pattern Recognition, New Orleans, LA, USA, 18–24 June 2022; pp. 12892–12901.
26. Wei, Y.; Liu, S.; Rao, Y.; Zhao, W.; Lu, J.; Zhou, J. Nerfingmvs: Guided optimization of neural radiance fields for indoor multi-view stereo. In Proceedings of the IEEE/CVF International Conference Computer Vision, Montreal, BC, Canada, 11–17 October 2021; pp. 5610–5619.
27. Rematas, K.; Liu, A.; Srinivasan, P.; Barron, J.; Tagliasacchi, A.; Funkhouser, T.; Ferrari, V. Urban radiance fields. In Proceedings of the IEEE/CVF Conference Computer Vision and Pattern Recognition, New Orleans, LA, USA, 18–24 June 2022; pp. 12932–12942.
28. Toshi, F.; Tonioni, A.; Gregorio, D.D.; Poggi, M. NeRF-supervised deep stereo. In Proceedings of the IEEE/CVF Conference Computer Vision and Pattern Recognition, Vancouver, BC, Canada, 18–22 June 2023; pp. 855–866.
29. Huber, P.J. Robust Estimation of a Location Parameter. *Ann. Math. Statist.* **1964**, *35*, 73–101. [CrossRef]
30. Kingma, D.P.; Ba, J. Adam: A method for stochastic optimization. In Proceedings of the International Conference Learning Representations, San Diego, CA, USA, 7–9 May 2015; pp. 1–15.
31. Yodokawa, K.; Ito, K.; Aoki, T.; Sakai, S.; Watanabe, T.; Masuda, T. Outlier and artifact removal filters for multi-view stereo. In Proceedings of the International Conference Image Processing, Quebec City, QC, Canada, 27–30 September 2015; pp. 3638–3642.
32. Kirillov, A.; Mintun, E.; Ravi, N.; Mao, H.; Rolland, C.; Gustafson, L.; Xiao, T.; Whitehead, S.; Berg, C.A.; Lo, W.; et al. Segment anything. In Proceedings of the IEEE/CVF International Conference Computer Vision, Paris, France, 2–3 October 2023; pp. 4015–4026.
33. Weibel, J.B.; Sebeto, P.; Thalhammer, S.; Vincze, M. Challenges of Depth Estimation for Transparent Objects. In Proceedings of the International Symposium Visual Computing (LNCS 14361), Lake Tahoe, NV, USA, 16–18 October 2023; pp. 277–288.
34. Kaya, B.; Kumar, S.; Oliveira, C.; Ferrari, V.; Gool, L.V. Multi-View Photometric Stereo Revisited. In Proceedings of the IEEE/CVF Winter Conf. Applications of Computer Vision, Waikoloa, HI, USA, 2–7 January 2023; pp. 3126–3135.

Journal of
Imaging

MDPI

Article

Fast Data Generation for Training Deep-Learning 3D Reconstruction Approaches for Camera Arrays

Théo Barrios*, Stéphanie Prévost * and Céline Loscos

LICIIS Laboratory, University of Reims Champagne-Ardenne, 51100 Reims, France
* Correspondence: theo.barrios@univ-reims.fr (T.B.); stephanie.prevost@univ-reims.fr (S.P.)

Abstract: In the last decade, many neural network algorithms have been proposed to solve depth reconstruction. Our focus is on reconstruction from images captured by multi-camera arrays which are a grid of vertically and horizontally aligned cameras that are uniformly spaced. Training these networks using supervised learning requires data with ground truth. Existing datasets are simulating specific configurations. For example, they represent a fixed-size camera array or a fixed space between cameras. When the distance between cameras is small, the array is said to be with a short baseline. Light-field cameras, with a baseline of less than a centimeter, are for instance in this category. On the contrary, an array with large space between cameras is said to be of a wide baseline. In this paper, we present a purely virtual data generator to create large training datasets: this generator can adapt to any camera array configuration. Parameters are for instance the size (number of cameras) and the distance between two cameras. The generator creates virtual scenes by randomly selecting objects and textures and following user-defined parameters like the disparity range or image parameters (resolution, color space). Generated data are used only for the learning phase. They are unrealistic but can present concrete challenges for disparity reconstruction such as thin elements and the random assignment of textures to objects to avoid color bias. Our experiments focus on wide-baseline configuration which requires more datasets. We validate the generator by testing the generated datasets with known deep-learning approaches as well as depth reconstruction algorithms in order to validate them. The validation experiments have proven successful.

Keywords: 3D vision; training database; deep learning; 3D reconstruction

1. Introduction

The principle of photogrammetric 3D reconstruction is to recover the depth of a scene by exploiting the parallax existing on images acquired from different viewpoints. More precisely, this means matching pixels from one image with others (co-homologous pixels, i.e., projections of the same 3D point in images). The search space for co-homologous pixels [1] varies according to the structured (aligned, planar) or unstructured (free position) configuration of the cameras in the acquisition system. These variations have a decisive influence on the process of reconstructing a 3D scene from images. In this paper, we focus solely on 2D camera array configuration, where the principles of simplified epipolar geometry [2] can be applied. Thanks to them, the search space is reduced to a single line following the pixel grid of the image, i.e., vertical for vertically adjacent cameras and horizontal for horizontally adjacent cameras.

In this configuration, the depth computation becomes a disparity computation (i.e., the computation of an offset of a number of pixels separating the co-homologous pixels of 2 successive images in one of the horizontal or vertical axes). The use of deep neural networks for photogrammetric 3D reconstruction had a significant impact on improving state-of-the-art performance in terms of speed, accuracy, and robustness of reconstruction in stereo and light field configurations. However, they require training datasets, of more or less significant size depending on the camera configuration, usually including ground truth information. While reconstruction methods for light field cameras can be trained on a small number of scenes (state-of-the-art methods can be trained with a few dozen scenes), this is not the case for stereo and wide-baseline multi-view stereo configurations, which require a high number of training scenes (several thousand) to be efficient. The main reason for this is the wide range of correspondence search space. The light field configuration has a disparity of approximately 10 pixels, while the stereo and camera array configurations have a disparity range of around 200 pixels. The latter configurations, therefore, require a larger amount of data to train the network.

Some contributions have attempted to work around this problem by proposing deep neural network training without the need for ground-truth data, with either unsupervised [3,4] or self-supervised training [5]. Other works propose virtual datasets that have by construction more accurate ground truth data, and for some of them, more data [6,7]. However, a lot of these datasets only have a few dozen images and are thus more suited for method evaluation rather than training.

In this paper, we propose a dataset generator, i.e., to create a high number of scenes, and render them in the form of images and disparity maps, from a user-chosen set of models and textures. We show that our approach allows for a fast generation of a training dataset with enough variety to improve the results of deep learning methods for disparity estimation. We also demonstrate that the proposed dataset is best used for first-step training before fine-tuning is performed with a state-of-the-art dataset.

After a review of different types of available state-of-the-art datasets in Section 2, we present our highly configurable generator and describe our training dataset and the protocol for our experiments in Section 3. The experiments in Section 4 compare the use of our dataset versus Li et al.'s dataset [7] for training. They highlight the relevance of our training dataset, and hence such a generator, by comparing use cases with two deep learning reconstruction methods [7,8], firstly, as a single source, secondly as a primary, and finally as a fine-tuning dataset. We conclude and address future work in Section 5.

2. Related Work

In this section, we distinguish three types of available data to review the state-of-the-art datasets/generators. The first is real data, where images are recorded through sensors, such as cameras, possibly with ground truth using depth cameras, or Lidar sensors. The second is hand-made virtual data, i.e., scenes that are manually created and rendered with 3D modeling software but where scene conception and lighting are decided by a human being. The third type is procedurally generated data, where scene conception is decided by an algorithm.

In Table 1, we summarize the features of discussed training datasets in this section. For a more extensive review, please refer to [9].

Table 1. Summary of discussed training datasets and ours. MVS: Unstructured multi-view stereo, LF: light field, #: number.

Reference	Nature	# Cameras	Structure	# Captures	Resolution	with GT
Mayer et al. [6]	both	2	stereo	≈25 k (200 real)	960 × 540	✓
Li et al. [7]	virtual	81	9 × 9 array	353	512 × 512	✓
Scharstein et al. [10]	real	2	stereo	6	512 × 384	✓
Scharstein et al. [11]	real	2	stereo	33	2864 × 1924	✓
Menze et al. [12]	real	2	stereo	400	1242 × 375	✓
Schops et al. [13]	real	4	MVS	38	6048 × 4032	✓
Honauer et al. [14]	virtual	81	9 × 9 LF	28	512 × 512	✓
Butler et al. [15]	virtual	2	stereo	1628	1024 × 436	✓
Dosovitskiy et al. [16]	virtual	2	stereo	≈22 k	1024 × 768	✓
Sabater et al. [17]	real	16	camera array	12	2048 × 1088	×
Ours (for experiments)	virtual	25	5 × 5 array	3978	1920 × 1080	✓

2.1. Real Datasets

Most of the real scene datasets were made for testing purposes rather than training. Before the emergence of machine learning techniques in stereoscopic reconstruction methods, real scenes were provided as benchmarks for method evaluations, as for example by Scharstein et al. [10,11]. More recently, several benchmarks were proposed for stereo reconstruction and unstructured multi-view stereo reconstruction, made of real scenes associated with their ground truth data, expressed in the form of a disparity or depth map [11–13]. Early deep neural network methods, such as [18], were trainable on the small number of scenes, offered by these datasets (around 20 scenes).

In 2015, Menze and Geiger [12] also proposed a set of 200 real training scenes for the purpose of stereo disparity reconstruction on car-embedded cameras. The scenes are exclusively driving scenes and serve the purpose of autonomous driving.

However, using real data involves handling the properties and imperfections of physical image sensors (optical and color distortions). Correspondingly, when depth is captured, it also means dealing with the inaccuracy of the depth sensor (noise), and sometimes its inability to provide ground truth values in certain areas (highly reflective, absorptive and transparent area, etc.). Moreover, due to their nature and size, none of these real datasets are used as standalone training datasets by current deep neural network methods. Nevertheless, the datasets can be also used for network fine-tuning, i.e., for adapting the weights of a pre-trained neural network to a specific context.

2.2. Hand-Made Virtual Datasets

Virtual datasets allow to have precise and complete data with ground truth. In the context of light field disparity reconstruction, Honauer et al. [14] proposed a benchmark and a hand-made training dataset with 25 scenes. This low number of scenes, compared to other configurations, is enough to train state-of-the-art methods for this configuration. Li et al. [7] proposed a training and a testing dataset for a 9×9 wide-baseline camera array with a disparity range of 50. The testing dataset is composed of 12 virtual hand-crafted scenes and the training dataset also contains eight hand-crafted scenes.

While most of these proposed datasets have very few scenes, some efforts were made in improving the scene variety by proposing datasets based on image sequences of animated scenes instead of still scenes [6,15]. This allows for the creation of a higher number of scenes than with hand-crafting scenes, within the same time span. However, the scenes generated by this method do not increase the variety of objects in the dataset.

2.3. Procedurally Generated Datasets

Procedurally generated scenes can be used to have a large amount of data, without the need for time-consuming human design. For the stereo configuration, Dosovitskiy et al. [16] proposed a training dataset with various chair models that are randomly positioned. Mayer et al. [6] proposed training and testing datasets with more variety in models based on the ShapeNet [19] taxonomy. Furthermore, textures for this dataset are randomized based on various existing and also procedurally generated images.

For camera arrays, Li et al. [7] proposed a similar process for generating a training dataset with a nearly photo-realistic rendering. This dataset contains 345 scenes, with images taken by a 9×9 camera array. While these images are very high quality, the relatively small number of scenes makes it only practical for training lightweight neural networks. The disparity range is set at 50 pixels disparity range. However, this range can be extended to 200 pixels if you consider the dataset as a 3×3 camera array, by taking the images on every fourth row and column. This dataset contains scenes with a realistic rendering, although, they are in small numbers and thus are only efficient for training lightweight neural networks—around 2 M weights for Li et al.

In summary, the state of the art lacks large datasets when it comes to 3D wide-baseline camera array reconstruction, and even more so when it comes to network training, as many do not have the necessary ground truth and/or do not have a sufficient quantity of data. Existing deep neural network methods rely on training on datasets of relatively small scale and thus need to adapt to these small scale datasets, limiting their efficiency. We thus propose a way to generate data suitable for training more heavyweight and data-sensitive neural networks.

3. Materials and Methods

3.1. Virtual Data Generator

3.1.1. Principle

The goal of our data generator is to be able to: (i) quickly generate a large number of training scenes, with a great variety, for a two-dimensional camera array of a user-defined size; (ii) render and save these scenes as different data (RGB, disparity) to provide a dataset useful for 3D reconstruction methods. To be used in deep neural network supervised training, these scenes must also have associated disparity ground truth. For this, the principle is to randomly associate 3D models with a texture each, and randomly position them in order to render scenes taken from a virtual $N \times M$ camera array with a regular baseline, equal in the horizontal and vertical direction. The random selection of models ensures diversity of scene content, while the random assignment of textures to objects ensures that the method will not learn to recognize an object by its color, like green grass, blue sky, etc., thus avoiding a shape-color association bias. Similarly, to avoid shape-object association bias, we apply random scaling to each object on each axis.

Although the generated images are non-realistic, and quite unintelligible for a human being, when taken as a whole as illustrated in Figure 1, their local geometry mimics the variety of shapes and colors that can be found in real scenarios. It is thus possible to train deep neural networks on such scenes.

Figure 1. Example of our generator output. An RGB rendering image of one scene (**left**) with its disparity map (**right**) encoded on three RGB channels (see Section 3.1.3).

3.1.2. Parameters

Our proposed dataset generator allows for several types of parameters to be configured through its configuration file:

– Camera array configuration.

The number of cameras on each row and column (*cam_grid_row*, *cam_grid_Col*), as well as the baseline (*grid_spacing_row*, *grid_spacing_col*), i.e., the space between two adjacent cameras can be configured, one for each axis. This is one of the main factors in the disparity range of the created dataset. The virtual camera array can also be selected in off-axis or parallel disposition with the focus point parameter (*focusPoint*), although this paper focuses on the parallel disposition. When the focus point is set to 0, the cameras are positioned in parallel, otherwise, their vision pyramid is off-center to focus on it.

– Camera parameters.

The configuration is the same for every camera on the array. The configuration of their intrinsic matrix is realized with the following parameters: image resolution in pixel (*width_pixel*, *height_pixel*), *near* and *far* parameters. The vertical field of view (*fov*) can also be parameterized allowing for datasets from several types of cameras. As their positions in the camera array are constrained, we do not propose extrinsic parameter selection. The extrinsic matrix is fixed as the identity matrix since objects will be placed based on the camera field of view. It is modified according to the camera's position in the grid. In addition, for future work and the adaptability of our generator, we already integrated the camera exposures (*exposures*). This will be useful for multi-exposed recording in a High Dynamic Range reconstruction context. If several values are entered, each viewpoint is rendered once with each exposure value.

The parameters in these two sections (Camera array configuration and Camera parameters) are used to generate the capture system of our generator (see line 4 of Algorithm 1).

Algorithm 1 CameraArrayDatasetGenerator ()
▷ *A function for generating a dataset with ground truth from a camera Array respecting the parameters described in the user configuration file*

Input: cfgFile ▷ *configuration file*
Output: *cam_grid_row* × *cam_grid_col* RGB images with their disparity maps (see Table 2)

```
1:  cfg        = cfgFile contents
2:  models     = Load the models from the model file folder
3:  texs       = Load textures from the texture file folder
4:  camArray   = Generate the camera arrays, initialized with the camera parameters and
               camera array configuration defined by the user in cfgFile
5:  Create the OpenGL Context with a size set at (cfg.width_pixel, cfg.height_pixel)
6:  for nbreCaptureToDo = cfg.number_of_frame_to_render;
7:                nbreCaptureToDo > 0; nbreCaptureToDo- = 1 do
8:      populateScene()
9:      for each cam from camArray do
10:         Render the view from cam with its disparity map and save the both in the
            cfg.output_dir folder
11:     end for
12: end for
```

Table 2. Parameter list with their name in the configuration file and their value for the experiment.

Parameter	Name in Configuration File	Experiment Value
camera array configuration		
nb Camera in a Row	cam_grid_row	5
nb Camera in a Column	cam_grid_col	5
baseline in meters in row and column	grid_spacing{_row, _col}	0.2
camera focus point (if off-center)	focusPoint	0
camera parameters		
image resolution in width	width_pixel	1920
image resolution in height	height_pixel	1080
Z near distance in meters	near	0.1
Z far distance in meters	far	1000
vertical field of view	fov	60
Camera exposures	exposures	[1.0]
Scene configuration		
Minimum and Maximum distances between object and camera array center (in meters)	object_range	[2, 500]
Number of different models loaded in the scene	n_models	51
number of repetition of model in a scene	n_textures	3
rate of hidden	visible	[0.3, 0.6]
scene number	number_of_frame_to_render	4000
Number of different textures in the directory		108

– Scene configuration.

The major part of the configuration is on the scene. The minimum and maximum distances of objects can be set (*object_range*). However, these minimum and maximum values are not hard limits, as they are only used to position the object centers themselves. Part of the objects can still be in front of the minimum distance or behind the maximum distance. As a side-effect, some of the scenes generated do not conform to the desired maximum disparity. Other configuration parts are the different numbers of models (*n_models*) and textures (*n_textures*) loaded and how many times a given model is loaded with a different texture on a scene (*n_textures*). Finally, we propose to set the probability of hiding each object on each new scene (*visible*). The probability can be set in a range of probabilities so that some scenes are more or less full than others. Lastly, the user can set the number of generated scenes (*number_of_frame_to_render*). The Algorithm 2 gives step-by-step the construction of one randomly generated scene. In our model folder, we put only the required untextured models for the training. Random selection from this folder is not needed as we process models iteratively. This position is reflected in the Algorithm 2.

Algorithm 2 populateScene ()

▷ *A function which randomly insert some randomly textured and distorted objects. The function random(x,y) generates a random value between x and y with a uniform distribution.*

Input: *cfg* ▷ *configuration Structure*
Input: *models* ▷ *set of the loaded models*
Input: *texs* ▷ *set of the loaded textures*

proba = random(cfg.visible[0], cfg.visible[1])
for (i = 0; i < *cfg.n_models* ; i++) **do**
    for (j = 0; j < *cfg.n_textures* ; j++) **do**
        model = Clone of *models[i]*
        tex = Random texture selection from *texs*
        $dispZMin = MIN(cfg.object\_range[0]^{cfg.rep}, cfg.object\_range[1]^{cfg.rep})$
        $dispZMax = MAX(cfg.object\_range[0]^{cfg.rep}, cfg.object\_range[1]^{cfg.rep})$
        if (random(0,1) − *proba*) > 0 **then**
            zpos = random($dispZMin$, $dispZMax$)
            Translate the *model* on the z-axis by $zpos^{\frac{1}{cfg.rep}}$
        else
            Hide the model
        end if
        Randomly translate of *model* on the x-axis and y-axis to place it in the frustum of the camera array
        Randomly scale the *model* on each axis
        Randomly rotate of *model* on each axis
        Add *model* to the 3D scene
    end for
end for

– Output.

Each rendered output (an RGB image with its disparity map) is saved in the defined output directory (*output_dir*).

3.1.3. Implementation Details

We chose to implement our dataset generator as a webGL application using html and javascript, with the electron API [20] and threeJS [21]. Indeed, on the one hand, ThreeJs is a well-known javascript 3D library in the computer graphic world, which has also the advantage of already having a large number of mesh loaders (obj, ply, fbx, gltf, etc.). On the other hand, the web nature of this application allows it to be cross-platform and makes it easy for others to reuse.

Meshes and textures are each in their own directory, which must contain at least two items. Given the nature of our mesh data, we are currently only using the obj loader, but integrating the other loaders should not pose any problems.

The output data (RGB image and disparity map) is rendered from each camera of the array, using OpenGL rasterization [22] without incorporating any complex lighting effects (no shadow, no transparency, etc.). We generate the color rendering and the disparity map with two shader passes. We compute the depth as explained in [23] and we deduce from it the disparity value with the properties of the cameras in the array. This disparity value δ is instantly encoded in the shader as described in the following storage section.

The Algorithms 1 and 2 describe the global pipeline and main steps.

– Storage.

We save images and disparity maps as PNG images. Files are named as follows: {*tag*}{*type*}{*position*}_{*exp*}.png where:

- *tag* is a 21 alphanumeric scene label, randomly generated each time a new scene is rendered. This label has around 4.8×10^{32} different possibilities.
- *type* is either *rgb* for pictures or *depth* for disparity maps.
- *position* is a single number identifying the position of the view on the array. It identifies the view in a top-to-bottom, left-to-right order, i.e., if the position is (i, j) on a $cam\_grid\_row \times cam\_grid\_col$ array, the *position* number will be $i.cam\_grid\_col + j$.
- *exp* is the exposure factor, a higher number means brighter images, and a lower number means darker images. The *exp* value is the exposure value for RGB images and 0 for ground truth disparity maps (file with *type = depth*).

The disparity values are encoded in a 32-bit fixed-point precision format, with possible values ranging from 0 to 8192. The disparity is encoded using the four channels of the image:

- Red channel encoding the coarsest part of disparity with a disparity step of 32.
- Green channel encoding disparity with a step that is 256 times smaller, i.e., $\frac{1}{8}$.
- Blue channel encoding disparity with a step of $\frac{1}{2048}$.
- Alpha channel encoding disparity with a step of $\frac{1}{524,288}$.

3.2. Creation of the Dataset

We propose a Full HD dataset for deep neural network training. It is taken from a 5×5 camera array in a parallel disposition. Image sizes are 1920×1080. We set the cameras *near* and *far* at 0.1 m and 1000 m, respectively, and the field of view at 60 degrees.

For models and textures, we took 51 models from the ShapeNet taxonomy [19], each from a different semantic class, and 100 texture-like images from the Pixabay [24]. In each scene, models are loaded three times with a random texture. Each textured model has a probability to be hidden varying between 30 and 60%.

Objects are positioned in a random place within the field of view of the camera array and at a distance between 2 and 500 m. To smooth disparity repartition the positioning is not uniform but based on the distance. The likelihood of an object being put at a given distance is $\frac{1}{\sqrt{x}}$ with x the distance to the camera array center. This means that objects are more likely to be put closer to the camera array than further. Combined with our rescaling making objects that are bigger, gives a smoother distribution.

The baseline for the camera array is set to 0.2 m for the horizontal and vertical axis. This gives this camera array a disparity range of 128 (from 0 to 128). We rendered 4000 images and the output datasets have a size of 185 GB and 526 GB, respectively. From this dataset, we remove the scenes with disparity outside the target range, which leaves us with 3978 scenes. The Table 2 summarizes the parameter settings. Since we do not experiment on multi-exposed camera arrays, the exposure value is always set to 1 in our experiments.

The generated dataset, which is also the one used in our experiments and following the configuration given in this article, is freely available to the community (see Data Availability Statement).

3.3. Protocol for Experiments

With the proposed camera array, we conduct experiments on a 3×3 configuration by taking views on every other row and column of our 5×5 dataset. For the training dataset, the disparity range is 0–256 in the 3×3 configuration. For experiments, we compare the results obtained by two deep neural network methods from Li et al. [7] and Barrios et al. [8]. It should be noted that our aim is not to compare these methods with each other, but only to compare the contribution of our dataset to these methods, without or with a fine-tuning.

- Considered training datasets

As shown in the related work section, no dataset with ground-truth exists in the literature suitable for training heavyweight and data-sensitive neural networks, in order to estimate disparities within a context of a wide-baseline camera array. As illustrated in the Table 1, only the Li et al. [7] dataset has ground truth, even though it was not originally designed for use in a wide-baseline scenario. The considered training dataset thus are the datasets described in this paper and the one proposed by Li et al. in [7]. Networks are trained with the same amount of iterations on either dataset. We then compare the results obtained by methods on Li et al.'s testing dataset, for the 3×3 configurations. We also propose experiments on training with one of these two datasets and refinement with the other. The refinement part consists of 10k additional iterations, with a learning rate fixed at 10^{-5} throughout the process for every network and dataset considered.

- Test metrics

For comparison, we use the metrics *bad x* that are used by Li et al. in their paper and also on the Middlebury Stereo website [25]. The metric *bad x* is the percentage of pixels for which the absolute error, when the resulting disparity is compared to the ground truth, is greater than x.

For comparison to ground truth, we use the testing dataset proposed by Li et al. in [7]. This dataset is originally a 9×9 testing dataset with a disparity range of 50 (from 0 to 50). For our experiments, we will use this dataset with a 3×3 array by using one in every four columns. The disparity ranges will thus be 200 for the 3×3 configuration. Due to the different configurations and disparity ranges, we adapt the metrics used by the Middlebury stereo benchmark [25]. We use for the 3×3 configuration, bad 0.5, 1, 2, and 4, similar to the main metrics on the reference.

- Tested methods

Two deep neural network methods are tested. The first one, proposed by Li et al. in [7], is a lightweight neural network with less than 2 M weights. It computes a disparity map mainly through convolution neural networks following the classical structure of disparity inference with deep neural networks [9].

The second tested method, proposed by Barrios et al. [8], is a neural network that computes disparity maps in two parts, following a similar structure. The first part computes a downsampled disparity map and the second step computes a residual disparity map from low to high resolution. The network has 5 M weights in total, with the second step counting more than 4 M weights.

These networks are trained on each dataset with the number of iterations or training time indicated on their respective papers, i.e., 1.5 M iterations for Barrios et al. [8] and 250 k iterations corresponding to the training time indicated on the paper [7].

4. Results and Discussion

4.1. General Results on the 3 × 3 Configuration

The results with the proposed metrics are shown in Table 3 for the 3 × 3 configuration. The results shown are the average results of the training performed.

Table 3. Comparison of results between training with either our dataset and the one proposed in Li et al.'s work [7]. "Bad x" metrics represent the percentage of pixels for which the difference to ground truth is higher than x. Lower is better. Bold values have the lowest errors for their respective method and metric.

Method	Training Dataset	Bad 0.5	Bad 1	Bad 2	Bad 4
Results with a 3 × 3 configuration					
Li et al. [7]	Li et al. [7]	23.95	12.58	7.70	5.85
	Ours	25.40	11.49	6.46	4.60
	[7] fine-tuned with ours	27.27	14.95	10.14	7.74
	Ours fine-tuned with [7]	**22.90**	**10.28**	**5.78**	**3.96**
Barrios et al. [8]	Li et al. [7]	16.60	11.46	8.22	5.75
	Ours	14.96	8.27	5.33	3.59
	[7] fine-tuned with ours	20.14	12.78	9.19	6.79
	Ours fine-tuned with [7]	**12.23**	**7.51**	**4.60**	**2.98**

On the 3 × 3 configuration, the results show that the use of our training dataset instead of that of Li et al. strongly improves results for the bad 2 and bad 4 metrics for both considered neural networks. For example, with Li et al.'s method, the bad 4 metric is 5.85 when trained with their dataset and 4.60 when trained with ours. While this 3 × 3 configuration is not optimal for Li et al., it remains a possible configuration for their method and our training dataset shows improvement. With the network from Barrios et al. using our training dataset improves the results from 5.75 to 3.59 for bad 4. With higher tolerance thresholds, these metrics can identify the outliers in reconstruction. Since a lower error rate means greater robustness, we can, therefore, conclude that using our training dataset instead of Li et al.'s significantly improves the robustness of deep neural networks. This is especially visible on images with thin elements, for example, on the image shown in Figure 2. In this figure, the thin elements are correctly reconstructed with a network trained with our dataset when it is not with networks trained on the dataset of [7]. This can be seen in the third column of the figure, where bars are more efficiently reconstructed when our training dataset is used.

When considering fine precision (bad 0.5), the results are different depending on the method that is considered. While training with our dataset improves the results obtained with the network from Barrios et al. [8] (with a bad 0.5 of 14.96 versus 16.60), they degrade the results with Li et al. (with 25.40 versus 23.95). This puts forward the main limitation of our dataset. As we chose rasterization as our method of rendering, some low textures and light effects are not taken into account. Our training dataset thus underperforms, even compared to the one in [7] when images contain low texture and light effects, as can be seen in Figure 3. The zone below the teddy arm on the right is not reconstructed correctly when the methods are trained with our dataset.

Moreover, optimal results are obtained when the network trained with our training dataset is fine-tuned with Li et al.'s dataset. This fine-tuning results in a better reconstruction rate than without fine-tuning regardless of the network, the training dataset, or the bad metric considered. These results are shown in the last column in Figures 2 and 3. For each scene presented, detailed numerical results, shown in Table 4, confirm the conclusion but also indicate that on some views, results can be degraded by the fine-tuning step.

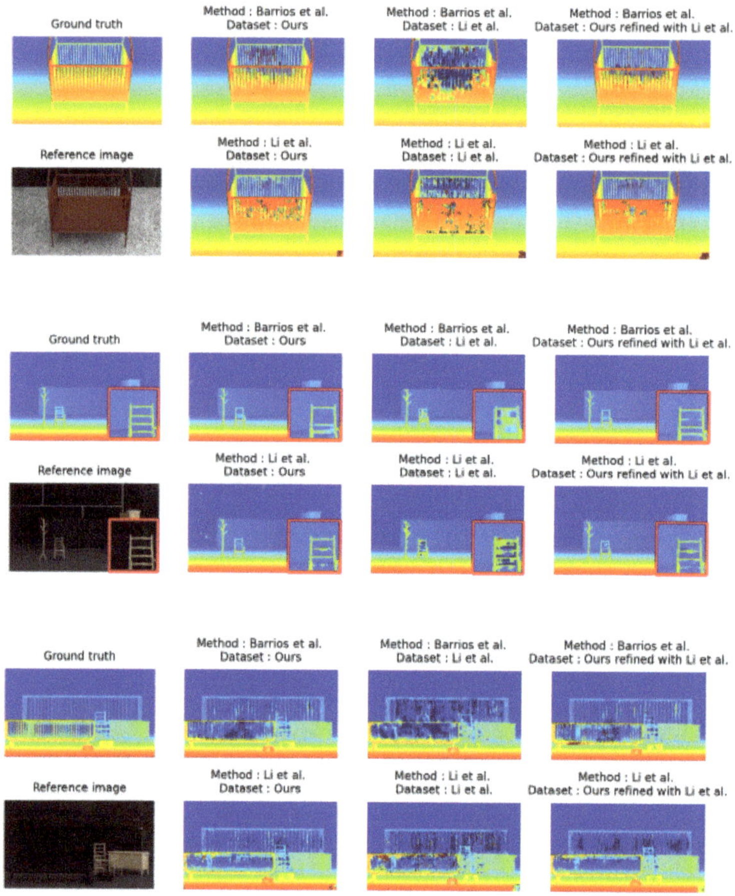

Figure 2. Difference of disparity maps between networks trained with ours and Li et al.'s with examples of images containing thin elements, taken from Li et al.'s [7] and Barrios et al.'s [8] test dataset. The red squares are zooms of a detailed part of the image.

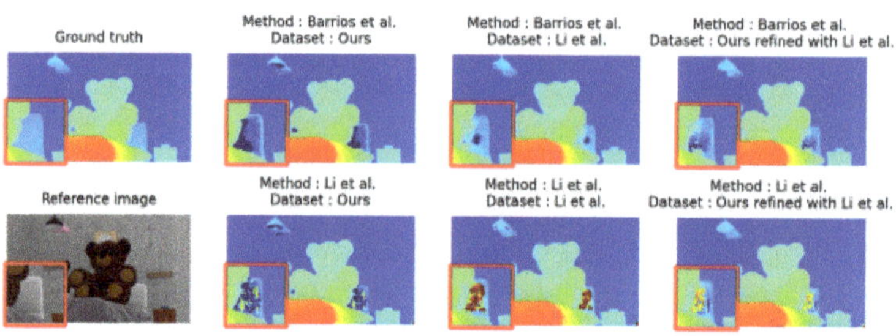

Figure 3. Difference of disparity maps between networks trained with ours and Li et al.'s with an example image some bright untextured elements. Images are taken from Li et al.'s [7] and Barrios et al.'s [8] test dataset. The red squares are zooms of a detailed part of the image.

Table 4. Comparison of results between training with either our dataset and the one proposed in Li et al.'s work [7] on some specific views. "Bad x" metrics represent the percentage of pixels for which the difference to ground truth is higher than x. Lower is better. Bold values have the lowest errors for their respective method and metric.

Method	Training Dataset	Bad 0.5	Bad 1	Bad 2	Bad 4
Results for view #1, "Cot"					
Li et al. [7]	Li et al. [7]	32.52	20.56	15.95	13.73
	Ours	26.72	14.07	9.23	**7.00**
	[7] fine-tuned with ours	29.28	19.14	15.45	13.32
	Ours fine-tuned with [7]	**29.47**	**15.14**	**10.13**	7.88
Barrios et al. [8]	Li et al. [7]	32.01	26.77	22.73	18.97
	Ours	21.96	14.67	10.01	6.86
	[7] fine-tuned with ours	33.10	27.94	24.58	20.96
	Ours fine-tuned with [7]	**21.51**	**13.35**	**8.10**	**5.20**
Results for view #2, "Furniture"					
Li et al. [7]	Li et al. [7]	12.72	5.59	2.97	2.14
	Ours	24.77	7.84	2.04	1.17
	[7] fine-tuned with ours	30.40	16.40	11.42	9.07
	Ours fine-tuned with [7]	**16.53**	**5.08**	**1.96**	**1.10**
Barrios et al. [8]	Li et al. [7]	7.10	3.93	2.47	1.52
	Ours	11.41	2.59	1.31	0.66
	[7] fine-tuned with ours	10.85	4.97	2.88	1.82
	Ours fine-tuned with [7]	**5.56**	**2.43**	**1.24**	**0.63**
Results for view #3, "Sidebars"					
Li et al. [7]	Li et al. [7]	38.98	31.84	26.46	22.87
	Ours	38.19	26.22	19.65	15.74
	[7] fine-tuned with ours	43.85	35.44	30.06	26.06
	Ours fine-tuned with [7]	**35.56**	**26.17**	**19.56**	**15.35**
Barrios et al. [8]	Li et al. [7]	42.08	36.69	30.97	24.75
	Ours	36.65	28.02	22.20	17.44
	[7] fine-tuned with ours	45.21	39.40	35.47	30.93
	Ours fine-tuned with [7]	**34.21**	**26.98**	**20.78**	**15.96**
Results for view #4, "Teddy Bears"					
Li et al. [7]	Li et al. [7]	22.98	8.23	3.85	2.83
	Ours	**21.38**	**6.37**	3.54	2.88
	[7] fine-tuned with ours	17.07	5.54	3.47	2.84
	Ours fine-tuned with [7]	22.86	6.74	**3.22**	**2.18**
Barrios et al. [8]	Li et al. [7]	10.51	5.66	3.29	2.05
	Ours	8.00	4.66	3.27	2.55
	[7] fine-tuned with ours	11.31	5.37	3.11	1.89
	Ours fine-tuned with [7]	**7.26**	**4.16**	**2.51**	**1.67**

As shown in detail in Table 5, the fine-tuning process improves results on some pixels and degrades them on others. For example, Li et al.'s neural network using our dataset in initial training with a fine-tuning with the dataset of Li et al. (see Table 5), allows 9.92% and 7.21% of pixels to go from bad 1 to bad 0.5 and vice versa. Those values are 4.13% and 2.50%, respectively, for the method of Barrios et al. We can, however, note that for both neural networks considered the amount of pixels improved is higher than the amount of pixels degraded (15.82% vs. 11.15% in the method of Li et al. and 7.71% vs. 5.63% for the method of Barrios et al.), whether this is in total (on the first column of the tables) or when considering evolution from any two categories (considering two opposed categories, e.g., from errors between 0.5 and 1 to error smaller than 0.5 compared with the inverse evolution).

However, when the training and fine-tuning datasets are swapped, i.e., training is conducted with the dataset from Li et al. [7] and fine-tuning with our dataset, the results are significantly degraded. The third rows of each section of Table 3 show that results with this choice of training are worse than even non-fine-tuned training. We can thus conclude that our training dataset is not relevant for fine-tuning purposes. As the results are significantly worse with this reversed training, we do not propose a detailed analysis of the evolution as we did in the previous case.

Table 5. Evolution in results between training with our dataset with and without fine-tuning on Li et al.'s dataset [7]. The numbers correspond to the percentage of pixels that change category with the addition of fine-tuning. Green values under the diagonal show the pixels whose category is improved with fine-tuning, and red values above the diagonal show the pixels whose category is degraded with fine-tuning. The white values on the diagonal show the pixel for which there is no change of category.

Evolution between Training with Our Dataset only (Row) and Training with Fine-Tuning (Column)						
Li et al. [7]		$0 \leq$ err < 0.5	$0.5 \leq$ err < 1	$1 \leq$ error < 2	$2 \leq$ err < 4	err ≥ 4
Category improved	$0 \leq$ err < 0.5	64.65	7.21	1.22	0.18	0.17
15.82	$0.5 \leq$ err < 1	9.92	3.42	1.07	0.18	0.15
	$1 \leq$ err < 2	1.95	1.46	1.35	0.38	0.23
Category degraded	$2 \leq$ err < 4	0.26	0.25	0.48	0.57	0.36
11.15	err ≥ 4	0.32	0.27	0.4	0.51	3.05
Barrios et al. [7]		$0 \leq$ err < 0.5	$0.5 \leq$ err < 1	$1 \leq$ error < 2	$2 \leq$ err < 4	err ≥ 4
Category improved	$0 \leq$ err < 0.5	81.41	2.50	0.70	0.27	0.17
7.71	$0.5 \leq$ err < 1	4.13	1.50	0.70	0.24	0.12
	$1 \leq$ err < 2	0.73	0.67	0.92	0.41	0.21
Category degraded	$2 \leq$ err < 4	0.28	0.23	0.40	0.51	0.31
5.63	err ≥ 4	0.27	0.22	0.33	0.45	2.32

4.2. Comparison with Data-Sensitive Networks

When considering solely the network from Barrios et al. [8], two things must be considered. First, when comparing training with either our dataset or Li et al.'s alone, Table 3 shows that using our dataset gives better results in every metric, whether it considers fine reconstruction or outliers. This is mostly due to the refinement step that is more data sensitive than the network from Li et al. and thus is not efficiently trained by Li et al.'s smaller-scale training dataset. This is also visible with the results in Table 6. When we only consider the downsampled disparity map of Barrios et al. [8] that was computed with a very lightweight network that has an overall structure similar to the one in [7], we observe the same behavior [7], i.e., better robustness but less overall precision when trained with our dataset compared to Li et al.'s.

Table 6. Comparison of results between training with either our dataset and the one proposed in Li et al.'s work [7]. These comparisons of error are conducted for the small resolution and high-resolution disparity maps obtained by the neural network proposed by Barrios et al. [8]. Bold values have the lowest errors for their respective method and metric.

Output from [8]	Training Dataset	Bad 0.5	Bad 1	Bad 2	Bad 4
Results with a 3×3 configuration					
Small resolution (downsampled)	Li et al. [7]	16.39	11.37	8.42	6.30
	Ours	17.15	10.26	7.42	5.43
	[7] fine-tuned with ours	20.96	12.86	9.43	7.26
	Ours fine-tuned with [7]	**14.61**	**9.60**	**6.78**	**4.89**

Table 6. *Cont.*

Output from [8]	Training Dataset	Bad 0.5	Bad 1	Bad 2	Bad 4
High resolution (upsampled)	Li et al. [7]	16.60	11.46	8.22	5.75
	Ours	14.96	8.27	5.33	3.59
	[7] fine-tuned with ours	20.14	12.78	9.19	6.79
	Ours fine-tuned with [7]	**12.23**	**7.51**	**4.60**	**2.98**

Second, fine-tuning is less efficient when conducted on the refinement step of this method, as proved by the results obtained by turning on or off fine-tuning on the last part of the network in Table 7.

Table 7. Comparison of quality of results between turning on or off fine-tuning of the refinement part on the method proposed by Barrios et al. [8]

Variant	Bad 0.5	Bad 1	Bad 2	Bad 4
No fine-tuning at all	14.96	8.27	5.33	3.59
Fine-tuning on every part	13.11	8.00	4.99	3.13
Fine-tuning except on the refinement part	12.23	7.51	4.60	2.98

5. Conclusions and Future Work

We introduced a dataset generator to automatically compose scenes and render them as a set of images and disparity maps with a large variety from a set of user-defined models and textures. The scenes that we generate are nowhere near realistic in terms of color and their composition (layout of objects). Nevertheless, they present geometric challenges that are found in realistic scenes and avoid any shape-color association bias. As we opted for the very fast but limited rasterization render method, some light effects are not present in our dataset and methods trained with it cannot process them correctly. However, we showed that a short fine-tuning step on a smaller dataset that does take these light effects into account not only resolves this problem but obtains overall more stable results.

Future work includes testing different disparity ranges, from the very short disparity range as in lightfield configuration to wider disparity ranges, like what was proposed in this work. The objective would be to assert the amount of data required to train methods depending on the target disparity range. Another future work possibility is to find a compromise between the speed of rendering and its quality and accounting for specific light effects by changing the rendering engine to more modern engines that can provide fast rendering with a higher visual quality, such as Unreal Engine [26] or NVIDIA Omniverse [27]. In addition, it would also be interesting to extend our experiments by using a testing dataset consisting of real data and ground truths obtained by LIDAR technology.

Author Contributions: Conceptualization, T.B.; Formal analysis, T.B.; Funding acquisition, S.P. and C.L.; Methodology, T.B. and S.P.; Project administration, S.P. and C.L.; Validation, T.B., S.P. and C.L.; Writing—original draft, T.B. and S.P.; Writing—review & editing, T.B., S.P. and C.L. All authors have read and agreed to the published version of the manuscript.

Funding: This work was funded by the ANR-ReVeRY projet (ANR-17-CE23-0020).

Institutional Review Board Statement: Not applicable.

Informed Consent Statement: Not applicable.

Data Availability Statement: The training dataset, generated from a given set of free-to-use models and textures collected on various websites, and used to lead these experiments, is downloadable at https://revery.univ-reims.fr/dataset/big/Dataset_paper.tar (accessed on 21 December 2023).

Acknowledgments: The experiments were conducted on the infrastructures of the CentreImage of the University of Reims (https://centreimage.univ-reims.fr/ (accessed on 21 December 2023)) and the ROMEO computing center (https://romeo.univ-reims.fr/ (accessed on 21 December 2023)). Special thanks to Julien Gerhards for his strong contributions to the code of our solution.

Conflicts of Interest: The authors declare no conflict of interest.

References

1. Prévost, S.; Niquin, C.; Chambon, S.; Gales, G. Multi- and Stereoscopic Matching, Depth and Disparity. In *3D Video: From Capture to Diffusion*; John Wiley & Sons, Ltd.: Hoboken, NJ, USA, 2013; Chapter 7, pp. 137–155. [CrossRef]
2. Prévoteau, J.; Lucas, L.; Rémion, Y. Shooting and Viewing Geometries in 3DTV. In *3D Video: From Capture to Diffusion*; John Wiley & Sons, Ltd.: Hoboken, NJ, USA, 2013; Chapter 4, pp. 71–89. [CrossRef]
3. Huang, B.; Yi, H.; Huang, C.; He, Y.; Liu, J.; Liu, X. M^3VSNet: Unsupervised Multi-metric Multi-view Stereo Network. *arXiv* **2020**, arXiv:2004.09722.
4. Zhou, C.; Zhang, H.; Shen, X.; Jia, J. Unsupervised Learning of Stereo Matching. In Proceedings of the 2017 IEEE International Conference on Computer Vision (ICCV), Venice, Italy, 22–29 October 2017; pp. 1576–1584. [CrossRef]
5. Yang, J.; Alvarez, J.M.; Liu, M. Self-Supervised Learning of Depth Inference for Multi-View Stereo. *arXiv* **2021**, arXiv:2104.02972.
6. Mayer, N.; Ilg, E.; Häusser, P.; Fischer, P.; Cremers, D.; Dosovitskiy, A.; Brox, T. A Large Dataset to Train Convolutional Networks for Disparity, Optical Flow, and Scene Flow Estimation. In Proceedings of the 2016 IEEE Conference on Computer Vision and Pattern Recognition (CVPR), Las Vegas, NV, USA, 27–30 June 2016; pp. 4040–4048.
7. Li, Y.; Wang, Q.; Zhang, L.; Lafruit, G. A Lightweight Depth Estimation Network for Wide-Baseline Light Fields. *IEEE Trans. Image Process.* **2021**, *30*, 2288–2300. [CrossRef]
8. Barrios, T.; Niquin, C.; Prévost, S.; Souchet, P.; Loscos, C. A Wide-Baseline Multiview System for Indoor Scene Capture. In Proceedings of the 19th ACM SIGGRAPH European Conference on Visual Media Production, London, UK, 1–2 December 2022; CVMP '22. [CrossRef]
9. Laga, H.; Jospin, L.V.; Boussaid, F.; Bennamoun, M. A Survey on Deep Learning Techniques for Stereo-based Depth Estimation. *IEEE Trans. Pattern Anal. Mach. Intell.* **2020**, *44*, 1738–1764. [CrossRef] [PubMed]
10. Scharstein, D.; Szeliski, R.; Zabih, R. A taxonomy and evaluation of dense two-frame stereo correspondence algorithms. In Proceedings of the IEEE Workshop on Stereo and Multi-Baseline Vision (SMBV 2001), Kauai, HI, USA, 9–10 December 2001; pp. 131–140. [CrossRef]
11. Scharstein, D.; Hirschmüller, H.; Kitajima, Y.; Krathwohl, G.; Nešić, N.; Wang, X.; Westling, P. High-Resolution Stereo Datasets with Subpixel-Accurate Ground Truth. In *Pattern Recognition*; Jiang, X., Hornegger, J., Koch, R., Eds.; Springer International Publishing: Cham, Switzerland, 2014; Volume 8753, pp. 31–42. [CrossRef]
12. Menze, M.; Geiger, A. Object Scene Flow for Autonomous Vehicles. In Proceedings of the Conference on Computer Vision and Pattern Recognition (CVPR), Boston, MA, USA, 7–12 June 2015.
13. Schops, T.; Schonberger, J.L.; Galliani, S.; Sattler, T.; Schindler, K.; Pollefeys, M.; Geiger, A. A Multi-view Stereo Benchmark with High-Resolution Images and Multi-camera Videos. In Proceedings of the 2017 IEEE Conference on Computer Vision and Pattern Recognition (CVPR), Honolulu, HI, USA, 21–26 July 2017; pp. 2538–2547. [CrossRef]
14. Honauer, K.; Johannsen, O.; Kondermann, D.; Goldluecke, B., A Dataset and Evaluation Methodology for Depth Estimation on 4D Light Fields. In *Computer Vision—ACCV 2016*; Lai, S.H., Lepetit, V., Nishino, K., Sato, Y., Eds.; Springer International Publishing: Cham, Switzerland, 2017; Volume 10113, pp. 19–34. [CrossRef]
15. Butler, D.J.; Wulff, J.; Stanley, G.B.; Black, M.J. A naturalistic open source movie for optical flow evaluation. In Proceedings of the European Conference on Computer Vision (ECCV), Florence, Italy, 7–13 October 2012; Fitzgibbon, A., Lazebnik, S., Perona, P., Sato, Y., Schmid, C., Eds.; Springer: Berlin/Heidelberg, Germany, 2012; Part IV, LNCS 7577, pp. 611–625.
16. Dosovitskiy, A.; Fischer, P.; Ilg, E.; Hausser, P.; Hazirbas, C.; Golkov, V.; Smagt, P.V.D.; Cremers, D.; Brox, T. FlowNet: Learning Optical Flow with Convolutional Networks. In Proceedings of the 2015 IEEE International Conference on Computer Vision (ICCV), Santiago, Chile, 7–13 December 2015; pp. 2758–2766. [CrossRef]
17. Sabater, N.; Boisson, G.; Vandame, B.; Kerbiriou, P.; Babon, F.; Hog, M.; Gendrot, R.; Langlois, T.; Bureller, O.; Schubert, A.; et al. Dataset and Pipeline for Multi-view Light-Field Video. In Proceedings of the 2017 IEEE Conference on Computer Vision and Pattern Recognition Workshops (CVPRW), Honolulu, HI, USA, 21–26 July 2017; pp. 1743–1753. [CrossRef]
18. Zbontar, J.; LeCun, Y. Stereo Matching by Training a Convolutional Neural Network to Compare Image Patches. *J. Mach. Learn. Res.* **2016**, *17*, 32.
19. Chang, A.X.; Funkhouser, T.; Guibas, L.; Hanrahan, P.; Huang, Q.; Li, Z.; Savarese, S.; Savva, M.; Song, S.; Su, H.; et al. ShapeNet: An Information-Rich 3D Model Repository. *arXiv* **2015**, arXiv:1512.03012.
20. Electron: Build Cross-Platform Desktop Apps with JavaScript, HTML, and CSS. Available online: https://www.electronjs.org/ (accessed on 21 December 2023).
21. three.js. Available online: https://threejs.org/ (accessed on 21 December 2023).
22. Segal, M.; Akeley, K. *OpenGL 4.6 (Core Profile)—May 5, 2022*; Chapter 14. The OpenGL® Graphics System: A Specification (Version 4.6 (Core Profile) - May 5, 2022) Editor (version 2.0): Pat Brown Copyright © 2006–2022 The Khronos Group Inc. Available online: https://registry.khronos.org/OpenGL/specs/gl/glspec46.core.pdf (accessed on 21 December 2023).
23. Available online: https://threejs.org/docs/?q=depth#api/en/textures/DepthTexture (accessed on 21 December 2023).
24. Available online: https://pixabay.com/ (accessed on 21 December 2023).
25. Available online: https://vision.middlebury.edu/stereo/eval3/ (accessed on 21 December 2023).

26. Epic Games. Unreal Engine. Available online: https://www.unrealengine.com (accessed on 21 December 2023).
27. NVIDIA. Available online: https://www.nvidia.com/en-us/omniverse/ (accessed on 21 December 2023).

MDPI AG
Grosspeteranlage 5
4052 Basel
Switzerland
Tel.: +41 61 683 77 34

Journal of Imaging Editorial Office
E-mail: jimaging@mdpi.com
www.mdpi.com/journal/jimaging

www.ingramcontent.com/pod-product-compliance
Lightning Source LLC
Chambersburg PA
CBHW061454210926
51850CB00008B/1708